RESHAPING AGRICULTURE'S CONTRIBUTIONS TO SOCIETY

PROCEEDINGS
OF THE
TWENTY-FIFTH
INTERNATIONAL CONFERENCE
OF AGRICULTURAL ECONOMISTS

Held at Durban, South Africa
16-22 August, 2003

Edited by
David Colman, University of Manchester,
England
and
Nick Vink, University of Stellenbosch,
South Africa

2005

Blackwell
Publishing

350 Main Street, Malden, MA 02148-5020, USA
108 Cowley Road, Oxford OX4 1JF, UK
550 Swanston Street, Carlton, Victoria 3053, Australia

First published 2005 by Blackwell Publishing Ltd.

Library of Congress Cataloging-in-Publication Data

International Conference of Agricultural Economists (25th : 2003 : Durban, South Africa)
Reshaping agriculture's contributions to society : proceedings of the twenty-fifth International Conference of Agricultural Economists, held at Durban, South Africa, 16–22 August, 2003 / edited by David Colman and Nick Vink.
    p. cm.
Includes bibliographical references and index.
ISBN-10: 1-4051-3328-7
ISBN-13: 978-1-4051-3328-9
1. Agriculture–Economic aspects–Congresses. 2. Agriculture–Social aspects–Congresses. 3. Agriculture–Environmental aspects–Congresses. 4. Farms–Congresses. 5. Food supply–Congresses. 6. Sustainable development–Congresses.
I. Vink, N. II. Colman, David. III. Title.

HD1405.I58 2003
338.1–dc22

2004066095

For further information on
Blackwell Publishing, visit our website:
http://www.blackwellpublishing.com

CONTENTS

Preface

As President of the IAAE until the end of its 25th conference, I wish to record my thanks to all who made this splendid event possible in Durban. South Africa provided a most stimulating environment for the IAAE Conference and we are grateful for the hospitality extended to the IAAE and its members. Up front, a very special thank you should be offered to the Government of South Africa, especially to the Minister of Agriculture, Thoko Didiza, and the Director General of the Ministry, Bongiwe Njobe for their unswerving support of the conference. This was not only delivered in terms of financial support, but very noticeably through their enthusiastic participation in the conference itself.

Organization

There are four principal groups centrally involved in an international event such as this, the Local Organizing Committee (LOC), the program organizing team drawn from officers and members of the IAAE, the sponsors, and the old and new members who made the commitment and raised the funds to attend the conference. Seven-hundred-thirty-five people from 75 countries registered for the conference, which was higher than expected.

The LOC, led by Professors Gerhard Coetzee and Johann Kirsten at the University of Pretoria did a magnificent job of coordinating all the many tasks to be undertaken in South Africa, and fronting the negotiations with IAAE about finance, facilities, travel advice, accommodation and all the myriad of matters entailed in bringing people from so many countries together. They arranged to enlist Event Dynamics, a professional conference management company to handle all of the many practical routine matters. That practical management led by Melanie Campbell and Helen du Toit at Event Dynamics, with David Goodwin in charge of lighting and visuals, produced facilities at the Durban International Conference Centre, which everybody who attended agreed were the

best experienced at any IAAE conference to-date. That, and the smooth organization and courteous handling of individual needs helped greatly in making this a memorable event. Generous sponsorship for the LOC by the Land Bank, ABSA, Santam, AFGRI, ECI-Africa and SAPPI enabled many extra services to be provided to delegates at the conference. A special word should also be added to thank Professor Johan van Zyl, who in his role as Vice-Chancellor of the University of Pretoria until 2002 had done so much of the preparatory work to bring IAAE to South Africa.

The very fact of this conference being in Africa, where agricultural economics has a particularly strong role to play, stimulated generous sponsorship from many sources, much of it to enable young professionals from Africa and other countries where professional salaries are too low to be able attend. Without the financial support of the Rockefeller Foundation, the United States Department of Agriculture, the UK Departments of Environment, Food and Rural Affairs and of International Development, German Technical Cooperation (GTZ), the W. K. Kellogg Foundation, and FAO and IFPRI, such high levels of participation from developing countries could not have been achieved. As it was, 124 delegates attended from African countries other than South Africa, and 131 from the host country itself. Another 88 came from other developing and transition countries, many with grant support.

The superb conference program was put together by a team coordinated by Vice-President Program David Colman, who chose the theme and was responsible for choosing the 15 plenary speakers, whose papers are published in this volume. Organizing other components of the program involves some very large jobs. We all owe David Colman a lot for his service to the IAAE. His contribution was central for making the conference a success.

The administration of selecting contributed papers was splendidly managed by Michele and Terry Veeman at the University of Alberta. A record 501 papers were submitted, of which 140 were selected after thorough review for presentation as contributed papers, with a further 158 transferred to Professor Kei Otsuka and his team at FASID/Tokyo Metropolitan University who reviewed the poster paper submissions. In addition to the contributed papers transferred for consideration, a further 103 papers were submitted specifically as posters. Out of the number selected 123 posters were eventually presented in Durban, with the majority in a highly professional format. These are very demanding jobs and my sincere thanks goes to all of the above for their generous contribution of effort and time, and also to the 93 contributed paper and 55 poster paper reviewers for their essential input. Wally Tyner at the University of Purdue did a fine job of organizing the 16 invited panels, and the numerous discussion groups and mini-symposia, and Peter Wehrheim of the Universities of Bonn did likewise with the computer program presentations, which are a steadily growing element of these conferences, and of such interest that extra space had to be allocated in the program.

At the conference itself the production of the daily news-sheet, Cowbell, is a pressurised job; thanks go to the joint editors, Nick Vink and Ruvimbo Chimedza-Mabeza, for their expert management.

Pre-conferences and learning workshops

A remarkable development of our conference structure has been the addition of extra or related events. For the Durban conference three learning workshops were organized for the day before the conference proper. One of these on *Water Reform, Institutions' Performance, Pricing and Resource Accounting*, was presented and planned by members of the International Water Consortium, who organized their own conference in South Africa to coincide with ours so that they could also present the learning workshop. 56 people, including presenters, attended this workshop and nearly all remained for the conference.

A second workshop was organized under the guidance of Susan Offutt and the Economic Research Service of the USDA, which provided additional financial support specifically for attendance at this. The workshop, which was attended by 95 participants, concerned *Food Security Measurement in a Developing World Context with a Focus on Africa*.

The third learning workshop on *Analytical and Empirical Tools for Poverty Research*, organized by Christopher Barrett from Cornell University, again attracted over 90 participants. As with the other workshops there were a large number of lecturers involved, and by general consent the workshops were all a great success, which have established the learning workshops as a key element of future conferences. My warm thanks goes to the organizers.

The conference

The opening ceremony was marked in dramatic and entertaining style by the Zulu group African Frenzy, with a generous greeting to the delegates and some stirring stories of local history and dancing. That warm, enthusiastically-received welcome, was then followed by a sincere and well-targeted address by the Minister of Agriculture, Thoko Didiza, in which she adjured the conference to address a whole range of pressing issues, and to support African countries developing policies to accelerate agricultural and rural development. The Minister's commitment to the conference was underlined by a second visit to meet some of the senior IAAE members present to discuss possible areas of collaboration with her staff.

Following this stirring opening it was down to the intense business of the conference. In my own address I underlined the continuing tendencies of farming populations (internationally and within countries) to divide into a mainstream group forging ahead and a very significant group who are left behind. More must be done to reduce the divergence in order to meet the reasonable and proper targets for poverty reduction set out at the beginning of this Millennium. Professor Bruce Gardner presented in the Elmhirst Lecture his innovative research on the extent to which agricultural growth might itself significantly be a crucial factor in reducing rural poverty. Both papers are published below, and are briefly reviewed by the editors of this volume.

Possibly 75% of all participants in the conference played an active role in either presenting papers, leading discussions and chairing sessions. The number of parallel contributed papers, discussion and panel group sessions was larger than at previous conferences, and it was impossible for anybody to attend all the papers that were of key interest. In part this was overcome

by issuing all participants with a CD containing all plenary and contributed papers plus poster paper abstracts. That means that those with restricted access to libraries, but with access to a reasonable computer have readily available permanent copies. The full program is published as an annex in this volume, as a permanent record of the conference, and as an acknowledgement of all who had an assigned role. I would personally like to thank them all.

There were several other innovations at the conference. The Theodore W. Schultz Prize (for the best contributed paper by, as senior author, a young member of the profession) was jointly awarded to two papers (i) Flora Nankhuni and Jill Findeis and (ii) Arnab K. Basu, Nancy H. Chau, and Ulrike Grote. They and the papers of the two excellent runners-up, Timothy Dalton and Ran Tao, were presented in plenary session to great acclaim, and all four will be published in the IAAE journal "Agricultural Economics." In parallel, the Nils Westermarck prize for the best poster paper was inaugurated at this conference and awarded to Matin Qaim. A powerpoint version of that poster and of the two runner-up posters, by (i) M. M. Waithaka, P. K., B. D. Salasya, K. D. Shepherd, S. J. Staal, and N. N. Ndiwa, and (ii) R. J. Armour and M. F. Viljoen, were also presented in plenary session. More computer displays were presented than ever before. These provide valuable opportunities to see new software applications and to meet their designers; they have proved to be particularly stimulating and attractive to new members of the IAAE attending their first conference. There were also changes made to the way in which the proceedings were wound up. Instead of the traditional large numbers of reports, a number of senior IAAE members were asked to summarize their assessment of the conference. Three of these synoptic reviews are published in this volume along with that of my successor as IAAE President, Prabhu Pingali. I wish him and all the colleagues on the Executive as enjoyable and interesting a tenure as I have had.

Honorary Life Memberships were voted to seven members who have given dedicated and distinguished service to IAAE over many years. They are Jock Anderson (Australia), Ian Behrman (South Africa), Ruvimbo Chimedza-Mabeza (Zimbabwe), Csaba Csaki (Hungary), Carl Eicher (USA), Laurent Martens (Belgium), and Kirit Parkikh (India).

It is also noteworthy that the agricultural economics community of Africa came together at the conference and established the foundations of an African Agricultural Economists Association. This may be the most significant impact of the IAAE Conference for years to come. The IAAE has made a commitment to support the African colleagues in the process of the building of this important organization.

It is impossible to complete this brief summary of events without commenting on the exceptionally friendly and informal atmosphere which pervaded this conference. That was highlighted by the way people let their hair down and danced at the superb conference dinner, with excellent bands and entertainers. Many participants, in their formally recorded assessments, and also informally, stated how much they value these opportunities to meet and have access to international colleagues from all across the spectrum. In addition to the many academics and students present, there were large numbers of agricultural economists and development specialists from government, international agencies, NGOs, and commercial companies. It is this mix which makes the IAAE conferences so highly-valued by its members, and it is an attribute I am sure will continue.

Joachim von Braun

Introduction: The 25th conference and the association

David R. Colman and Nick Vink

Members of the International Association of Agricultural Economists have diverse interests and come from many different types of organization and, of course, many countries. Membership of IAAE is individual rather than organizational, and is motivated by a genuine wish to learn and achieve sustainable development through agricultural and rural progress in all regions. The theme of the 25th conference, *Reshaping Agriculture's Contribution to Society* was chosen to embrace the broadest possible set of contemporary issues covering these concerns, with four specific subthemes (1) strategies for reducing poverty, (2) efficiency in food and farming systems, (3) food safety and security, and (4) environmental management. Interestingly, in the last three of these areas the broad societal agendas in developed and developing countries diverge.

The theme was deliberately chosen to highlight the contrasts between the changing social contract with agriculture in richer nations and the harsher set of problems in developing and transitional countries. With regard to the efficiency of farming systems, deepening rural and farming poverty in developing countries and the failure of technology to alleviate these conditions are high on the agenda: while in developed countries concern is more that the rapid march of technological change is causing large reductions in full-time family farming and its replacement by corporate farming, thus changing the social impact of farms in rural areas. Thus the issues of efficient farming differ in key respects. In developed countries many contemporary issues relate to healthier food (e.g., eating less, rejecting genetically modified foods), whereas in many developing countries food security and obtaining sufficient food is still of primary concern. In the area of environment, degradation of resources and water shortage are at the top of the developing agenda, while in many rich countries the increasing concern is for sustaining (and restoring) the environmental quality of farmed areas rather than with sustaining high levels of production. Thus there

are changing agendas for the agriculture sector, which differ by region.

For a conference held in Africa it was natural that the lead subtheme should be *strategies for reducing poverty*, with the unstated corollary that this was poverty among farming families and in rural areas. Happily this theme was also chosen for the two opening papers at the conference, the Elmhirst lecture and the presidential address. The president chooses the Elmhirst lecturer, and Professor Joachim von Braun duly invited Professor Bruce Gardner to undertake this task, which he did in his paper *Causes of Rural Development*. The paper reviews the literature on causes of agricultural growth and its links to rural income growth, and involves a large-scale econometric exploration (only partially reported in the paper) of a panel of data for 71 countries over the period 1980 to 2001. From the standpoint of agriculture's role in reducing rural poverty in general and that of farmers in particular, Gardner's findings and conclusions are somewhat negative. His analysis points to continuing divergence in growth of agricultural GDP per worker between the poorest countries and the higher growth rate in OECD and East Asian countries. In the OECD, at least, this is partly due to the decline in the number of workers, but no simple explanation emerges as to why the poorest countries have not been able to take advantage of the international spread of knowledge and technology to reduce the performance gap in growth, whether that be in cereal yields or GDP per worker. Gardner also fails to find compelling support for the argument that agriculture is the key dynamo for economic development and reduction in rural poverty in the poorest countries and areas. Of course, strong agricultural growth does certainly have a positive effect in this regard, but Gardner concludes it is growth in nonagricultural sectors that is the most crucial factor in raising farmers incomes and reducing rural poverty. Here he draws on his past work in the United States, where it is growth in off-farm income that has supplemented farming income and closed the gap with nonfarm incomes. In East Asia similar opportunities have been created by nonagricultural growth, and, in terms of rural (as opposed to agricultural) poverty, new employment opportunities outside agriculture are probably the key factor. This leads Gardner to produce a very challenging conclusion, namely "But what I am coming to believe is that rural income growth and poverty alleviation are not sub-fields of agricultural economics." This is not a position that sits easily with the conventional view that in poor countries, with a still large proportion of the employed engaged in agriculture, agriculture growth is a crucial motor for general development, or that increasingly its agenda (or that of the agricultural economics profession) should be seen as one for poverty reduction. The panel data approach adopted, in which the statistical weight attached to OECD and East Asia data is high, may not be best suited to testing the basic hypothesis, and Gardner himself declares unease with his conclusions and commits himself to further research.

The president, Joachim von Braun, himself followed this by addressing *Agricultural Economics and Distributional Effects.* One of the key aspects of this paper was to emphasise what he terms "bifurcations," or the contrasting duality of so many tendencies in agricultural, rural, and general economic development. The central divergence highlighted concerned income distribution, with widening inequality between the rich and poor in the last two decades, and the unsatisfactory implications this has in causing poverty (particularly rural) to persist on a large scale in many countries. He observes a bipolar political situation where, in countries with rich consumers, there is an ever smaller number of large farms using advanced and increasingly sustainable science in a competitive environment of integrated markets, while, at the other extreme in poor countries, numerous small farms with little connection to science use unsustainable technologies in fragmented, noncompetitive markets. These inequalities and contrasting trajectories are extensively explored, to be subsequently picked up in varying measure in the ensuing plenary sessions, and in the many other paper sessions at the conference.[1] Thus, one key

---

[1] A selection of contributed papers was published in Volume 31, issue 2/3 of *Agricultural Economics*, while virtually the full set is available as working papers on the IAAE website.

question for the conference was—can the traditional objectives of agricultural policy, to promote food security and efficient production, be reshaped to focus more on general poverty reduction.

Strategies for reducing poverty

As befits a conference taking place in Africa, the first plenary session was concerned with rural poverty and agriculture's role in alleviating it. The three papers in the session addressed quite different, but highly relevant questions. The paper by Christopher Barrett, is titled "Rural Poverty Dynamics: Development Policy Implications." In it Barrett picked up the "bifurcation and duality" themes of the president's address. The paper explores the trajectories of poverty, and argues that there are thresholds. Below certain thresholds, individuals or families are trapped in poverty and may be on a downward path. Others may be in poverty but in a "recovery zone," where they have a high likelihood of escaping upward without policy intervention, or with minimal intervention. The more fortunate lie above thresholds of well-being and asset ownership, where they need not be defined as poor or in need of policy assistance on account of poverty; of course, this does not rule out policies that will accelerate their upward dynamic where that conduces to the general good. The policy implications arise from budgetary pressures that require targeting policies to only those that need help to escape their predicament, and to policies that are effective. Thus, in summary, Barrett argues for "cargo net" policies to assist the chronically poor escape the poverty tray, and "safety net" policies to prevent those transitorily poor (because of diverse shocks) from falling into chronic poverty.

The paper by Alain de Janvry and Elisabeth Sadoulet addresses what is required for "Achieving Success in Rural Development: Towards Implementation of an Integral Strategy." This is done very much from a Latin American perspective. They argue that 80% of the success there has been in reducing the share of rural in total poverty has been due to migration from rural areas, rather from increasing rural incomes, thus re-stating the notion that nonfarm income growth has been the most important force. It is argued that the impact of general economic growth (largely centred on urban areas) has had little direct impact on rural incomes, but has stimulated migration and the shift of much poverty to urban areas. To address the task of increasing rural incomes requires what they call in "integral approach," which shares features of the earlier "integrated rural development" programs of the 1970s and 1980s, but which is driven and motivated more by the poor themselves than by central government or international decision makers. Key to this is devolving governance to local-civil society, investing in the social capital of rural people, and, echoing Barrett's paper, putting in place safety nets to prevent descent into poverty traps. Whether pursuing the integral approach will reverse the neglect of rural development lamented by a succession of IAAE conferences and speakers remains to be seen.

Simon Maxwell's paper rather controversially casts doubts on whether that neglect will be reversed in his paper "Six Characters (and a few more)in Search of an author: how to rescue rural development before it is too late." He argues that there is probably insufficient consistency in the rural development strategies of major policy institutions, or "characters" (The World Bank, EU, FAO, and IFAD), the characters, to convince the major donors and international community to commit a larger proportion of a diminished share of their reluctantly granted aid funding to rural development, even though there can be agreement about the central properties of such a strategy. Maxwell identifies these as growth, empowerment (as with de Janvry and Sadoulet), and security—so all three papers in this session emphasise safety nets. Anticipating the second plenary session Maxwell argues that the evidence is not consistent with support for agriculture, especially for small-farm agriculture as a priority strategy for reducing rural poverty. More widely, the conference noted the need for agriculture to connect to high-valued and rapidly growing markets, something that small subsistence farming by definition has no competitive advantage in.

Efficiency in food and farming systems

Despite its broad title, this plenary session of four papers was primarily concerned with the prospects for small farms. Given the preoccupation in the past of agricultural economists, and indeed of agricultural policy, with the survival of the family farm, this is an issue

that remains a key one, but with much less sentiment than before. It is also an issue that relates to both developed and developing countries, but whereas in the former the future of small farms is more connected to cultural, lifestyle, and environmental issues rather than with supply concerns, in developing countries it is very much linked to the rural poverty issue and the millenium goals for reducing it; in developing countries many of the rural poor have some engagement in small-scale agriculture.

This is very much highlighted in the paper by Shenggen Fan and Connie Chan-Kang, "Is Small Beautiful: Farm Size, Productivity and Poverty in Asian Agriculture." In the two most populous of the five countries studied, India and China, average farm sizes are steadily declining as farms are subdivided in circumstances of growing rural populations. The paper reveals productivity improvements, which at least partially offset the reduction in farmed area, which now averages less than 1.4 hectares in India and less than 0.4 hectares in China. The productivity gains are in part produced by a shift from crop production to livestock, vegetables, and higher-valued products, as well as by technological and efficiency improvements. But a vast number of the farms are now too small to meet family needs, and increasingly families have to rely on off-farm work. Since the near landless are very likely to be poor, this explains why the problem of rural poverty can no longer be tackled primarily by strategies for agricultural growth.

This point is reinforced by Peter Hazell's paper "Is There a Future for Small Farms?" Hazell sets out the reasons why we must still care about small farms. Too many people in developing countries are still dependent (even if only partially) on their farming output and income, for them to be ignored, and will be for many years to come. Nevertheless, as Hazell recounts, there are many factors working to their disadvantage as competitive production units. Thus, at the end of his paper he restates the question—Is There a Future for Small Farms? the unstated implication being that they will remain marginalized and under pressure in a majority of poor countries.

Ulrich Koester's paper "*A Revival of Large Farms in Eastern Europe?—How Important are Institutions?*" considers an entirely different issue about small farms, by examining the persistence and growth of very large enterprises in Russia and East Germany. In Russia the number of farm enterprises of more than 5,000 hectares. grew slightly between 1995 and 2000, while the number of private farms fell, although they increased in average size to 7.5 hectares. This is counter to the expectation that, with liberalization of markets and the economy, private farming grows at the expense of former state farms and cooperatives. What Koester traces out is the institutional and attitudinal factors working in favor of larger-scale new capitalists and to the disadvantage of smaller private farmers.

The fourth paper in this session by D. S. Prasada Rao and Tim Coelli, "Catch-up and Convergence in Global Agricultural Productivity," reports a large-scale econometric exercise to examine trends in productivity and efficiency growth in 97 countries from 1980 to 1995. Its results appear to confound some of the pessimism about the performance of agriculture. For example, although South American countries performed on average worst over this period, there was a little annual growth in total factor productivity and efficiency. Perhaps surprisingly, African countries performed slightly better on both these indicators than Latin America and also had positive technical change. Globally, the overall annual average growth in total factor productivity in agriculture is recorded at 2.7%, with Asia leading the way. One significant conclusion of the research is that there was no slowdown in productivity, efficiency, and technical improvement over the decade and a half, although there were poor years. That is, the argument that progress slowed after the Green revolution is not supported, although an extension of the analysis beyond 1995 will be welcome when the data can be assembled. The authors also conclude that those countries starting from the lowest productivity base did exhibit some catch-up, although this is not readily apparent from the results presented in the paper.

Food safety and security

The four papers in this session addressed widely divergent aspects of food safety, a relatively new area of research interest for agricultural economists.

In the first paper, Jean Kinsey asks: "Will Food Safety Jeopardize Food Security?" In answering this question, she started with definitional issues, as befits a new focus area for research. In her view, food safety is not only about safe food, but also about the safe

consumption of food. Using this definition allows the analyst to add the modern problems of overeating to the more traditional focus on undernutrition and unsafe food (as a result of the presence of microbes, pesticide residues, or foods that have unknown but suspected health consequences such as irradiation or genetically modified foods), i.e., to add chronic problems to the more familiar list of acute problems associated with food safety.

While the cause-and-effect relations for some manifestations of unsafe food (principally food-borne illnesses due to microbes) are well known, others, such as the relationship between food, diet, and chronic diseases and delayed illnesses is less well established, often because of difficulties in measurement. Nevertheless, Kinsey argues that some of these relationships have been established, such as that between obesity and type 2 diabetes, and between obesity and 20–40% of cancers found among U.S. adults. Obesity is, therefore, becoming a major health care issue. In this regard, Kinsey provides truly startling data on the prevalence of obesity in the U.S. population, and estimates that the cost to society of obesity is between 6.2 and 13.5 times higher than the cost of microbial contamination.

In returning to the question posed in the title of the paper, Kinsey argues that the evidence shows that: "... poverty, hunger and being overweight exist simultaneously, and that being overweight jeopardizes health, which jeopardizes the ability to work and be productive, which in turn jeopardizes the ability to earn income to buy healthy food. Therefore, safe consumption of food is compatible and consistent with food security in all parts of the world . . . . If the food available is not safe or its consumption does not improve health, it does not contribute to food security."

The second paper in this session by Michael LeBlanc, Betsey Kuhn, and James Blaylock, entitled "Poverty Amidst Plenty: Food Insecurity in the United States," addresses a similar issue in a very different manner. They start their argument by providing different measures of the prevalence of poverty in the United States, showing that (a) it differs by race, sex, and household head, and (b) that only a small proportion of poverty is chronic ("... the designation persistently poor falls disproportionately on blacks, on the elderly, and on those living in rural areas and in the South"). By extension, most poverty is transitory, with the

evidence showing that poverty spells begin because of factors such as a decline in the earnings of the household head or of other family members, and changes in family structure resulting from the breaking up of a marriage or the birth of a new child. In contrast, poverty spells end when the earnings of the household head or of other family members increase, or when changes in the structure of the family bring in new earners or lead to a smaller household.

In this regard, food assistance programs provide a critical source of relief for food-insecure households. The authors survey these programs for the United States, showing their effect on poverty. For example, the Food Stamp program reduces poverty as measured by the poverty gap index, by about 16%. Nevertheless, they conclude that these programs cannot lead to the elimination of poverty; they also conclude that economic growth alone will not result in the elimination of poverty. In the long run, they argue, only improved returns to labor will have this result.

The third paper in this session differs markedly both in its geographic focus (Europe as opposed to the United States) and in its subject matter from the first two. Johan Swinnen, Jill McCluskey, and Nathalie Francken provided a lively presentation of the way in which the media (in their case newspapers in Belgium) shape opinions on food safety issues in their aptly titled paper "Food Safety, the Media, and the Information Market." Their point of departure is to question the underlying assumption prevalent in the literature on the economics of asymmetric information, namely that information provision is neutral. However, they argue that the providers of information "... have an internal incentive to select certain information items and certain forms of information over others in their information distribution activities." On the supply side, this imperative is driven by the need for profits as well as the preferences of the media organization, while market structure and the preferences of consumers affect the decision making of a media company. On the demand side, they argue that readers of newspapers use information "... to reduce the variance of his/her estimate of truth ..." while the amount of time a reader will spend reading will be determined by equality in the net marginal benefit of reading, work, and leisure.

From this model, the authors derive a set of hypotheses about the behavior of readers and media owners.

First, they show that it is rational for consumers to be imperfectly informed, partly because information is costly, partly because the opportunity costs of acquiring information are positive even if the information were free, and partly because the story may have an ideological bias that diverges from that of the reader. Second, they show that "... the generally recognized tendency of the popular media to publish mostly negative aspects of news items is driven by the demand of their audience, rather than by inherent preferences of the media itself." Third, they argue that if consumers are heterogeneous, the media will be heterogeneous if entry barriers to publishing are not too high. Finally, they argue that the timing of publication of a story depends on a rational comparison of the rewards to getting it out before the competition and the penalty of damaging the reputation of the company because it published incorrect information.

In the final paper in this session, Vittorio Santaniello takes on the subject of "Agricultural Biotechnology: Implications for Food Security." After defining biotechnology and highlighting the critical role played by intellectual property rights in its development and dissemination, he provides a survey of the most critical recent and expected breakthroughs in the development of new products focused on the needs of developing country farmers, and of the literature that has arisen from analyses of these experiences in countries as diverse as Brazil, India, Mexico, Colombia, South Africa, and Argentina.

Santaniello argues further that, while there are positive signs that agricultural biotechnology can foster greater food security for the rural poor in developing countries, the benefits will only reach farmers if the innovations meet farmers' needs and if there is an adequate national research system in place and well-functioning markets for seeds, fertilizer, and other inputs.

Environmental management

The four papers in this session are neatly balanced between two conceptual views, on the meaning of the concept of "sustainability" (Bromley) and on the causes of under-investment in public goods (Lopez); and two applied papers, one on Africa (Ehui and Pender) and one on the Philippines (Rola and

Coxhead). The subject matter at hand, the methods employed and the applications are too diverse to draw any general conclusions from these analyses, but nonetheless they provide interesting insights into the state of the art with respect to environmental issues and the way in which they influence agriculture and the rural areas of the world.

In his inimitable style, Dan Bromley shows in his paper entitled "The Poverty of Sustainability: Rescuing Economics from Platitudes" how the use of the concept of sustainability has evolved to become a term that "...conveys nothing of substance." This is largely because sustainability is focused on natural and man-made capital—a construct that is dependent on institutions ("...to whom does the capital belong? Who may control its use?...")—rather than on the institutional arrangements themselves. In his words: "If we are to understand sustainability we must be concerned with the ways in which humans relate to each other—and why those particular interactions produce particular implications for the natural environment." As a result, "...the ecological dimension of sustainability cannot be considered and understood apart from the social dimension". Of course, this leads to the need to analyze the reasons why specific rules exist in the first place, and what their effect is on people's choices, and a need to understand that sustainability depends on "...*constant change in social and economic institutions, and not in their preservation.*" (Emphasis in the original)—hence the need for an "evolutionary environmental economics", something which is "...impossible in the equilibrium models and metaphors of contemporary economics."

Bromley summarizes his argument as follows:

Sustainability can be rescued from platitudes and incoherence by rediscovering the evolutionary predecessors of the ordinalist revolution in economics...I used to believe that conversations about sustainability were conversations about what is worth saving for the future...I now insist that sustainability is best thought of as *looking for those aspects of our natural and constructed settings and circumstances for which we can, at the moment, mobilize the best reasons to make sure that they are passed on to future persons.* This is not a process in which we seek to maximize time paths of consumption or welfare into the infinite future. It is, instead, a process

in which we search for the best reasons to bequeath a particular endowment bundle to those who will follow. And that task is precisely the subject matter of a properly constituted evolutionary economics....

Ramon Lopez poses three key questions as an introduction to his "Under-Investing in Public Goods: Evidence, Causes, and Consequences for Agricultural Development, Equity and the Environment." The first is "...why have the environmental effects of agriculture been so negative in most developing countries?" He argues that this question cannot be approached using the conventional externality toolbox, and that the performance of the sector and underlying political economy factors need to be taken into account.

The second question rests on the empirical observation that, even in middle-income countries, agricultural growth has benefited those working directly in the modern agricultural sector as well as urban dwellers via the market for unskilled labor, but not the rural poor, who often constitute the majority of the rural population. As Lopez argues: "It is thus paradoxical that rapid agricultural growth, whenever and wherever has occurred, has been good to reduce poverty in non-rural areas but it has been less powerful in promoting higher incomes among the poorest segments of the rural population. The second question is now natural: Why has agricultural growth not benefited these groups?"—this despite the removal of macroeconomic distortions in many parts of the world. This leads to his third question: "Why has such slow growth continued even in countries that have removed anti-agriculture macroeconomic biases?"

Lopez then argues that the reasons for the weak rural poverty-reducing impact of agriculture, and the resultant negative impact on the environment, "...are associated with a more fundamental distortion in the allocation of public expenditures that leads to a chronic under-supply of public goods. Investments in public goods get crowded out from government budgets by massive expenditures in subsidies to the wealthy and other expenditures in private goods that play no role in ameliorating market imperfections. In turn, the under-supply of public goods is at least in part related to political economy forces." In his view, these political economy forces consist of the ability of rural elites to capture the benefits of most public expenditure in a manner that provides them with government-funded private goods. Poor farmers therefore face a "double crowding out" as the limited government budget provides insufficient public goods to the rural poor, while the government crowds out private provision of private goods for the poor with its subsidies to the elites.

In their paper entitled "Resource Degradation, Low Agricultural Productivity and Poverty in Sub-Saharan Africa: Pathways out of the Spiral," Simeon Ehui and John Pender report on detailed research in East Africa aimed at finding ways out of the spiral of rural poverty that afflicts the subcontinent. Their research shows that strategies need to be tailored to the circumstances and needs of the different parts of the subcontinent, although there are common elements that include: ". . . assurance of peace and security, a stable macroeconomic environment, provision of incentives through markets where markets function, development of market institutions where they do not, and public and private investment in an appropriate mix of physical, human, natural and social capital."

Finally, Agnes Rola and Ian Coxhead, in their paper "Economic Development and Environmental Management in the Uplands of Southeast Asia: Challenges for Policy and Institutional Development" use a case study from the uplands of the Philippines to explain the factors responsible for recent changes in economic behavior and institutional arrangements in these areas. These changes include a shift from more traditional farming practices (long-phase forest fallow rotations regulated by customary law) to more intensive commercial operations. The latter are the result of in-migration, leading to the displacement of traditional institutions that governed land and resource use by both de facto and de jure means. One adverse result has been that there are now fewer constraints on natural resource use, and hence more environmental degradation. Proposed solutions for a win-win scenario rest on (i) better accountability requires genuine decentralization, (ii) addressing the externality problem requires watershed-based institutions and policies, and (iii) market-based mechanisms that can support sustainable upland management, but only if the appropriate institutions are in place.

Agricultural economics and distributional effects

Joachim von Braun*

Abstract

The paper examines the main issues surrounding distributional effects in the domains of natural resource management and land policies, agricultural technology and research policies, agricultural market and trade policies, and consumer-oriented policies, including standards, subsidies, and labeling. Agriculture is drifting into an ever more drastic bifurcation at a global level and within many countries. Correcting that bifurcation will require large investments in rural areas and rural people, in institutions, and in information and biological technologies accessible by the poor in the world's smallholder sector. Large and growing national and international inequalities related to agriculture and rural areas threaten peace, growth, and sustainable development.

JEL classification: Q18

Keywords: income distribution; economic development; agricultural technology; agricultural markets; natural resource management; land ownership and tenure; agriculture in international trade; agricultural policy; food policy; bifurcation

1. Why revisit distributional effects?

Agricultural economists have always been concerned with enhancing the productivity and efficiency of agriculture, and rightly so, as these are essential for increased wealth and human welfare. Efficient allocation of resources drives the spatial distribution of economic activity and the rents earned by factors of production, including labor (von Thünen, 1826, 1850). But the founders of our association were also concerned about the effects of agricultural change on income distribution (Ashby, 1930), because then much of world poverty was concentrated in rural areas as is the case now. Over the past seven decades, however, the representation of distributional effects at our conferences has been rather uneven and may be more a product of zeitgeist—the spirit of the times—than of the actual nature and scope of the issues.[1]

The approach toward achieving equity through agricultural policies during the twentieth century has been largely to allow growth and markets to generate an (Pareto-) efficient equilibrium, and to rely on redistribution policies to take care of adverse distributional consequences. Kuznet's curve thinking reinforced the neoclassical concept that economic growth would eventually (after rising initial inequality) result in more equal income distribution. To a large extent, the issue of income distribution was "out in the cold"—as more of an adjunct to economics than a research priority in its own right (Atkinson, 1997). After decades of economic growth in much of the world, it is witnessing a dramatic splintering of income equality, both internationally, and intranationally, and a declining progress in the reduction of poverty and hunger (Kanbur and Lustig, 1999; Pinstrup-Andersen and Pandya-Lorch, 2001). Globally, the incomes of the world's richest 1% of earners are equivalent to those of the poorest 57%, and international inequality, which had remained rather stable with the Gini coefficient of world income distribution of about 0.46 between 1950 and 1985, has increased dramatically by 17%

International Food Policy Research Institute (IFPRI), 2033 K Street, NW, Washington, DC 20006–1002, U.S.A.

[1] An account of distribution-related themes, including poverty and development themes, in IAAE conference volumes shows the following years as most prominent: 1947—47%; 1955—43%; 1982—65%. Since 1985, there has been a low level of papers covering distributional effects: 1985—14%; 1988—9%; 1994—22%; 1997—18%;

2000—13%. (These figures are drawn from the papers published in the IAAE proceedings volumes.)

(to 0.54) over the past decade (Milanovic, 2002, 2003). The average per capita income in the industrialized nations was 9 times the sub-Saharan African average in 1960; the disparity has since doubled to 18-fold (UNDP, 2001). In addition, with rapidly expanding communications networks, the recognized standards of comparison relative to other world citizens changed for billions of people—especially in rural areas—and rendered relative deprivation increasingly relevant for welfare perceptions.

Economic growth is necessary to reduce poverty and inequality, but it is not sufficient (Chen and Ravallion, 2000; Ravallion, 2003).[2] In the absence of appropriate policies, institutions, and public investments, the highest income earners capture the lion's share of the benefits of economic growth. Agricultural growth reaches the poor hardly better than nonagricultural growth does when income distributions are highly skewed (Gardner, 2000; Timmer, 2002). These facts, combined with a rapid regional and global change in the characteristics of agriculture, particularly in the past two decades, collectively demand a reexamination of the interaction between agricultural policy and distributional effects, as well as a fresh look at agricultural economics' potential contributions to this field. This paper argues that:

- We ought to rethink the nature of the relationships between growth and distribution in agriculture and the rural space, because of fundamental changes in structures and dynamics of agriculture and food systems, driven by new technologies and institutions;
- We are confronted with large and growing national and international inequalities that are related to agriculture and rural areas, which threaten peace, growth, and sustainable development; and in view of these observations;
- Our profession is probably underresearching distributional effects.

Three categories of distributional effects are relevant for this discussion: (1) variance (distributional equality or inequality, typically measured by Gini coefficients);

(2) absolute deprivation (i.e., poverty, including consideration of how far below an accepted cut-off point a subset of population may fall),[3] and (3) relative deprivation (the patterns of distances within the distributions that may affect people's aspirations and perceptions).[4] All three categories are subjects for economic study in their own right, and they will selectively be referred to throughout the paper.

Examining the major domains of agricultural policy making provides an instructive overview of the main issues surrounding distributional effects and agricultural economics. These include:

1. Natural resource management and land policies;
2. Agricultural technology and research policies;
3. Agricultural market and trade policies; and
4. Consumer-oriented policies, including standards, subsidies, and labeling.

Policies related to public goods cut across these four domains. The vector of the relationship between growth and distributional outcomes—subject to initial conditions and context—can be depicted in their interaction with these policy domains, as shown in Figure 1.

I will selectively discuss these domains from global, regional, and household perspectives; explore the scope for new areas of research; and consider the interactions of agricultural policies and their distributional effects, as these domains are interlinked and partly reinforce and partly counterbalance each other.

2. Transformation of world agriculture with bifurcations

The institutional and political context within which agriculture operates has changed rapidly in the past two decades. The bipolar global political system has come to an end, as have the devastating experiences with

---

[2] Chen and Ravallion (2000), drawing on 265 national sample surveys spanning 83 countries, found that although there was a net decrease in the overall incidence of consumption poverty over 1987–98, it was not enough to reduce the total number of poor by various definitions.

[3] See the important contributions by Amartya Sen (1997) to this category.

[4] See Stark and Wang (2000). The idea is that a comparison of the income of i (an individual, a household, a family) with the incomes of others who are richer in i's reference group results in i's feeling of relative deprivation. The associated negative utility impinges, for instance, on migration behavior. Stark shows that the relative deprivation of an individual (or, for that matter, of a household or a family), whose income is y, is where $F(x)$ is the cumulative distribution of income in y's reference group.

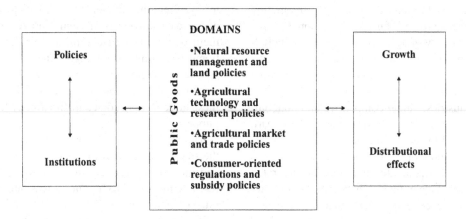

Figure 1. The conditioning of agricultural growth and distributional effects.

central planning of economies and societies (which often included disastrous experimentation with agriculture). Other political changes include the introduction of democratic systems, the strengthening of the rule of law, and—very relevant for rural areas—moves toward decentralization, devolution, and privatization (von Braun and Grote, 2002). In many countries, national governments are devolving authority to subnational and local governments or ceding roles to the private sector, civil society, or—especially in the case of natural resource management—user groups (Meinzen-Dick et al., 2002; Birner, 2003). As national governments in developing countries have reduced their economic and social roles, nongovernmental organizations (NGOs) have helped sustain vital social safety net programs and played an important role in local development activities. However, NGOs are limited in their ability to take on the task of complete public goods provision (Paarlberg, 2002).

Agricultural economics research must address changes in political systems, as well as the greater diversity of actors, the more complex context in which food systems and food policy operate, and the capacity of local organizations to take up new roles. Such research must engage stakeholders as active participants and not simply as objects of study. Hedley (2001) pointed out these important issues in his International Association of Agricultural Economics presidential address.

The agriculture and food system is increasingly changing from a relatively large and distinct sector of the economy into a more pervasive, integrated system,

in which resource use and ecosystems functions are linked to consumers via extended food and service chains with multiple market and nonmarket institutions shaping the system. Essentially, a development is underway from a linear relationship between farmers, markets, agro industry, and consumers toward systems of interaction between and among these four, with policy making and institutional innovations cutting across the system in more complex fashions. These developments proceed to a different extent and at different speeds in different parts of the world; and when technology and education investments are low, the transformation of agriculture proceeds slowly at best (Schultz, 1964), which is one reason for bifurcations.

Viewing the domain of agricultural economics as defined by a narrowly delineated agricultural sector puts an unnecessary constraint on the scope and relevance of the profession. Even the economics of traditional and subsistence agriculture must be studied beyond the farm level, or much of the value-added of household processing and ecosystem functions would simply not be accounted for. The economics of small- and large-scale modern agriculture, in which much of the value-added is created farther down in the value chain, artificially reduces itself when defining "agriculture" as isolated from that chain. Thus, our profession should not define its scope by a narrow statistical concept of agricultural sector production but embrace the whole food and agriculture system.

The increased challenge confronting agricultural economics is the bifurcation of fundamental agricultural developments:

- At the farm level around the world, the distinction between large and small farms is ever more stark.
- The sustainability of agricultural ecologies, including water management, appears to be on the decline in many parts of the world. Sustainable and nonsustainable systems exist in parallel.
- In food industries, global competitive and noncompetitive industries move farther and farther apart, as exemplified by the technologies and institutions used in local and global food industries.
- Markets that are nationally and globally integrated coexist much longer than expected with nonintegrated (subsistence) and local exchange systems.
- The share of consumers who are poor remains high and is increasing in some regions of the world.

The number of people operating at a marginal level (the third column of Table 1) is larger than the number operating at a dominant level (the second column). Yet the economic weight of global agricultural systems depends much more on the small number of dominant actors. Agricultural economics may be driven unduly by economic weight rather than by relevant population shares. The bulk of our profession's research efforts focus on the structures and actors in the second column—the small subset made up of large farms, sustainable agro-ecologies, users of advanced science, integrated markets, competitive industries, and rich consumers—and much less on the large subset made up of small farms, nonsustainable agro-ecologies, users disconnected from science, fragmented markets, noncompetitive industries, and poor consumers.

The distributional effects of agricultural change cut across these bifurcations in new ways. Agricultural policy at a global, national, and local level is confronted with new, far-reaching distributional effects both in

Table 1
Stylized bifurcations of world agriculture

Agricultural domains	Dominant	Marginal
Farms	Large	Small
Agro-ecologies	Sustainable	Nonsustainable
Technologies	Using advanced science	Little connected to science
Markets	Integrated	Fragmented
Agro-industry	Competitive	Noncompetitive
Consumers	Rich	Poor
People directly affected	Few	Many

terms of regional and intertemporal distribution effects of policies. Institutional change is an important element of this. Understanding and predicting distributional effects and providing guidance for efficient and equitable policies will require agricultural economists to expand their toolbox. New approaches to modeling these effects are needed and are already being actively pursued.

The remainder of this paper will assess changes in the four major domains of agricultural policy making and their distributional effects, and will describe the research implications for our profession. At the outset it must be stated, however, that analyses of distributional effects in agriculture are impaired by lack of sound and comparable statistical information, which may be a consequence of the lack of demand for such data by policy makers and researchers.

3. Distributional effects of natural resource management and land policies

Access to land and natural resources, it could be argued, is of decreasing importance for agricultural distribution effects because of the growing technology and knowledge content of agricultural production and processing. The flipside of this hypothesis suggests that access to land and natural resources remains of great relevance to poor farmers who have little access to technology to date. In low-income countries, the distribution of land still matters greatly for income distribution and for poverty reduction (Binswanger et al., 1996; Carter, 2000; de Janvry et al., 2001).[5] The world now contains about 460 million farms. Table 2 depicts their estimated size distribution (not controlled for quality of land). Eighty-five percent of the world's farms are smaller than 2 hectares, and of these farms smaller than 2 hectares, 90 percent are in low-income countries and 10 percent are in middle and high-income countries. These small farms require special attention by agricultural and rural development policies to facilitate pro-poor growth.

Changes in *farm size inequality* reveal that bifurcation is increasing. In developing countries average farm

[5] In India landlessness is the best predictor of poverty: 68% of landless laborers are poor, compared with 51% of scheduled castes and tribes and 45% of illiterate households (World Bank, 1997).

Table 2
An approximation of world farm size distribution, late 1990s

Farm size (hectares)	% of all farms	Number of farms (millions)
<1	73.2	333.95
1–2	11.7	53.29
2–5	8.9	40.28
5–10	3.0	13.77
10–20	1.5	7.12
20–50	0.8	3.72
50–100	0.4	1.67
100–1,000	0.4	1.98
>1,000	0.1	0.30
Total	100.0	456.07

Sources: Estimates based on FAO World Agricultural Census, 1990; Supplement to FAO World Agricultural Census (various years) and various country statistics.

Table 3
Gini coefficients of income distribution and land distribution for selected countries (different years in the 1990s)

Country	Index of Gini coefficient of land concentration	Index of Gini coefficient of income
Nepal	45.0	37.0
Ethiopia	47.0	44.0
Thailand	47.0	41.0
Philippines	55.0	46.0
India	58.0	38.0
Turkey	61.0	41.5
Peru	86.0	46.0
Paraguay	93.0	58.0
Japan	59.0	25.0
United Kingdom	67.0	37.0
Germany	68.0	30.0

Sources: World Bank (2002); Lipton (2001); IFAD (2002).

size is generally shrinking, whereas in high-income countries it is increasing. The changes in average size come with a mixed pattern of inequality: in a sample of 32 developing countries with information over time, only 11 countries show decreasing inequality. Furthermore, in many countries a relatively small proportion of large and growing farm units coexist with a large number of small farm units, which often provide important shares of household income and are managed by households whose members hold multiple jobs.[6]

Land reforms have long been major mechanisms to redistribute assets. Some weak positive correlation between land distribution and income distribution remains in low-income countries, but the relationship has become less relevant in middle- and high-income countries (Table 3). Land reforms remain dependent on the distribution of political power and typically happen in politically volatile circumstances. The past two decades have seen large policy changes in relation to land distribution and landownership as elements of larger societal transformations in many parts of the world—for instance, in China, Vietnam, the former Soviet Union and Eastern Europe, Southern and East Africa, and parts of Latin America. Many of these processes are far from complete, and agricultural economics research could deliver large benefits by providing guidance on productive and sustainable land use, land market, and land tenure systems.

Gini coefficients of landownership are typically higher than the Gini coefficients of income distribution.[7] This difference occurs because people with little or no land have other income sources, and (especially in middle- and high-income countries) these sources are more significant than the distribution of land.[8] Moreover, even where land is not a major source of income, land reforms that provide at least some landownership—even homestead sites—can be important for improving the security, status, and bargaining power of asset-poor households (Hanstad et al., 2002). Even in generally land-rich sub-Saharan Africa, the relationship between land distribution per capita and consumption levels remains very close, and this relationship is also found in parts of South Asia.[9] However, many poor rural households are unable to gain sufficient access to land when such access could be their best option for escaping poverty (Bardhan et al., 1998; de Janvry et al., 2001; Binswanger et al., 1995).

Levels of *initial* land distribution have proved to have a significant impact on economic growth. Deininger (2003) finds a strong relationship between initial land equality in developing countries and economic

[6] This applies, for instance, to large parts of Eastern Europe, especially Russia (von Braun and Qaim, 1999).

[7] In a sample of comparable landownership Gini coefficients and income Gini coefficients of countries, the former are on average 0.60 and the latter are 0.40.

[8] See Gardner (2002) for the United States.

[9] In East Africa off-farm income of farm households is lower than farm income even in farms smaller than 0.2 hectare per person (Jayne, 2001).

Joachim von Braun

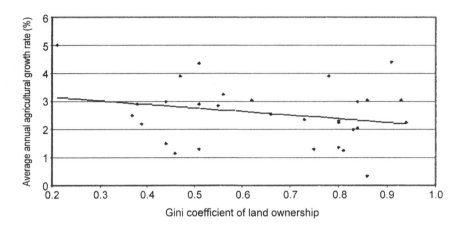

Figure 2. Inequality of landownership (1980s) and average annual agricultural growth rate for selected countries, 1989–2000.

growth—especially in Asia—between 1960 and 2000. However, in this more recent sample, changes in farm size inequality, as measured in Gini coefficients of farm size, and growth show at best a weak negative relationship (Figure 2). Furthermore, there is no apparent relationship between *changes* in Gini coefficients of land and *changes* in Gini coefficients of per capita income (Figure 3). This may be explained by a reduced relevance of land distribution for income distribution in many countries (but certainly not everywhere) in recent decades.

The *institutions* accompanying land (and other resources) seem to matter more for distributional outcomes than the mere distribution of the resources themselves. The distribution of land and other assets matters not only regionally and between households, but also within the household (Haddad, 1999). Research has shown that women's lack of landownership reduces their productivity as farmers by restricting their access to credit, extension advice, and decision making opportunity. Moreover, where women have independent rights to land or are recognized as co-owners with

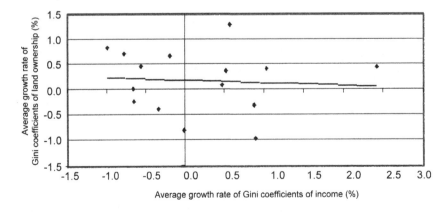

Figure 3. Rates of change in inequality of land ownership and income for selected countries, 1970s–90s.
Note: Selected countries: Argentina, Brazil, Egypt, Ethiopia, India, Indonesia, Korea, Lesotho, Mexico, Nepal, Pakistan, Peru, Philippines, Thailand; countries were selected such that time periods covering land Gini changes overlapped substantially with income Gini changes, but often not identical periods between 1970 and 2000.
Sources: Lipton (2001); IFAD (2002); Deininger and Squire (1998); WDI (2002); and Adams (2002).

their husbands of land they also have more bargaining power within the household, which has been shown to increase the proportion of household income spent on food, education, and the welfare of children (Quisumbing et al., 1995; Meinzen-Dick et al., 1997).

Quality of land is, of course, not homogeneous or constant over time. Soil degradation reduces agricultural productivity and affects about 25% of the world's agricultural land. Between 5 and 12 million hectares of arable land are lost each year as a result of salinization, flood-induced erosion, or nutrient mining. These factors also reduce productivity on an estimated additional 20 million hectares annually. Water and wind account for 80% of all erosion. Slow-onset disasters caused by soil fertility destruction are possible in some regions. Research is needed on policies for landscapes and land use that protect the world's soil fertility, promote integrated nutrient management, assure that poor farmers have information about plant nutrient use in various production systems, and foster efficient and effective plant nutrient markets (Scherr, 1999).

Access to water at low private cost in irrigated agriculture remains a major factor in the distributional outcomes of agricultural policies in the domain of natural resources. Initial distribution in irrigation settings tends to be rather equal because of the key role of public institutions in providing irrigation and distributing irrigated land. As land changes hands over time, this situation can change. Yet the rather equal distribution of irrigated agriculture implies also fairly equal distribution of access to irrigation water in these settings. Increasingly, however, water resources in developing countries are being privatized, which may reduce the equal distribution. In addition, a growing urban/rural bifurcation with regard to water usage has profound implications for the distributional effects of water policy. Increasing urbanization in developing countries has dramatically increased the demand for water and may jeopardize rural communities' access to water for agricultural purposes and thus undermine income-generating opportunities for the rural poor (IFAD, 2001).

Climate change and related policies could have a variety of important implications for agriculture, but the distributional effects of climate change, mediated through agricultural adaptation or the lack thereof, are barely understood today. It is likely, however, that climate change has had (and will continue to have) its most severe impact on the tropical and developing regions of the world, where its effects may include decreased water availability, increased risk of natural disasters such as flooding, and an increase in health hazards such as tropical and waterborne diseases (FAO, 2002). Complex multidisciplinary modeling might improve our understanding of the impacts of global and regional climate change, including its impacts on agricultural inequalities. Agricultural economists should engage more actively in these modeling activities to add innovation based on sound understanding of agriculture. Research is needed to better explain how technology, trade, and insurance can help facilitate global and local adaptation to climate change. The challenge is to provide the information needed to design effective insurance schemes and to offer policy options for ensuring that poor farmers have access to climate forecasting and other tools that can help manage risks.

Long-term food security depends on the availability and efficient use of diverse *plant genetic resources*. Although this is a global policy issue, policies on conserving and using plant genetic resources are partly national and partly local and involve interplay between public and private actors. Multidisciplinary research—with economic, legal, ecological, and technological expertise—is needed to devise sustainable and fair solutions. Researchers should examine appropriate governance options for these globally important resources and identify sustainable, efficient, and equitable outcomes for low-income countries, farmers, and consumers. The chronic underinvestment in these global public goods contributes to growing inequalities, because large farmers, who are linked to the growing private seed industry, are much less dependent upon public investments in seed improvement.

Policies concerning the *multifunctionality* of agriculture encompass land use, water use, genetic resources, and the management of forests and landscapes. The policies and regulations related to multifunctionality entail costs and benefits, but the distribution of net benefits today and for future generations is difficult to assess. Many of the direct costs of multifunctionality are paid by the general taxpayers. Implicit consumer taxation (through price protection) and shifting of the cost of regulations to farmers are also significant. On the other hand, the benefits of rural amenities, including long-term natural resource protection, biodiversity, and landscape beauty, are society-wide in nature and only are partly captured by the actual "users" of multifunctionality, including ecosystems services. This remains a complex area where new agricultural economics approaches to natural resource valuation can make a contribution (Randall, 2002).

It is possible to interpret agriculture's contributions to growth and its distributional effects as part of multifunctionality. Whereas multifunctionality has been seen partly as a mechanism for protecting agriculture in the Organization of Economic Cooperation and Development (OECD) countries, in a different sense it is also relevant to poor people in developing countries. Much of its relevance hinges on protecting customary access and use rights to land, water, and other biological resources. Although individual property rights related to land and access to water are increasingly well defined in world agriculture, governments often do not recognize the property rights of communities. Community rights to regulate uses of private land or to manage common property require collective action in order to be established and managed. Because of the heavy dependence of the poor on common property resources and the environmental importance of such resources, agricultural economics and sociological research on the formation of such institutions, including those for managing biodiversity, watersheds, and variable landscapes, will continue to have high payoffs (Meinzen-Dick and Bakker, 2001; Otsuka and Place, 2001; Meinzen-Dick et al., 2002).

Agricultural economics research today is challenged by new developments in the domain of land and natural resource policy, and the distributional implications thereof. The efficiency and equity implications that result from the effects of agricultural and natural resource policy are increasingly complex, especially because land and natural resource policies have joint effects for private and public goods. Further, agricultural systems are less stable in developing countries, where agriculture has a precarious and complex relationship to the natural environment, and where institutional ability to manage the interface between environment and agriculture is often weakly enforced (Lopez, 2002). To assess and predict these effects is a challenge. By confronting this challenge with an expanded understanding of institutional economics approaches, as well as with multidisciplinary research that helps explain long-term returns and risks for the resource base, our discipline may be led into new methodological and empirical research of great relevance.

4. Distributional effects of agricultural technology and research policies

Extraordinary new technologies—in areas such as molecular biology (including genetic engineering, tissue culture, and marker-assisted breeding), information, communication, and energy technology—are revolutionizing global productivity. Many of these technologies have a direct bearing on agriculture and food systems. The development of new technology in food and agriculture can be a political matter, however, because of perceived risks and perceptions of nontransparent benefits, the scope for regulation, and a naturally incomplete knowledge base in the new and fast-changing sciences. Acceptance and adoption of new crop technologies, including genetically modified crops, will depend upon transparent and convincing advantages for consumers and producers. In many cases, including those involving genetic engineering, consumer organizations, the media, and retail industries—rather than farmers, agro industries, and food processors—are playing an increasingly important role in technology policy.

The impact of new technologies on distribution will depend largely on the existence of policies that ensure that technologies are developed and disseminated and can actually be adopted by poor producers. With the emergence of more vibrant public agricultural research institutions serving the smallholder sectors in many countries, the gap between those with much access to technology and those with little access may have narrowed for an intermediate period in the 1970s and 1980s. With the breakthroughs in information and communications technology and biotechnology in the 1990s, however, as well as the institutional innovations along the food chain, the gap has widened again. Although agriculture and the rural economy have the potential to "leap frog" over several generations of technological advances, they are currently on the wrong side of the digital divide. Agricultural economics is challenged to take a fresh look at the growth and distributional effects of the advanced technologies and their interaction in agriculture and agriculture-related industries.

In the past two decades, *information and communications technologies* (ICTs—i.e., the telephone and the Internet) have probably changed food and agricultural systems more than biotechnology has so far, even though the latter dominates the agricultural technology debates. Access to information and the ability to use it efficiently are critical for allocating resources, whether labor, capital, or natural resources,[10] under market or nonmarket conditions, and for access to public goods. As tools to help access and use information more efficiently, ICTs impinge on all of the determinants of distribution, and by providing new channels of information at reduced costs they can be expected to reduce inequality. Unlike roads, ICTs generate network externalities, which means that their returns can increase over time. The concept of network externalities is critical for the distributional effects of ICTs in rural areas (Torero and von Braun, 2004). More specifically, ICTs contribute to

- Lowering the costs of market participation for farm households and small rural enterprises;
- Reducing costs and improving quality of public goods provision (such as research-extension linkages in agriculture, and education and health services);
- More effective use of existing social networks or their expansion; and
- New institutional arrangements and consequent strengthening of peoples' rights.

It is important to keep in mind that access to information through ICTs is a question not only of *connectivity,* but also of the *capability* to use the new tools and of *content* or relevant information in accessible and useful forms. Although all these three "c"s are critical, it is connectivity that matters most for the poor, given that it is a precondition for the others. In terms of public access, the incidence of phone use by rural households is already substantial in rural areas of six developing countries studied, ranging from 33% in Laos to 69% in Peru (Torero and von Braun, 2003). Technological change in ICTs brings down unit prices and stimulates new products; this is partly facilitated by foreign and domestic research and development policies, as well as by the prevailing regulatory framework. Digital divides within a country—between poor and nonpoor, urban and rural, educated and uneducated—seem more prominent in developing countries. Whether

rural households benefit from increased access to a telephone depends on whether the gain from having a telephone closer by is higher than the direct and indirect costs of the phone use. Based on household data analyses of rural phone programs in Bangladesh and Peru, Chowdhury (2002) finds that households in Bangladesh have a net benefit of US$0.11 to US$1.59 per call, and Peruvian households a net benefit US$1.45 to US$2.91 per call, depending on the applicable communication alternatives (such as sending a letter, travel to distant other phone, messenger). The poorest quartiles tend to benefit more. Moreover, participation in land and labor markets increases by at least 8% in the Bangladesh sample because of access to a phone, and high-value produce from small farms (eggs, milk, poultry) is channeled to markets at reduced transaction costs (Chowdhury, 2002).

Some observers have suspected that many *agricultural technological innovations* for farmers favor large-scale agriculture, but a closer look at the net distributional benefits suggests more scale-neutral distributions. This closer look requires a careful assessment of both the direct effects on agricultural incomes of farmers and workers and the indirect effects of technology on declining prices of food (Hazell and Ramasamy, 1991; Hazell and Haddad, 2001). The Green Revolution did, in fact, have positive distributional effects for small landowners in Asia. The initial and well-known lag in adoption by small farmers is not necessarily indicative of the sustained income distribution outcomes of technology. The conditions that facilitated the positive income distribution effects of green revolution technology were the relatively even distribution of land, the complementary roles played by research institutions, and investments in basic infrastructure. Negative distributional effects of green revolution technology occurred where poor agricultural smallholders shifted from tenant to nonagricultural rural households, unless there were accompanying increases in nonfarm employment. Moreover, the differences in irrigation resources often led to widening disparities between villages and subregions. An assessment of household- and community-level effects without interregional assessment gives a distorted picture of the distributional outcomes of the technologies.

Attempts to narrowly target agricultural technology to low-income and resource-poor farmers show mixed

---

[10] A large share of cell phone conversations among small farmers in Bangladesh are on land issues (Chowdhury, 2002).

results.[11] Technology and its benefits are often captured by those who have access to input and output markets and who can mobilize the investment resources required for technology use. This situation has, in many instances, led to less than proportional benefits for poor and women farmers. Again, the final outcome of these effects for distribution is a question of consumption and investment patterns resulting from the overall income streams. These patterns include human resource investments in, for example, education and health, and other expenses that may enhance the long-run, even intergenerational, distribution effects of the benefits of technology (Hazell and Ramasamy, 1991; Kerr and Kolavalli, 1999). Thus, it is necessary to look at the assets required to benefit from a technology. Beyond scale neutrality, it is necessary to consider whether water control, large labor or cash inputs, market access, and education are needed, as each of these requirements are likely to exclude some of the poor. People with fewer assets may have fewer options, but policies can help them overcome the barriers.[12]

The distributional impact of technological change ultimately depends on the particular context of policies, markets, and institutions and on interregional connectedness through infrastructure (de Janvry and Sadoulet, 2002). This leads to *diverse patterns by world regions*. In large parts of Africa, society-wide technological gains in agriculture remain limited because of a lack of interconnectedness, low efficiency, and inequity in taxation and public investment policies. Still, even minimal access to, for instance, irrigation technology can have a decisive effect for coping with drought and famine. In Latin America, many of the poorest rural hinterlands have remained poor because of a lack of access to markets and public services, such as education, health, and human capacity building that are needed complements of technology for pro-poor

growth. A sizable share of the rural poor have remained excluded, and that exclusion may have even increased in the past two decades, despite many initiatives for administrative, fiscal, and political decentralization to foster progress in rural areas. In large parts of Asia, technological change has directly and indirectly led to rural growth combined with poverty reduction, including for landless rural people. In much of Eastern Europe and the former Soviet Union, after the economic transformation of the late 1980s and early 1990s, new technology only reached the large-scale sector. Agricultural technology and related services hardly reached the smallholder sector (where it existed, in countries such as Poland) and the sizable household agriculture sector.

A long-run view of technological change must also take into account the distributional effects of *agricultural research investments*. These research investments go beyond technology and include institutional innovations and the structure of the scientific system catering to agriculture. The benefits of agricultural research investments are large and undisputed, but their actual levels and distributional effects remain under discussion, as became very evident at the last IAAE conference in Berlin (Alston and Pardey, 2001). From a global perspective, it is important to recognize that there is an ever-greater disparity between private and public research and between developing and developed country agricultural research. Developed countries spend about 47% of the US$22 billion spent globally on public agricultural research, and spend more per farm and per unit of output than do developing countries, where spending is dominated by a few large countries including Brazil, China, and India (Pardey et al., 2002). Further, the growth of private investment in biotechnology has exacerbated the inequities in agricultural technology and research between developed and developing countries.

As biotechnological investments in agriculture have raised the private sector share in overall agricultural research expenditures over the past decade, the products from these investments have hardly reached smallholders. Most biotechnological research has yet to target cassava, sorghum, pearl millet, pigeon peas, or groundnuts, which are five of the most crucial crops in most of the developing world (Qaim et al., 2000). Transgenic crops are for the most part soybeans, corn, and cotton, which are found predominantly in the United States

---

[11] Attempts to target irrigation and rice crop technology to women farmers in West Africa did not succeed as well as expected because of the complexities of community and intra-household institutions of power, cost sharing, and lack of legal enforcement of contracts (von Braun et al., 1988). In Bangladesh, targeted agricultural technologies (such as improved vegetables) and ICTs (such as cell phones leased to low-income women) had an empowering and distribution-enhancing effect at the local level (Bayes et al., 1999; Hallman et al., 2002).

[12] Micro finance, for instance, can overcome barriers to credit for the poor (see Zeller and Meyer, 2002), and marginal lands can catch up if properly considered in research and investment strategies (Pender and Hazell, 2000).

(68%), Argentina (23%), Canada (7%), and China (1%) (Juma, 2001). The "fruits" of biotechnology thus tend to be contained within similar geographic areas, and exclude the subtropical and tropical regions of the world, which are also the poorest.

The growing private sector influence and the decline of the public sector in agricultural research investments underscores the bifurcation in world agriculture, and entails severe distributional consequences for much of the developing world. One of the challenges for agricultural economists is to identify institutional and incentive systems for transferring innovations from the private sector to the public sector, where these innovations can serve the poor. At the same time economists should develop institutional designs and economic incentives that would make it more attractive for the private sector to generate technologies for which effective demand could be forthcoming among the millions of smallholders. Currently, the private sector investments in technology, and biotechnology in particular, may be creating interesting innovations that never reach the public. They are discarded halfway through their testing and realization because of missing markets (i.e., lack of short-term commercialization potential) and deficient public–private partnerships. If this hypothesis is correct, the global knowledge system, especially in the private sector related to agriculture, is not functioning efficiently under a social cost perspective. The reasons are ill-designed intellectual property rights systems (especially concerning biotechnology) and codes of conduct in industry and in the public sector that are not sufficiently reliable to overcome these barriers.

New institutional research may help overcome these failures. From the perspective of seeking efficient and equitable solutions to allocating resources within and between the public and private research sectors, agricultural economists should devote more research attention to corporate governance related to technology and to the scientific innovation policies of the international agricultural companies. Complementary institutions, such as the Consultative Group on International Agricultural Research (CGIAR), must ensure that the benefits of agricultural technology and research extend widely around the world and especially to the rural poor (Pardey and Beintema, 2001; de Janvry and Sadoulet, 2002). To help provide distributional benefits, research is needed to identify—through stakeholder consultations and other means—appropriate policies in areas such as intellectual property rights, biosafety, and food safety regulations, seed systems, facilitation of access to new technologies, and the allocation of public and private research funds. The bifurcations in agriculture identified in Table 1 make it more and more difficult to facilitate technology and science transfers between the two branches of world agriculture. However, the private sector should find it in their interest to take a more long-term perspective toward market building and inclusion of the world smallholder sector, especially as smallholders are challenged to become more diversified and attempt to capture opportunities in high-value products in the fruit and vegetable sector and the meat and dairy sector.

5. Distributional effects of agricultural trade, market, and aid policy

Perhaps no other subject has commanded such controversy with regard to distributional effects as agricultural price and trade policy. Trade in goods and services and relatively open labor movement across borders could, in principle, be major driving forces for equality with growth. That they are not is largely a consequence of trade barriers in agriculture, such as domestic and regional market and trade policies, including the large subsidies connected to market intervention in OECD countries as well as hindered market access and export subsidies.

Although negotiating parties have made attempts in the Uruguay Round and in the current Doha Round of World Trade Organization (WTO) negotiations to move toward lower tariffs, quotas, export subsidies, and domestic support for agriculture between regions and between free-trade blocs (such as ASEAN, EU, MERCOSUR, and NAFTA), the world as a whole has yet to lower its barriers enough to bring about truly liberal agricultural trade. The failure to achieve this is heavily rooted in structural inequalities and in the political economy of agricultural protectionism.

It is likely that reducing barriers to agricultural trade, especially for developing countries, will deliver overall global benefits and thereby have a positive impact on global income equality. But while the international welfare effects of lowering subsidies, tariffs, and quotas in the global agricultural industry are positive, the magnitude of the potential gains remains the subject of much study and debate (Diaz-Bonilla et al., 2002;

Joachim von Braun

Hertel et al. 2003; Martin et al. 2003; OECD, 2002b; Wobst, 2002). Several sticky issues and adverse trends make it difficult for developing countries to capture benefits from agricultural trade, including the failure of industrialized countries to open up their markets for agricultural goods from developing countries, the use of nontariff barriers such as nontransparent requirements regarding food safety and standards that poor countries cannot meet, and high tariffs for high value-added and processed commodities; and quota and market conduct play an important role for some products (Herrmann and Sexton, 2001). This is not only an issue of rich nations versus poor nations. Developing countries' trade restrictions on agriculture often offset the limited gains from international agricultural trade. These countries would also benefit from their own liberalization (Gulati, 2002). High-value and processed food exports from developing countries have expanded rapidly and are a major source of revenue, and for these exports developing countries stand to benefit as much from lowering their own agricultural trade barriers as from the lowering of OECD barriers (Rae and Josling, 2003). Focusing only on sectoral trade policy reforms does not give a complete picture, because in Latin America, for instance, real domestic prices of farm tradables fell after the initiation of reforms in several countries as a result of currency appreciation, reinforced by a fall in world prices (Valdes and Foster, 2003).

Despite the large potential gains for developing countries from agro-trade liberalization and thus improvements in distribution between nations, this is not necessarily indicative of reductions in *income inequality within countries.* An analysis of the distributional effects of agricultural policies within and among OECD countries shows that the distribution of support is similar to the distribution of output. The largest farms, and hence the most prosperous, are the main beneficiaries (OECD, 1999). Direct payments are more equally distributed than market price support. For instance, the Gini coefficients of direct payments are 0.56 in the EU and 0.61 in the United States, whereas the Gini coefficients for market price support are 0.74 in the EU and 0.98 in the United States (OECD, 1999). These policies of agricultural support are neither cost effective nor equitable.

For *middle- and low-income countries* a diverse pattern is observed. Chile, for example, dramatically reduced its incidence of poverty in the 1990s while its relative income distribution remained remarkably static (Animat et al., 1999). The distributional effects of agricultural liberalization, at least in the short run, can be either negative or positive, and tends to be more positive for net exporters than for net importers. Ultimately, understanding the broader effects of trade on income distribution implies a need for further research on the household-level impacts of these changes.[13] For instance, agro-trade liberalization would drive up food prices and probably labor costs, which would be adverse for low-income people, who already devote a high share of their earnings to food, and for their employment. In Brazil, Hertel et al. (2003) estimated that for agricultural households, the impact of liberalization by OECD countries would reduce poverty by 7.6%. Yet for nonagricultural households OECD liberalization would increase poverty by 2.5% (Hertel et al., 2003). In another computable general equilibrium (CGE) analysis of the distributional effects of trade liberalization in 14 Latin American countries, it was found that although poverty fell in 13 of the 14 countries, the impact on distribution was more ambiguous, in that inequality rose in 5 countries and fell in 9 (Morley and Piñeiro, 2003).

The story of the distributional effects of agricultural trade and market policy is complicated by an increased demand for quality and production process information and related standards. Rising *quality control and food safety standards* in the agro-food industry can pose problems for agricultural exporters in developing countries, thereby contributing to the widening gap in income distribution, especially on an international level. As Hazard Analysis and Critical Control Points (HACCP) and other food safety and quality control standards increasingly become the global norm, developing countries must meet this demand in order to compete in increasingly consolidated and competitive agro-food markets in wealthy countries. Yet compliance with standards imposes transaction costs, as well as significant risk, on developing-country producers. Further, the debate over genetically modified (GM)

---

[13] Comparative static assessment of trade liberalization effects provides limited insights for distributional outcomes. Economy wide modeling that increasingly includes distributional effects has progressed over the past decade and brings new insights (see Lofgren et al., 2003, for instance).

crops complicates the agenda for agricultural liberalization. The different labeling policies adopted or planned by the EU, Japan, and the United States hinder developing-country exporters from tapping trade opportunities, complicate the global food aid system, and thus have potential adverse distributional effects.

Agricultural trade and market development should also be viewed in relation to the rural nonfarm economy. *Commercialization and market integration* of the millions of smallholder farms remain a central task in overcoming rural poverty and the bifurcations in agriculture (von Braun and Kennedy, 1995; Kherallah et al., 2002). The substantial reduction in international transport costs, as a result of new transport and storage technologies and ICTs, over the past decade is an important advance. The urbanization of rural areas and the decreasing cost of capital relative to labor are transforming market institutions at a more micro level and changing the nature of farming in many countries. Explicit and implicit capital subsidies as well as infrastructure investments tend to be biased against small farmers and less-favored areas. Although many rural people depend on agriculture for their livelihoods, many more do so indirectly by working in small-scale rural enterprises providing goods and services for farm families or in agro industries that add value to primary agricultural produce.

The appropriate use of these linkages between agriculture and rural industrialization, as well as rural–urban linkages in an open trade context supported by public goods that facilitate smallholder productivity growth, have proven essential for pro-poor growth processes in, for instance, Japan, South Korea, and Taiwan (Hayami, 2000). These broader externalities of markets, together with public goods and nonmarket institutions, have important developmental and distributional effects that are not yet well understood under different rural conditions. Ultimately, the distributional impacts of agricultural trade policy depend primarily on the structure of the macro economy, the structure of markets, and the structure of poverty, employment, and income distribution within the economy. Initial conditions and change then determine the outcome, especially for small farmers. In Latin America small farmers were often excluded from reform benefits because of their difficulties adjusting to an open trade regime with higher price risks, oligopsonistic buyers in the food industry demanding increasingly larger

volumes and higher standards, the trend toward greater capital intensity, and the reduction in agricultural subsidies (Valdes and Foster, 2003).

Official development assistance (ODA) directed toward agriculture could facilitate increases in global agricultural productivity and trade and add to global and national-level equality. Aid flows, however, are not well targeted at the poorest countries, and even less so at the rural poor. Moreover, development aid to agriculture and rural development has declined continuously in the past two decades. Agriculture and rural development aid totaled US$5.9 billion in 2001, compared with US$12.1 billion in 1979–1980 (OECD DAC, 2002a). A major impediment to agriculture and rural development is the limited public investment in infrastructure and research. The decline of aid for infrastructure is inhibiting the potential gains from market integration and trade, just at the time when countries with location advantages could benefit from reductions in international transport costs.

At an international level, food aid remains an important instrument for distribution, especially during crises. A comprehensive reassessment of food aid in its various types would be useful, given changed market and food security circumstances in many countries (Barrett, 1998). The distributional effects of food aid at a global level are significant, but the record of response in times of international price increases remained disappointing in the 1990s: when prices increased, food aid declined. The overall allocation of food aid across recipient countries has not changed much since the Uruguay Round. Only a small share of concessional food aid goes to low-income food-deficit countries. Results-based criteria for food aid are needed, including for emergency aid, and trade distortions should be minimized.

6. Distributional effects of consumer-oriented regulations and subsidy policies

The world food system is rapidly and fundamentally moving toward industrialized food processing, long-distance marketing, and retail business dominance (Peters and von Braun, 1999). The global food processing industry and the retail business sector both dwarf the agro industries that focus on inputs and crop technologies. Driven by new technologies (especially in transport and information) and by the already mentioned changing demand patterns, this trend has far-reaching implications for consumers and governments. This trend partly bypasses small-scale food industries, low-income consumers, and smallholder farmers. Global food retailing with supermarket outlets is a well-known trend in high-income countries, a recent one in Latin America and the Caribbean, and a strongly emerging one in Asia and even Africa (Reardon and Berdegue, 2002). Research on institutional arrangements at global, national, and regional levels, such as antitrust standards, codes of conduct, and the means to enforce them, is increasingly called for.

At the same time, the demand for *food safety standards* is increasing and human and environmental health concerns are rising, posing a new challenge to our profession (Unnevehr, 2001). Issues of food safety, food security, and trade in agriculture are more than technical matters. They are a conflict over science, evidence, and values; over the future of agriculture and food cultures; over solutions to the hunger and malnutrition problems of the poor; and over trade, market shares, and competition. Research is needed on how to ensure that all links in the global food system function efficiently under this new more quality-demanding food system. Rising food safety standards have distributional implications that are not yet well understood.

Food safety policies, though targeted toward the export sector in developing countries, can have a positive impact on domestic food safety by facilitating spillovers of food safety policies to domestic markets and processing industries. To foster these spillovers, however, governments of low-income countries must invest in capacity strengthening. Without these investments, the spillovers will not come about and there will be a consolidation of the current bifurcation in the global food system in which one system with high standards caters to rich consumers and one with low standards caters to the poor.

Diet change is moving rapidly with urbanization, combined with rising prosperity in some regions, changing dietary preferences, and increasing time costs shifting the pattern of food demand toward processed food. Poor city dwellers' food security depends hardly on food production, but on income security. Research must provide solutions that can address rapid demographic shifts and assure sustainable livelihoods for

people in urban and rural areas alike, as urban–rural linkages change (Virchow and von Braun, 2001). The media and information may influence patterns and trends of food demand today more than price changes, posing a challenge for traditional food demand analysis. Changing lifestyles in combination with diet change are creating new health problems, including the symptoms of obesity, a phenomenon that contributes to chronic diseases in growing segments of the population worldwide and imposes large health costs (World Health Organization, 2003). The World Health Organization (WHO) has emphasized the need for major international initiatives to address the diet problems of low-income countries, in particular because of the failed transition from hunger to health. Sixty percent of mortality in the developing world is now related not to infectious disease, but to chronic disease, much of which stems from bad diets. Further, more than 1 billion people in the world are chronically malnourished, and approximately 2 billion people have deficient diets, especially in micronutrients such as vitamin A, iron (especially in women and children), and zinc. This severe deficiency leads to deterioration of public health, shortens lives, and makes people less productive. Thus, to study the distributional effects of policies that impact on diets, research must include health and nutrition effects. Research needs to explain how food policy interacts with these health crises and the failed diet transitions. Agricultural economists are well positioned to address food-related health economics issues.

Studies on the potential contribution of income growth to overcoming absolute poverty, as represented by under-nutrition, are not encouraging. Even under optimistic scenarios of 2.5% per capita growth, under-nutrition will not decline by more than about a quarter in a sample of developing countries until 2015 (Haddad et al., 2003). Other policy measures are needed. These measures include more or less targeted consumer-oriented subsidies and targeted investments in human capital.

Analyses of *consumer-oriented food subsidies* have shown mixed effects on income distribution and agriculture. Few of the programs have benefited the poor more than the nonpoor (Pinstrup-Andersen, 1988). Often—but not always—the interventions adversely affected agriculture through distorted prices and market interference (von Braun, 1988). A meta-analysis of the efficacy of targeting interventions in 47 developing countries finds the median program transfers 25% more to the target group than would be the case with a universal allocation, but more than a quarter of targeted programs are regressive (Coady et al., 2002). Neither targeted nor untargeted programs on the consumption side seem to make up for the noted inequalities in a systematic way. A more encouraging alternative seems to be conditional transfer programs, such as food for education programs (Ahmed and del Ninno, 2001) or Mexico's PROGRESA (Skoufias and Parker, 2001), which build human capital while transferring benefits.

A further alternative is strengthening insurance and self-insurance mechanisms in rural areas, including crop, health, and old age insurance (pensions). Innovations in community-based health insurance are emerging in many rural areas of middle- and low-income countries and seem to be promising, even under harsh conditions, such as in Ethiopia (Asfaw, 2003). Building social security systems in rural areas from the bottom up may be an option.

Transfer and subsidy policies have not been able to redress the noted bifurcations. Despite much improved knowledge on who and where the poor are, and a much better understanding of the potentials of targeted interventions, inequality and relative deprivation have, in general, widened. Part of the reason may be that many countries simply discontinued compensating transfer programs in the 1980s and 1990s rather than reforming them to achieve improved efficiency and coverage. Agricultural economics research, in conjunction with human resources and health economics research, needs to address the scale and design issues of these programs to come closer to conclusion for policies.

7. Rural public goods and distributional effects

To a considerable extent, rural public goods policies that shape institutions, governance, and public investment cut across policies related to all four of the above discussed domains of agricultural policies. Policies on public goods can also stimulate positive externalities and optimal combinations of actions in the different policy domains.

Different *public investments* can have differential impacts on growth, distribution, and poverty reduction. These effects may also vary by region. Evidence has shown that in the past public investments have delivered greater benefits to farmers in more favorable areas, such as irrigated areas in India and coastal and central areas in China. Villagers in unfavorable areas, such as rain-fed areas in India and the western region in China, did gain important indirect benefits through increased employment, migration opportunities, and cheaper food (David and Otsuka, 1994), but these factors were rarely sufficient to prevent further widening of income differentials. In recent studies, Fan et al. (1999, 2002) have tried to quantify these differential effects of public spending in agricultural research, irrigation, infrastructure, education, and antipoverty programs, using time series and cross-regional data. In both China and India, investments in agricultural research have the largest returns in promoting agricultural production and productivity. Their effects on poverty, however, are often smaller. The largest poverty reduction effects come from improved rural roads in India and rural education in China.

Distributional effects also relate to *redistribution of power* through a change in institutions, governance, and decentralization. Governance affects the allocation as well as the efficacy of public spending. Both allocation and efficacy in turn have different effects on efficiency and income distribution. Many cross-country studies have attempted to link governance, returns to public investment, and economic growth (Isham et al., 1995; Kaufmann et al., 1999). In general, these studies found that good governance had a positive effect on economic growth.

Few studies have quantitatively examined the link between governance and public goods provision at the local (or community) level. Using a recent village survey conducted over a significant period of time, Zhang et al. (2002) compared two different modes of governance. They found that the presence of elections negligibly affects the level of a local community's revenue but significantly shifts taxation or levies from individuals to enterprises. Elections alone do not necessarily improve the allocation of public expenditures. Only when decision-making power is shared is the share of public investment higher. There are more questions to be answered. For example, does local governance affect not only the financing and allocation of public spending, but also the efficacy of public spending and final development outcomes such as growth, income distribution, and poverty?

In the 1990s, decentralization and local government reform became preferred development strategies. Decentralization involves transferring rights and responsibilities from higher to lower levels of government. These rights and responsibilities may be political, administrative, and fiscal. On the one hand, for public goods that have large spillover effects, such as agricultural research, a more centralized mode is probably more appropriate. But even within agricultural R&D, the types of research or technology development that are centralized or decentralized can have a large impact on growth as well as on poverty reduction. On the other hand, in a more decentralized mode, local needs and preferences can be reflected through local participation, and local accountability and efficiency can be improved thanks to a better understating of local knowledge and conditions. Therefore, for some public goods, such as health, education, local roads, and to some extent agricultural extension, a more decentralized mode may be more appropriate. Strong local capacity and participation, however, are a necessary condition for these investments to have high returns.

Estache and Sinha (1995) found a strong relationship between fiscal decentralization and government spending on infrastructure, which has a strong growth-promoting effect. Enikolopov and Zhuravskaya (2003) emphasized the preconditions for decentralization to have a positive impact on economic growth, quality of government, and public goods provision, including weakness or strength of the party system, and whether local- and provincial-level executives are appointed or elected. There is no clear-cut relationship, however, between decentralization and economic growth, and poverty reduction (von Braun and Grote, 2002), and cross-country analyses provide little policy guidance on how to improve current governance structures at community levels—the levels at which change can be implemented most feasibly and is most relevant for rural change. It depends on the nature of public investment. Few studies have quantified the effects of decentralization on the efficiency and distributional effects of public spending across different regions within a country. In particular, there is a lack of empirical evidence to analyze the conditions and types of rural public goods provision and public spending that should be decentralized. This type of research can be fruitfully related to agricultural economics issues.

8. Conclusions

1. World agriculture at the beginning of the twenty-first century is confronted with a *dilemma*: the global integration of agriculture and its potential benefits for poverty reduction and income distribution are not forthcoming to a satisfying degree. No comprehensive set of policies is emerging to address the matter. The contributions of expected growth will not correct the problem in the foreseeable future. The Millennium Development Goals, including the one to cut under-nutrition in half by 2015, challenges the situation in appropriate ways, but follow-up so far is not promising.

2. Agriculture is drifting into an ever more drastic *bifurcation* at a global level and within many countries. This bifurcation undermines growth potential and potentially fosters political conflicts. Correcting that bifurcation will require large investments in rural areas and rural people, in institutions, and in information and biological technologies accessible by the poor in the world's smallholder sector. The societal risks of perpetuated inequality must not be underestimated. Poverty—being largely rural—was until recently of little risk for world security. Today, virtually all of the poor know of potential lifestyles elsewhere on the globe. Relative deprivation can no longer be ignored.

3. *Land and natural resource policies* need further attention, but the institutions related to them may be more important than size distributions. Rights and access to land are ill defined in many countries and require reform. Inequalities induced by differences in access to agricultural land, resources, and technology can, in principle, be balanced by taxation, subsidies, public investments, and transfer policies, including consumer subsidies and investments in human resources (education), but that is not happening to a significant extent.

4. *Technology policies* remain central for distributional equity in agriculture. Information and communications technology and biological technologies have enormous potential to address scale economy problems in rural development and for growth in low-income farming communities. However, these potentials remain far from being tapped. Investment in public research that could foster these positive effects is lacking.

5. Market-oriented global redistribution through *open market and trade policy* is a potentially efficient and effective approach. Protectionism in OECD countries and market interference in developing countries prevents agricultural trade from playing its key role in ensuring favorable distributional effects. This calls for a coordinated and coherent correction.

6. In order to improve equity and efficiency, national governments must provide *public goods*, including internal peace, rule of law, and public investment in education, health, nutrition, and infrastructure. A massive scaling-up of investment in enhancing productivity in rural areas of developing countries is needed, accompanied by social services and innovative insurance institutions. Many of the key ingredients in agricultural growth in developing countries—research and related services that facilitate implementation—are in the public goods domain.

7. Taking these steps requires governments to make difficult choices. The capability to make these choices depends upon the quality of *governance of the food and agriculture system*. Providing these public goods can help accelerate private investment, since private investors generally avoid rural areas and countries characterized by weak justice systems and arbitrary and corrupt public administration.

8. *Important linkages* exist among the four domains discussed here (resources, technology, markets, consumer policies). These linkages are generating new distributional effects. Policy coherence and trade-offs need to be considered for stimulating positive aggregate distributional effects. But in view of the complex food and agriculture systems, the call for coherence among policy domains, such as land policies and technology policies or market and consumer subsidy policies, is becoming exceedingly difficult to respond to in the policy process. These choices cannot be made efficiently if only a top-down approach is adopted. Externalities of agricultural policies require more attention in research.

9. *Agricultural economics* is part of the solution, but only if our profession sufficiently directs itself to research-based problem solving. Agricultural economists have a fair degree of freedom in making choices regarding research priorities. Today

more than ever before, the research agendas of agricultural economists are potentially more relevant for society, development, security, and peace. As a well-established global association, *the network of IAAE* is of tremendous value for agriculture-related decision making through its common spirit of professional ethics and its ambition to contribute to people's well-being with regard to agriculture, food, and rural areas. A *renewed focus on the distributional effects* of agricultural policy is part of such service to society.

Acknowledgements

I gratefully acknowledge the excellent research assistance by Mary Ashby Brown (IFPRI) during the development of the paper and most helpful comments on an early draft by Rajul Pandya-Lorch, Sherman Robinson, Ashok Gulati, Shenggen Fan, Ruth Meinzen-Dick, and Agnes Quisumbing.

References

Ahmed, A., and C. del Ninno, "Food for Education Program in Bangladesh: An Evaluation of Its Impact on Educational Attainment and Food Security," FCND Discussion Paper no. 138 (International Food Policy Research Institute: Washington, DC, 2001).

Alston, J., and P. Pardey, "Reassessing Research Returns: Attribution and Related Problems," Twenty-fourth International Conference of Agricultural Economists, published contributed paper from Proceedings volume, G. Peters and P. Pingali (Eds.) (Ashgate: Burlington, VT, 2001).

Animat, E., A. Bauer, and K. Cowan, "Addressing Equity Issues in Policy Making: Lessons from the Chilean Experience," in V. Tanzi, K. Chu, and S. Gupta (Eds.), *Economic Policy and Equity* (International Monetary Fund [IMF]: Washington, DC, 1999).

Asfaw, A., "Costs of Illness, Demand for Medical Care, and the Prospect of Community Health Insurance Schemes in the Rural Areas of Ethiopia," in F. Heidhues and J. von Braun (Eds.), *Development Economics and Policy Series*, vol. 34 (Peter Lang: Frankfurt, 2003).

Ashby, A., "Agricultural Economics as Applied Economics." Contributed paper from the Proceedings of the Second International Conference of Agricultural Economists, Ithaca, NY (George Banta Publishing Co.: Menasha, WI, 1930)

Atkinson, A., "Bringing Income Distribution in from the Cold," *Economic Journal* 107 (1997), 297–321.

Bardhan, P., S. Bowles, and H. Gintis, "Wealth Inequality, Wealth Constraints, and Economic Performance," in A. Atkinson, F. Bourguignon (Eds.), *Handbook of Income Distribution*, vol. 1 (Elsevier: Amsterdam, 1998).

Barrett, C., "Food Aid: Is It Development Assistance, Trade Promotion, Both, or Neither?" *American Journal of Agricultural Economics* 80 (1998), 566–572.

Bayes, A., J. von Braun, and R. Akhter, "Village Pay Phones and Poverty Reduction: Insights from a Grameen Bank Initiative in Bangladesh," ZEF Discussion Papers on Development Policy no. 8 (Center for Development Research: Bonn, Germany, 1999).

Binswanger, H., K. Deininger, and G. Feder, "Power, Distortions, Revolt and Reform in Agricultural Land Relations," in J. Behrman and T. Srinivasan (Eds.), *Handbook of Development Economics 3A*, (Elsevier: Amsterdam, 1995).

Birner, R., *Devolution and Collaborative Governance in Natural Resource Management: Theory and Empirical Evidence from Developing Countries* Habil. Thesis (Institute for Rural Development; Georg-August University of Gottingen, 2003).

Carter, M., "Land Ownership Inequality and the Income Distribution Consequences of Economic Growth," (World Institute for Development Economics Research [WIDER]; Working Paper no. 201, Helsinki, Finland, 2000).

Chen, S., and M. Ravallion, "How Did the World's Poorest Fare in the 1990s?," World Bank Working Paper no. 2409 (World Bank: Washington, DC, 2000).

Chowdhury, S., *Institutional and Welfare Aspects of the Provision and Use of Information and Communications Technologies in the Rural Areas of Bangladesh and Peru* (Peter Lang: Frankfurt, 2002).

Coady, D., M. Grosh, and J. Hoddinott, "Targeting Outcomes Redux," FCND Discussion Paper no. 144 (International Food Policy Research Institute: Washington, DC, 2002).

David, C., and K. Otsuka, *Modern Rice Technology and Income Distribution* (International Rice Research Institute, University of Los Banos, Philippines, IRRI, 1994).

Deininger, K., *Land Policies for Growth and Poverty Reduction* (World Bank: Washington, DC, 2003).

De Janvry, A., G. Gordillo, E. Sadoulet, and J. Platteau, *Access to Land, Rural Poverty, and Public Action* (Oxford University Press: Oxford, 2001).

De Janvry, A., and E. Sadoulet, "World Poverty and the Role of Agricultural Technology: Direct and Indirect Effects," *Journal of Development Studies* 38, no. 4 (2002), 1–26.

Diaz-Bonilla, E., S. Robinson, M. Thomas, and Y. Yanoma, "WTO, Agriculture, and Developing Countries: A Survey of Issues," TMD Discussion Paper no. 81 (International Food Policy Research Institute, Washington, DC, 2002).

Enikolopov, R., and E. Zhuravskaya, "Decentralization and Political Institutions," Centre for Economic Policy Research Discussion Paper series 3857 (CEPR: London, 2003).

Estache, A., and S. Sinha, "Does Decentralization Increase Spending on Public Infrastructure?," Policy Research Working Paper no. 1457 (World Bank: Washington, DC, 1995).

Fan, S., P. Hazell, and S. Thorat, *Government Spending, Agricultural Growth, and Poverty: An Analysis of Interlinkages in Rural India* Research Report no. 110 (International Food Policy Research Institute: Washington, DC, 1999).

Fan, S., L. Zhang, and X. Zhang, *Growth, Inequality, and Poverty in Rural China* Research Report no. 125 (International Food Policy Research Institute: Washington, DC, 2002).

FAO. *The State of Food and Agriculture* (Food and Agriculture Organization of the United Nations: Rome, 2002).

FAO. "World Agricultural Census (1990) and Appendices (1990–97)," (Food and Agriculture Organization of the United Nations: Rome, various years).

Gardner, B., "Economic Growth and Low Incomes in Agriculture," *American Journal of Agricultural Economics*, 82, November (2000), 1059–74.

Gardner, B., *American Agriculture in the Twentieth Century: How It Flourished and What It Cost* (Harvard University Press: Cambridge, MA, 2002).

Gulati, A., *The Subsidy Syndrom* (Oxford University Press: Delhi, 2002).

Haddad, L., "The Income Earned by Women: Impacts on Welfare Outcomes," *Agricultural Economics* 20, no. 2 (1999), 135–141.

Haddad, L., H. Alderman, S. Appleton, L. Song, and Y. Yohannes, "Reducing Child Malnutrition: How Far Does Income Growth Take Us?," *The World Bank Economic Review* 17, no 1 (2003), 107–131.

Hallman, K., D. Lewis, S. Begum, and A. Quisumbing, "Impact Of Improved Vegetable and Fishpond Technologies on Poverty in Bangladesh," Paper presented at the International Conference on Impacts of Agricultural Research and Development: Why Has Impact Assessment Research Not Made More of a Difference? San José, Costa Rica, February 4–7, 2002.

Hanstad, T., J. Brown, and R. Prosterman, "Larger Homestead Plots as Land Reform? International Experience and Evidence from Karnataka," *Economic and Political Weekly*, July (2002), 20.

Hayami, Y., "Toward a New Model of Rural–Urban Linkages under Globalization," in S. Yusuf, W. Wu, and S. Evenett (Eds.), *Local Dynamics in an Era of Globalization* (Oxford University Press: London, 2000).

Hazell, P., and L. Haddad, *Agricultural Research and Poverty Reduction*, 2020 Vision for Food, Agriculture and the Environment Discussion Paper no. 34. (International Food Policy Research Institute: Washington, DC, 2001).

Hazell, P., and C. Ramasamy, *Green Revolution Reconsidered* (The Johns Hopkins University Press: Baltimore, 1991).

Hedley, D., "Presidential Address: Considerations on the Making of Public Policy for Agriculture," in G. H. Peters and P. Pingali (Eds.), *Proceedings of the Twenty-Fourth International Conference of Agricultural Economists – Tomorrow's Agriculture: Incentives, Institutions, Infrastructure and Innovations* Held in Berlin, Germany 13–18 August, 2000 (Ashgate: Burlington, VT, 2001) 3–11.

Herrmann, R., R. Sexton, "Agricultural Globalization, Trade and the Environment," in C. Moss, et al. (eds.), *Market Conduct and Its Implications for Trade Policy Analysis: The European Banana Case* (Kluwer Acad: Dordrecht, 2001)

Hertel, T., P. Preckel, J. Cranfield, and M. Ivanic, "OECD and Non-OECD Trade Liberalization and Poverty Reduction in Seven Developing Countries. Agricultural Trade and Poverty: Making Policy Analysis Count," Papers from the Global Forum on Agriculture, May 23–24 2002 (OECD: Paris, 2003).

IFAD, *Rural Poverty Report: The Challenge of Rural Poverty* (Oxford University Press: Oxford, 2002).

Isham, J., D. Kaufmann, and L. Pritchett, *Governance and Returns on Investment—An Empirical Investigation* Policy Research Paper no. 1550 (World Bank: Washington, DC, 1995).

Jayne, T. S., *Smallholder Income and Land Distribution in Africa: Implications for Poverty Reduction Strategies* MSU International Development Paper no. 24 (Department of Agricultural Economics and the Department of Economics, Michigan State University: East Lansing, MI, 2001)

Juma, C., "Modern Biotechnology. 2020 Vision for Food, Agriculture, and the Environment on Appropriate Technology for Sustainable Food Security," Policy Brief no. 4, P. Andersen (eds.) (International Food Policy Research Institute: Washington, DC, 2001).

Kanbur, R., and N. Lustig, "Why Is Inequality Back on the Agenda?," Paper presented at the Annual World Bank Conference on Development Economics (World Bank: Washington, DC, 1999).

Kaufmann, D., A Kraay, and P. Zoido-Lobaton, *Governance Matters* Policy Research Paper no. 2196 (World Bank: Washington, DC, 1999).

Kerr, J., and S. Kolavalli, "Impact of Agricultural Research on Poverty Alleviation: Conceptual Framework with Illustrations from the Literature," EPTD Discussion Paper no. 56 (International Food Policy Research Institute: Washington, DC, 1999).

Kherallah, M., C. Delgado, E. Gabre-Madhin, N. Minot, and M. Johnson, *Reforming Agricultural Markets in Africa* (The Johns Hopkins University Press: Baltimore, 2002).

Lipton, M., Personal communication (2001).

Lofgren, H., R. Sherman, and M. El-Said "Poverty and Inequality Analysis in a General Equilibrium Framework: The Representative Household Approach," in F. Bourguignon and L.A. Pereira de Silva (Eds.), *The Impact of Economic Policies on Poverty and Income Distribution: Evaluation Techniques and Tools* (World Bank: Washington, DC; Oxford University Press: Oxford, UK, 2003).

Lopez, R. "The Economics of Agriculture in Developing Countries: The Role of the Environment," in B. Gardner and G. Rausser, (eds.) *Handbook of Agricultural Economics vol. 2A* (Elsevier: Amsterdam, 2002).

Martin, W., D. van der Mensbrugghe, and V. Manole, "Is the Devil in the Details? Assessing the Welfare Implications of Agricultural Trade Reforms," Contributed paper presented at the IATRC conference June 23–26 (Capri, Italy, 2003).

Meinzen-Dick, R., and M. Bakker, "Water Rights and Multiple Water Uses: Issues and Examples from Kirindi Oya, Sri Lanka," EPTD Discussion Paper 59 (International Food Policy Research Institute: Washington, DC, 2001).

Meinzen-Dick, R., A. Knox, F. Place, and B. Swallow, *Innovation in Natural Resource Management: The Role of Property Rights and Collective Action in Developing Countries* (The Johns Hopkins University Press: Baltimore, 2002).

Meinzen-Dick, R., L. Brown, H. Feldstein, and A. Quisumbing, "Gender and Property Rights: Overview," *World Development* 25, no. 8 (1997), 1299–1302.

Milanovic, B., "True World Income Distribution, 1988 and 1993: First Calculations Based on Surveys Alone," *The Economic Journal*, January (2002), 51–92.

Milanovic, B., "The Two Faces of Globalization: Against Globalization as We Know It," *World Development* 31, no. 4 (2003), 667–683.

Morley, S., and V. Piñeiro, "Does Globalization Hurt the Poor?," *Evidence from CGE Simulations in Latin America*, draft manuscript (International Food Policy Research Institute: Washington, DC, 2003).

OECD, "Distributional Effects of Agricultural Support in Selected OECD Countries," Document no: 84646 (Organization for Economic Cooperation and Development: Paris, 1999).

OECD, Development Assistance Committee (DAC) Statistical Annexes (2002a).

OECD, "Agricultural Trade and Poverty: Making Policy Analysis Count. Papers from the Global Forum on Agriculture, 23–24 May," (OECD: Paris, 2002b).

Otsuka, K., and F. Place (Eds.), *Land Tenure and Natural Resource Management: A Comparative Study of Agrarian Communities in Asia and Africa* (The Johns Hopkins University Press: Baltimore, 2001).

Paarlberg, R., *Governance and Food Security in an Age of Globalization* 2020, Vision for Food Agriculture and the Environment Discussion Paper no. 36 (International Food Policy Research Institute: Washington, DC, 2002).

Pardey, P., and N. Beintema, *Slow Magic: Agricultural R&D—A Century after Mendel* (International Food Policy Research Institute: Washington, DC, 2001).

Pardey, P., B. Koo, and B. Wright, *Endowing Future Harvests: The Long-Term Costs of Conserving Genetic Resources at the CGIAR Centers* (International Food Policy Research Institute University of California, Berkeley: Washington, DC and Berkeley, CA, 2002).

Pender, J., and P. Hazell (Eds.) *Promoting Sustainable Development in Less-Favored Areas* 2020 Vision for Food, Agriculture, and the Environment, Focus no. 4. (International Food Policy Research Institute: Washington, DC, 2000).

Peters, G. H., and J. von Braun (Eds.). *Food Security, Diversification and Resource Management: Refocusing the Role of Agriculture?* Proceedings of the Twenty-Third International Conference of Agricultural Economists, Ashgate, (1999).

Pinstrup-Andersen, P. (Ed.), *Food Subsidies in Developing Countries: Costs, Benefits, and Policy Options* (The Johns Hopkins University Press for the International Food Policy Research Institute: Baltimore, 1988).

Pinstrup-Andersen, P., and R. Pandya-Lorch, "The Unfinished Agenda–Perspectives on Overcoming Hunger, Poverty, and Environmental Degradation," (International Food Policy Research Institute: Washington, DC, 2001).

Qaim, M., A. F. Krattinger, and J. von Braun (Eds.), *Agricultural Biotechnology in Developing Countries: Towards Optimizing the Benefits for the Poor* (Kluwer Academic Publishers: Boston, Dordrecht, London, 2000).

Quisumbing, A., L. Brown, H. Feldstein, L. Haddad, and C. Peña, *Women: The Key to Food Security* Food Policy Report (International Food Policy Research Institute: Washington, DC, 1995)

Rae, A., and T. Josling, "Processed Food Trade and Developing Countries: Protection and Trade Liberalization," *Food Policy* 28 (2003), 147–166.

Randall, A., "Valuing the Outputs of Multi-functional Agriculture," *European Review of Agricultural Economics* 29, no. 3 (2002), 289–307.

Ravallion, M., *Growth, Inequality, and Poverty: Looking Beyond Averages*, World Bank Working Paper no. 2558 (World Bank: Washington, DC, 2003).

Reardon, T., and J. Berdegue, "The Rapid Rise of Supermarkets in Latin America: Challenges and Opportunities for Development," *Development Policy Review* 20, no. 4 (2002), 317–334.

Scherr, S., *Soil Degradation: A Threat to Developing Country Food Security by 2020?* 2020 Vision for Food, Agriculture, and the Environment Discussion Paper no. 27 (International Food Policy Research Institute: Washington, DC, 1999).

Schultz, T., *Transforming Traditional Agriculture* (Yale University Press: New Haven, 1964).

Sen, A., *On Economic Inequality* Second Edition (Oxford University Press: Oxford, 1997).

Skoufias, E., and S. Parker, "Conditional Cash Transfers and Their Impact on Child Work and Schooling: Evidence from the PRO-GRESA Program in Mexico," Food Consumption and Nutrition Division Discussion Paper no. 123 (International Food Policy Research Institute: Washington, DC, 2001).

Stark, O., and Y. Wang, "A Theory of Migration as a Response to Relative Deprivation," *German Economic Review* 1, no. 2 (2000), 131–143.

Timmer, P., "Agriculture and Economic Development, in B. Gardner and G. Rausser (Eds.), *Handbook of Agricultural Economics* vol. 2A (Elsevier Science: Amsterdam, 2002).

Torero, M., and J. von Braun (Eds.), "Information and Communications Technology (ICT) for Economic Development and Inclusion of the Poor," draft manuscript (2004).

Unnevehr, L., *Food Safety and Food Quality* 2020 Focus 8 (Shaping Globalization for Poverty Alleviation and Food Security), Brief No. 7 (International Food Policy Research Institute: Washington, DC, 2001).

Unnevehr, L., and D. Roberts, "Food Safety and Quality: Regulations, Trade, and the WTO," Paper presented at the IATRC Conference, June 23–26 (Capri, Italy, 2003).

UNDP, *Human Development Report 2001: Deepening Democracy in a Fragmented World* (United Nations Development Program: New York, 2001).

Valdes, A., and W. Foster, "The Breadth of Policy Reforms and the Potential Gains from Agricultural Trade Liberalization: An Ex Post Look at Three Latin American Countries" (mimeo, 2003).

Virchow, D., and J. von Braun (Eds.), *Villages in the Future—Crops, Jobs and Livelihood* (Springer, Heidelberg, New York, 2001).

von Braun, J., "Implications of Consumer-Oriented Food Subsidies for Domestic Agriculture," in P. Pinstrup-Andersen (Ed.), *Food*

Subsidies in Developing Countries: Costs, Benefits, and Policy Options* (The Johns Hopkins University Press for the International Food Policy Research Institute: Baltimore, 1988).

von Braun, J., and U. Grote, "Does Decentralization Serve the Poor?," in E. Ahmad and V. Tanzi (Eds.), *Managing Fiscal Decentralization* (Routledge: London, 2002).

von Braun, J., and E. Kennedy, *Commercialization of Agriculture, Economic Development, and Nutrition* (The Johns Hopkins University Press: Baltimore, 1995).

von Braun, J., and M. Qaim, "Household Action in Food Acquisition and Distribution under Transformation Stress," in M. Hartmann and J. Wandel (Eds.), *Food Processing and Distribution in Transition Countries: Problems and Perspectives* (Wissenschaftsverlag: Kiel, Germany, 1999).

von Thünen, J. H., "Der isolirte Staat in Beziehung auf Landwirthschaft und Nationalökonomie, oder Untersuchungen über den Einfluss, den die Getreidepreise, der Reichtum des Bodens und die Abgaben auf den Ackerbau ausübe." Hamburg ["The isolated state in relation to agriculture and the national economy, or an analyses on the influence of grain prices, value of land, and taxes on farming"] (1826).

von Thünen, J. H., *Der isolirte Staat, vol II, Der naturgemässe Arbeitslohn und dessen Verhältnis zum Zinsfuss und zur Landrente*. Hamburg ["The isolated state in relation to agriculture and the national economy: the natural wage and its relationship to interest rate and returns to land"] (1850).

Wobst, P., "The Impact of Domestic and Global Trade Liberalization in Five Southern African Countries," TMD Discussion Paper no. 92 (International Food Policy Research Institute: Washington, DC, 2002).

World Bank, *India: Achievements and Challenges in Reducing Poverty*, World Bank Country Studies (Washington, DC, 1997).

World Bank, "World Development Indicators, CD-ROM version," (The World Bank: Washington, DC, 2002).

World Health Organization, *Diet, Nutrition, and Prevention of Chronic Diseases: Report of a Joint WHO/FAO Expert Consultation* Geneva, January 28–February 1, 2002. (World Health Organization of the United Nations (WHO): Geneva, 2003).

Zeller, M., and R. Meyer, *The Triangle of Microfinance: Financial Sustainability, Outreach, and Impact* (The Johns Hopkins University Press: Baltimore, 2002).

Zhang, X., S. Fan, L. Zhang, and J. Huang, "Local Governance and Public Goods Provision in Rural China," EPTD Discussion Paper no. 93 (International Food Policy Research Institute, Center for Chinese Agricultural Policy [CCAP], and Chinese Academy of Sciences: Washington, DC and Beijing, 2002).

Causes of rural economic development

Bruce L. Gardner*

Abstract

Underlying factors in the growth of agriculture as a sector and of rural incomes in developing countries are investigated, using data from 85 countries during 1960–2001. Hypotheses about growth are derived from both the general growth literature and the empirical literature on past agricultural growth in the United States and other industrial countries. The growth of agriculture as a sector is surprisingly independent of the growth of income per capita for those who work in that sector. Neither is necessary nor sufficient for the other. Agricultural economics is in many circumstances not the key discipline in understanding the economics of rural income and poverty.

JEL classification: O13

Keywords: agricultural development; productivity; rural income; economic growth

We can be most helpful in locating the bottlenecks and constraints to growth and suggest means to their alleviation. In this, we sometimes have to operate at the frontier of professional knowledge, and often against the common wisdom of governments, but this is where the progress is to be made.

—Mundlak, 1999, p. 46

1. Introduction

This paper follows Yair Mundlak's recommendation, which concluded his Elmhirst lecture, to identify sources of and constraints upon economic growth in agriculture. The subject matter is approached not from the perspective of research on international agricultural development, but rather as a follow-up to studies of the development of U.S. agriculture. The story of U.S. agriculture serves as a possible source of lessons for countries where sustained growth in the real incomes of rural people has not yet occurred. Moreover, U.S. research bears on the question of how the relatively poorest farm people have fared in the growth process

* *College of Agriculture and Natural Resources, University of Maryland, College Park, MD, USA.*

(Gardner, 2000), a topic that fits well with the emphasis of President von Braun's address.

The paper starts by revisiting some fundamentals of agricultural development economics, the literature concerning it, and data measurement issues. In the second major section, empirical evidence is reviewed on the growth of agriculture as an industry. The third section turns to welfare consequences of agricultural growth as measured by real household incomes. The final section discusses conclusions about the causes of sector growth and real income growth.

2. Models of growth

Since World War II, a huge literature has emerged on economic growth, with special attention to agriculture. Since that time most of the poor countries of the world have become less poor, and agriculture in practically all of them has become more productive (in terms of output per worker, output per acre, or multifactor productivity growth). However, the success has varied widely from country to country, from one period to another, and across regions within countries. How well does the accumulated literature further an understanding of these variations, and what might have been done to improve the performance of the worst-performing countries? The central analytical task is to identify the causes of growth.

It might be expected that the most helpful writings would be those that cover the complexity and range of the societies being studied, and thus finding a multiplicity of causes, each having different weights in different countries and at different times. However, to date no such comprehensive approach has proven fruitful. Instead, economists' contributions have typically proceeded by over-simplification, either by a model fixing on only a few key causal factors, which are taken to be applicable over a range of countries and circumstances, or by focusing on a single country and dissecting events through an analytical description (as opposed to econometric hypothesis testing).

Many of the key conceptual contributions can be classified according to two polarities of approach: microeconomic versus macroeconomic, and theoretical versus empirical. No economist is purely in any of the four camps that these polarities generate—micro-theoretical, micro-empirical, macro-theoretical, and macro-empirical. Yet many have emphases that place their main contributions in one or another area. The macro-theoretical approach got a big initial boost in the 1950s from growth models treating output in the economy as generated by a neoclassical production function, with capital as an input created by savings. Agriculture as a sector in a general equilibrium context was treated in two-sector models in both comparative static versions (notably Simon, 1947) and the many dual-economy models that followed.[1]

Macro-empirical contributions until recent years have had a case study flavor, accumulating analytical description without a formal model. Mancur Olson (1982) is a good example. More recently the creation of panel data covering countries over time has made possible an econometric macro-empirical research, e.g., Barro and Sala-i-Martin (1995). This literature has austere but fruitful theoretical underpinnings, leading to ideas of "convergence" that have not yet been exploited sufficiently in investigating agricultural growth. An outstanding example of the micro-empirical approach is T. W. Schultz (1964). Many recent papers

on household behavior in poor countries are heavier on micro-theory. Yet in both Schultz and, for example Singh et al. (1986), there is an intimate integration of theory and empirical observation.[2]

All the approaches have generated hypotheses about causes of growth that will be discussed below in the context of rural economic development.

2.1. Measures of growth

One of the services of the models is to provide a conceptual basis for our choices of variables to measure in quantitative terms and test econometrically. Agricultural output growth is a measure that arises naturally from the estimation of a production function. However, output can grow for reasons that provide little or no support for a rising standard of living—for example, output could rise under population pressure simply because of a larger farm labor force or clearing of additional land. For many purposes a better indicator of growth is agricultural gross domestic product (GDP) (value added) per worker—what the sector generates for each productively engaged person over and above the cost of inputs from outside of agriculture. Agricultural GDP per worker readily translates to a potential living standard measure, namely real income per household. With respect to causes of growth, underlying production theory says that either output or value added per worker can grow for the same two principal reasons: investment (including investment in human capital) and technological progress. The question then becomes why investment and technological progress occur, or fail to occur.

Matters get more interesting analytically as well as better attuned to actual situations when the link between agricultural value added and rural household incomes is broken. Farms produce nonagricultural products and farm household members earn incomes from nonfarm sources. Then the causes of growth may well be different for agricultural growth and rural income growth. Nonetheless, a flourishing agricultural sector can still be important instrumentally as a means of achieving rural income growth. One of the key empirical questions about economic growth in

---

[1] "Comparative static" models investigate economic change through one-time shocks in exogenous variables, as opposed to dynamic models that investigate time paths of investment or other growth-generating endogenous variables.

[2] Of course, many economists' works don't fall so easily into any of these categories. Hayami and Ruttan (1985), Mundlak (1999), and Timmer (2002), for example, draw on all the approaches.

rural areas is how crucial agricultural sector growth is in the process.

While no measure of either agricultural sector growth or rural income growth is perfect, some useful indicators exist for many countries over a substantial period. The most promising way to learn about the causes of growth is to compare the record of such indicators across countries as associated with variables hypothesized to be causes of growth. In order to carry out such comparisons most meaningfully, uniformly constructed cross-country data are needed. The massive undertaking of constructing such a data set for agriculture has been taken on principally by the Food and Agriculture Organization of the United Nations (FAO). The World Bank's *World Development Indicators* (available on-line) combines the FAO data with other sectoral and macroeconomic information for the years 1961–2001. These data are the main source of statistical information in this paper. Measures of agricultural sector growth include cereal yields, crop and livestock output indexes, and agriculture's contribution to GDP (sectoral value added).

3. Agricultural sector growth

"Getting agriculture moving" is a slogan that encapsulates the problem as it appeared to agricultural economists early in the post–World War II period as population pressures were seen as requiring faster expansion of food production than looked likely to occur in many low-income countries, where traditional agriculture is the rule. Traditional agriculture is characterized by poverty or subsistence-level living standards, with famine an ever-present threat, and hope of transformation to a higher standard of living for the rural population as a whole remote. What has to happen for a country's agriculture to break out of that situation? In the early 1960s, T. W. Schultz formulated his answer, beginning with what is *not* likely to work: improved efficiency within existing resource and technological constraints is not the answer, nor is investing more, given those constraints. Clearing more land is an investment that tends to be too costly for the returns to generate sustained growth and "additional irrigation is on approximately the same footing as land" (Schultz, 1965, p. 45). The high-payoff sources of growth are to be found elsewhere, notably in "improvements in the quality of agricultural inputs," virtually all of which must come from outside of agriculture rather than being generated within it (p. 46). Here Schultz has in mind not only fertilizer, tractors, and improved crop genetics, but also schooling and other means to improve the skills of farm people.[3]

The thinking of all who take an interest in agricultural development has to be influenced by the high returns to agricultural research in many countries, notably in the "Green Revolution." There exist inspiring cases where agriculture has flourished as a result. What is the evidence of success from these developments? Three indicators are: cereal yields, multifactor productivity, and agricultural GDP per worker.[4] Acceleration in yield is an indicator that a technology/investment shock is generating streams of output from given land inputs; but it is partial in that yield increases themselves do not imply improved profits because the land-augmenting inputs may cost too much. Multifactor productivity takes into account all the measured inputs and so is conceptually a better indicator of what a country obtains from a given set of resources committed to agriculture. But despite recent progress, obtaining accurate cross-country comparisons of multifactor productivity over time remains a major problem; and multifactor productivity is still not a sufficient indicator of the returns to farm-origin land, labor, and invested capital that constitute the basis for farm household income growth (because, for example, product buyers may reap the bulk of productivity gains through lower product prices). Agricultural GDP subtracts the costs of purchased inputs from outside agriculture, and indicates the net gains available for the purposes of improved incomes of farm people. Yet the data are sparser and require often-dubious assumptions (e.g., in estimating capital service flows) for multifactor productivity and agricultural GDP.

---

[3] Education is expected to be productive by improving the basic skills of people, not by changing their outlook to be less traditional or via other cultural changes. Indeed, Timmer attributes to Schultz "the demise in the late 1960s of community-action programs," which focused on cultural/institutional transformation (Timmer, 2002, p. 1516).

[4] For purposes of evaluating investments in research that generate technological change, a more appropriate bottom line is the rate of return to investment. It is estimates of these rates being extraordinarily high that have sealed the case for the benefits of international agricultural research (see Alston et al., 2000 and Evenson, 2001 for comprehensive reviews).

3.1. Cereal yields

Data on growth rates of cereal yields (kilograms per hectare) are available for 86 countries during two time periods, 1961–80 and 1981–2001.[5] The two periods are referred to as "Early Green Revolution" and "Late Green Revolution" by Evenson and Gollin (2003). The yield data are instructive in showing that progress indeed has occurred worldwide. Yields increased during 1961–2001 in all but 9 of the 85 countries covered; 8 of these were in Africa (Angola, Botswana, Chad, Congo, Mozambique, Rwanda, Sierra Leone, and Sudan). The sub-periods show a slowdown in yield growth in recent years, a phenomenon that some have viewed with alarm (see e.g., International Fund for Agricultural Development, 2001, Chapter 4). However, the slowdown is not huge, and does not occur in many countries. Yields in 15 of the 31 African countries, and in 10 of 18 Latin American ones, grew faster in 1981–2001 than in 1961–80. The countries that fit best the notion of yield slowdown were the industrial country (OECD) group.[6]

Countries can be ranked within each regional group by the rate of yield growth in 1961–81. It is notable that countries with the highest yield growth in 1961–81 are typically observed to have slower yield growth in 1981–2001, while the departures from this generalization tended to be countries with lower growth rates in 1961–80. However, the correlation coefficient between the growth rates in the earlier and later periods is only −0.06, not statistically significant, for the whole

sample of 85 countries. Figure 1 shows the time series of yield growth for countries that had particularly high or low rates in 1961–80. China and India are examples where yields grew fast and then continued to do so; while yields in Belize and Swaziland grew fast and then stagnated. In Angola and Mozambique yields declined sharply and then rebounded. There are no cases where yields declined at a high rate and then continued to decline. The closest approximation is Haiti, where a 0.4% rate of yield decline in 1961–81 was followed by a 0.7% rate of decline in 1980–2001.

Figure 1 indicates a substantial divergence of yields over time. This is partly a matter of the small sample selected, but when all the 85 countries covered are charted, it also shows divergence of yields over time.[7] This divergence of yields across countries is a surprise from the viewpoint that underlies a lot of recent work in the theory of economic growth. The basic idea is that any economy's output is generated by technology and economic actors following neoclassical principles. The application to crop yields is that with the same technology available everywhere, countries with lower yields will have a higher marginal return to new inputs. Therefore, use of such inputs is expected to increase at a higher rate in lower-yield countries and to increase their yields faster than those of initially high-yield countries. So stated, this idea is unattractive for the historical evolution of most of world agriculture because of differences in climate and other natural resources, and because the same technology is not available everywhere. The issue then arises of the international transfer of technology. On this subject we do have plausible dating of at least one important element of the cross-country story: the international integration of agricultural research under the Consultative Group on International Agricultural Research (CGIAR) in about 1970. Thus, if agricultural research is an important element of the story, there ought to be more convergence in the years following 1970 than before. A more complex version of this story is what Evenson and Gollin (2003, p. 758) call "broader and deeper impacts" of CGIAR research on more crops in more countries after the mid-1970s. Yet the yield data give no indications of yield convergence, even in the 1990s.

---

[5] These data are available in Table 1 of the version of the paper delivered at the conference (www.iaae-agecon.org).

[6] The World Bank (2002) states that "the yield growth experienced since the 1970s has slowed sharply in the 1990s due to diminishing returns to further input use, the rising cost of expanding irrigation, a slowdown in investment in infrastructure and research (in part induced by declining commodity prices), and resource and environmental constraints" (p. 47). This story is generally plausible but not supported with evidence, and what seems most fundamentally dubious is the initial claim of a sharp slowdown in yield growth. Of the 31 African countries, 16 had yields that increased more rapidly in the 1990s as compared to the 1980s than in the 1980s compared to the 1970s; and similarly the trend rate of growth during the 1990s was greater than the trend rate during 1970–89 for 16 of the 31. The 11 Asian countries in the sample are similarly split. But China and India, the two biggest, do conform to the idea of a yield slowdown in the 1990s. Nonetheless, we have about as many instances of yields accelerating in the 1990s as of yields decelerating.

[7] See Figure 1(a) in the conference version of the paper (www.iaae-agecon.org).

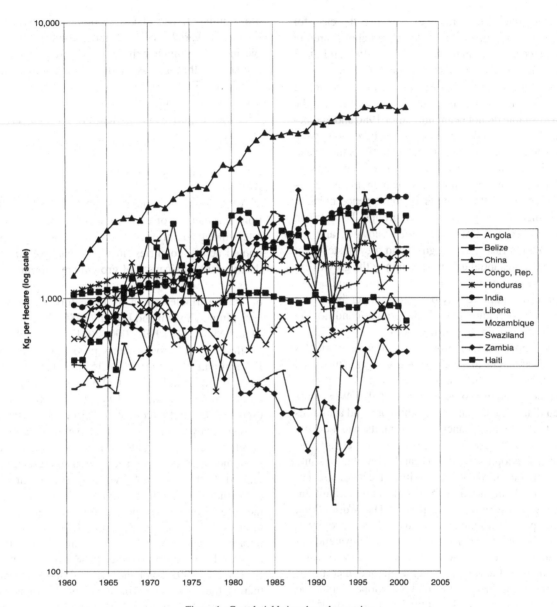

Figure 1. Cereal yields in selected countries.

3.2. Agricultural GDP per worker

The convergence hypothesis might be more likely to apply to real agricultural GDP per worker. The idea is that, at an initial point in time, countries with a lower agricultural GDP per worker will have a higher marginal return to capital investment under the classical laws of production. Therefore, more investment will occur in lower-GDP countries and their agricultures will grow faster than those of initially high-GDP countries. Notwithstanding questions of accurate measurement, the FAO/World Bank indicator series provide a substantial, consistently constructed panel of cross-country comparisons over time. Data on the 1980–2001 growth rates of agricultural GDP per worker for the 79 countries that have sufficient data for this purpose are available.[8] These rates, as was also the case for the cereal yield growth rates discussed above, are not calculated from changes between the 1980 and 2001 endpoints on the grounds that random year-to-year variation makes the calculated rate too sensitive to the choice of endpoint years (and for some countries the data available do not begin until after 1980 and end before 2001). Instead, log-linear trend regressions were estimated and the slope for each country is the "trend growth rate" for that country. The time series for several groups of countries are shown in Figure 2(a)–(d). In none of these charts is convergence evident, nor is there convergence between groups. The African countries started lowest and grew slowest, and the OECD countries started highest and grew fastest. Yet there are substantial differences among the growth experiences of countries within each group and across all groups, and more may be learned about the causes of growth by finding out which "growth-conditioning variables" explain those differences.

The importance of growth-conditioning variables became apparent to scholars of both agricultural and general economic growth as thwarted expectations of technology-led rural prosperity mounted. T. W. Schultz noted that while advances in technology and availability of capital for financing new inputs had become ever more widespread, "it has became increasingly evident that adoption of the research contributions and efficient allocation of the additional capital are being seriously thwarted by the distortion of agricultural incentives" (Schultz, 1978, p. vii). The World Bank (2002, p. 47) summarizes a range of recent opinion about constraints to growth, noting the prevalence of problems created by micro and macro policy discrimination against agriculture, inefficient and uncompetitive marketing institutions, underdeveloped labor and financial markets, weak political and property rights institutions, and world price-depressing policies of the OECD countries.

Which variables are the most important? Are there some that do not matter much in fact even though in principle they might have been expected to? Are there some conditions that are so important that, if they prevail, they are sufficient for real income growth? Two quite different approaches to answering such questions

are prominent. The first is econometric, pooling data on similar variables for as many countries as possible and attempting to explain the differences in growth statistically through association with candidate causal variables. The second is qualitative and narrative, essentially the accumulation of case studies by scholars with wide experience in agriculture across a variety of countries.

3.3. Cross-country regressions

An econometric approach that has immediate attraction as a method of explaining differences in growth rates among countries is to use time series regressions, pooled across countries, in which changes in candidate variables as causes of growth are correlated with rates of growth in real agricultural GDP per worker, or other variables taken as indicators of growth in agriculture. Hayami and Ruttan (1985, Chapters 5 and 6) and Mundlak (1999, 2000, 2001) have explored in depth the use of cross-country production functions.[9] Hayami and Ruttan (1985, p. 157) explained an agricultural output index as a Cobb–Douglas function of inputs for 43 countries in 1960, 1970, and 1980, and used the results to account for growth in output per worker. They found that output per worker in less-developed countries could effectively be increased by input increases along with education and research, and viewed the findings as "essentially encouraging" because they showed the possibility of progress even in the face of population pressure with limited agricultural land availability. Mundlak (2000) worked with improved data, especially for capital and investment, increased the country coverage to 88, extended the data coverage to 1992, explained agricultural GDP rather than output, and generalized the model to incorporate incentives (prices and risk) and constraints from the economic and physical environment as well as the usual input quantities. He also used country-specific "within" as well as "between" time period estimators to minimize identification problems that plague cross-sectional production function estimates.[10] He found

---

[8] Table 2 in the conference version of the paper (www.iaae-agecon.org).

[9] Other recent studies similar in approach include Fulginiti and Perrin (1993), Frisvold and Ingram (1995), and Craig et al. (1997). For a summary of findings from earlier such studies, see Hayami and Ruttan (1985), p. 149.

[10] See Deaton (1995, pp. 1824–1827) for a succinct presentation of the problems and remedies.

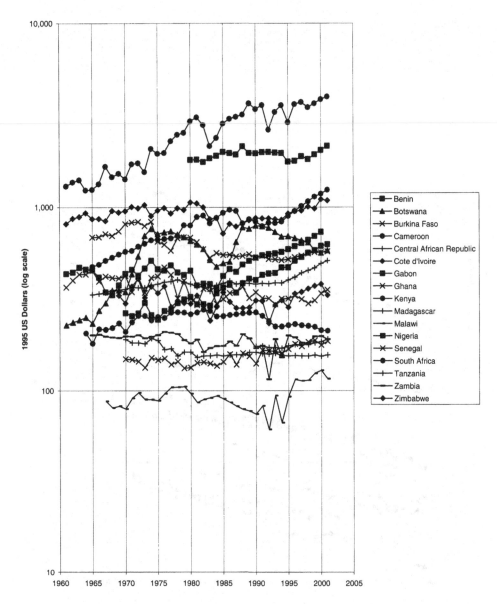

Figure 2(a). Real agricultural GDP per worker, sub-Saharan Africa.

input quantities to be important largely as expected, largely in line with earlier findings, and found increases in capital especially important in generating increased agricultural GDP. Variables over which a country can have some control as a matter of policy, notably agricultural prices and schooling, were estimated to have quite small effects (the schooling results stand in sharp contrast to the findings of Hayami and Ruttan). Yet, as Mundlak (2001) notes, these policy variables may influence investment and adoption of improved inputs,

and the regressions already include the input levels. To sort out the full contributions of such variables as causes of growth one needs either a more complete structural model of input and output supply and demand, or else reduced-form equations in which growth is estimated as a function of exogenous or policy variables only.

Attempts to test the convergence hypothesis, such as Barro and Sala-i-Martin (1995), have generated an essentially reduced-form approach that may be

Bruce L. Gardner

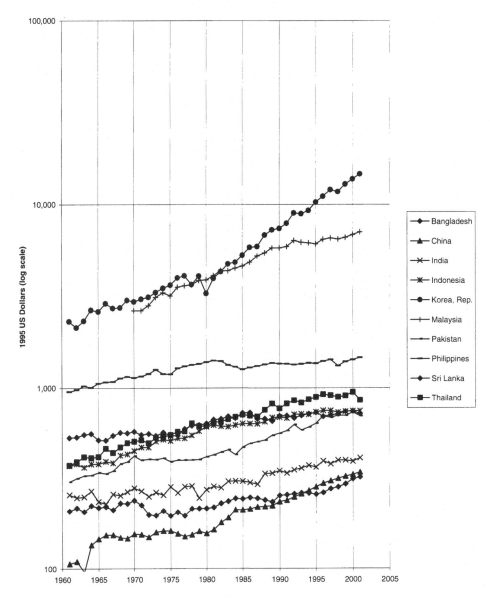

Figure 2(b). Real agricultural GDP per worker, Asia.

helpful. The simplest equation for estimating convergence is

$$G_{t,0} = \alpha + \beta y_0 \qquad (1)$$

where $G_{t,0}$ is the rate of growth of agricultural GDP per worker between time 0 and a later time, t; y_0 is the log of the initial level of agricultural GDP per worker; and α and β are parameters to be estimated. The estimate of β indicates change in the growth rate resulting

from a 1% higher level of y_0. Analogously, other initial-year levels of other variables hypothesized to influence growth can be added to equation (1). What we give up with this simple approach is the capability to estimate the dynamics of growth—how changes in causal variables affect growth and its timing—and the capability to estimate structural parameters of production or supply relationships. However, the econometric problems of sorting out causal effects from trending time

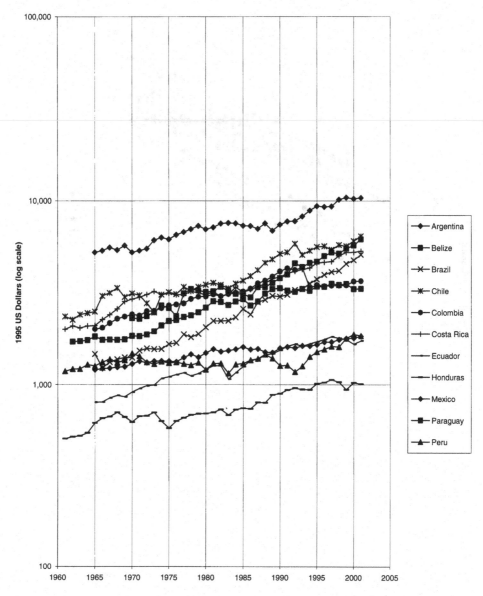

Figure 2(c). Real agricultural GDP per worker, Latin America.

series, and the predominance of measurement errors or other random fluctuations in year-to-year changes, are likely to preclude estimating dynamic relationships anyway.

Econometric objections to equation (1) have been raised.[11] One is the likelihood of bias toward a negative value of the estimated β because initial measured levels are temporarily low or high just by chance, owing to measurement error or transitory single-year events. When this occurs convergence is, according to equation (1), likely to be observed even if in fact no real convergence occurs. A second problem is that if variables omitted from the equation are positively correlated with income growth but negatively correlated with initial income, the estimated β will be biased toward a negative value. To address this criticism,

[11] For details, see Quah (1996) or Nerlove (2000).

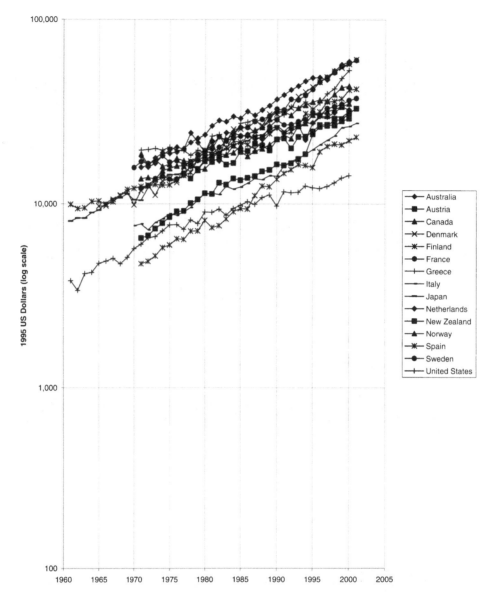

Figure 2(d). Real agricultural GDP per worker, OECD countries.

initial-year values can be added for likely omitted variables, thus estimating "conditional" convergence. This approach is what was suggested above, adding additional variables hypothesized to be causes of growth. The first problem is one that may be intractable given likely measurement error as well as random fluctuation in the agricultural GDP data. It means that there cannot be much confidence in what is really measured by the estimated β.

The general form of the linear regressions to be estimated is

$$G_{t,0} = \alpha + \beta y_0 + \gamma X_0 + \delta \Delta Z_{t,0} + \varepsilon \qquad (2)$$

where X_0 is a vector of initial values of hypothesized causal variables that may be endogenous, $\Delta Z_{t,0}$ is a vector of rates of change between 0 and t (changes in natural logs) of hypothesized causal variables that change over time and are exogenous to $G_{t,0}$, ε is an independent and identically distributed error term, and γ and δ are vectors of parameters that provide estimated impacts of the variables on growth. The observations are of a cross section of countries (country subscripts on the variables and error term are suppressed).

Explaining growth in the framework of equation (2) places the emphasis differently than the estimation of cross-country production functions. The key variables in production functions are input quantities and, in the dynamic context, investment in capital. But these are endogenous variables and not appropriate either as X_0 or $\Delta Z_{t,0}$ variables in equation (2). The use of equation (2) is rather to explore quantitatively the influence of factors that the literature on economic development has given attention to, typically in a descriptive or qualitative fashion. For example, a nation's rural infrastructure, human capital, market institutions, and political framework are aspects of the initial conditions that may be conducive or inimical to growth. Policy changes, educational improvements, or world market changes, if they are exogenous, are examples of possible $\Delta Z_{t,0}$ variables.

Attempts to consider multiple routes to growth, and as a result of that consideration narrow the focus to key factors, typically take a case study, analytically descriptive approach for one or a few countries rather than trying to systematically compare many countries in a cross-country regression. Examples of thoughtful studies of this genre include Pearson et al. (1987), Lele (1989), and Eicher (1999), among many others. While such studies have country-specific objectives, they can be helpful in specifying cross-sectional regressions because their findings for particular cases suggest hypotheses that can sometimes be tested in the cross-country context—the main constraint being whether reasonably believable data can be found to embody the hypothesis.

It is striking in such case studies that the factors that end up being the focus of interest are typically governmental actions. The underlying reason is that the countries considered tend to be those in which economic growth has been weak, as is the case in so much of developing country agriculture for most of recorded history. If a country is mired in stagnation and poverty, one has to look for major changes or shocks to the system, and governments (albeit sometimes foreign governments) are the instruments at hand to provide public

goods such as research, or to remove public bads such as monopoly, abuse of power, or legal disorganization. Given that, Mundlak's admonition quoted at the beginning of this paper, namely for economists to operate "against the common wisdom of governments" may seem surprising. The problem is that governments are often not willing or able to undertake the recommended policies. After all, governments are not usually entities exogenous to the status quo that can be used to shock the economy, but rather are an integral part of the status quo.

Consider the following conditioning factors for agricultural GDP growth, in all of which government has some role: providing macroeconomic and political stability; establishing reliable property rights and incentives; fostering productivity-enhancing new technology; and enabling access to competitive input markets (including credit) and output markets, without exploitive taxation.

The effects of policies or the institutional situation can be tested by introducing variables in equation (2) for initial-period values of variables representing policies or institutions. Unfortunately, because of lack of data this approach is often infeasible. The World Bank development indicators include estimated annual inflation rates, which were used to construct an indicator of macroeconomic instability, the variability (standard deviation) of the rate of inflation over the 1980–2001 period. With respect to political stability and obstacles to investment, indicators from O'Driscoll et al. (2003), Transparency International, and Freedom House (2003) are used that are intended to measure, respectively, commercial freedom, regulatory propriety (lack of corruption), and the overall state of repression in a country.

Factor market constraints, apart from those that stem from general economic and political conditions of the country, are often country-specific. Factor quantities themselves are used as explanatory variables in analyses like Mundlak's (1999, 2000), but as mentioned earlier what one really wants to know is why factor quantities increase or decrease. Prices of specific factors in each country's agriculture vary across countries but they are endogenous variables, consequences as much as causes of agricultural growth.

Product prices are often good candidates as causal variables, essentially treating agricultural sector growth as a matter of estimating supply functions, but in the present analysis we are explaining a single period of time, with all variation in the observations being cross-sectional, so all countries are, to a first approximation, operating under the same world market conditions. Where countries differ in the product market is in the role of subsidies, trade barriers, or other governmental interventions in the markets. Unfortunately, while there have been major efforts to quantify the support provided to agriculture to permit cross-country comparisons among the OECD countries, such measures hardly exist for most less-developed countries of the world. Product prices in these countries are undoubtedly affected by subsidies of agriculture in the OECD countries, and some countries, because of their product mix or location, are affected differently than others. Adverse effects upon growth would appear most directly in a lower agricultural GDP, or slower growth of agricultural GDP, in the most vulnerable countries. In order to obtain evidence of the importance of differential product price experiences on differences in the rates of agricultural GDP growth, the rate of growth of FAO's crop production index, a quantity indicator, should also be considered.[12] The ranking of countries from high to low growth rates in 1980–2001 is quite similar for the two indicators, and not generally lower for agricultural GDP. Because low or declining prices reduce agricultural GDP directly, they affect crop output only indirectly through supply response; the similarity of the two columns makes it seem unlikely that differential price experience is a dominant force in these rankings—though this hypothesis certainly could use better confirming or disconfirming evidence.

In searching for causes of the transformation of agriculture from economic stagnation to economic growth, it is natural to be drawn to investigation of the experiences of countries where this has happened. This draws attention more to the industrial countries and less to the developing world. Tomich et al. (1995), among others, have drawn lessons from developed countries where agricultural sector growth has occurred. The story economists are most familiar with is as follows: scientists, engineers, and tinkerers, in both the private and public sectors, apply their knowledge to problems of agriculture; extension services and other sources of

information place new knowledge in farmers' hands; and with sufficient property rights and price incentives to call forth the necessary investment, farmers adopt new technology and generate more output and income from their resources. Most of the gains may accrue to buyers of farm products rather than their producers, as increased output drives down prices, but nonetheless this is the paradigm of growth.

Historians have unearthed evidence on what was going on during the period when U.S. agriculture entered its period of strong and sustained productivity growth. This evidence, which has been largely neglected by economists, includes facts about farmers' attitudes and preferences, the intellectual and exemplary contributions of visionary individuals, and the establishment of institutions and forms of economic organization conducive to growth. The idea of the farmer as an ignorant, intellectually ossified follower of traditional practice and fearful of change, and constitutionally unable to forgo consumption in order to invest, was an influential view in the first half of the twentieth century. In this context investment in new technology would require a cultural transformation. Nonetheless, Griliches (1957) brilliantly showed that profitability was sufficient to explain the pattern of adoption of hybrid corn. But rural sociologists also staked a claim to cultural/social explanations such as community leadership and informational networks (e.g., Ryan and Gross, 1943; Havens and Rogers, 1961). Danbom (1979) describes the efforts by many promoters of progress in agriculture, notably President Theodore Roosevelt in the first decades of the twentieth century, to instill in farm people a mentality conducive to commercialization of their enterprises, investment, and the adoption of innovative technology. Historians like Clarke (1994) have also given a broader interpretation of agricultural support programs, particularly the New Deal programs of the 1930s, emphasizing how they altered farmers' outlook in ways that promoted investment and adoption of new technology.

Broader modes of thought are of course not new in the theory of development. Hagen (1962) is exemplary of ideas in the 1950s that obstacles to development are largely traditional rural village institutions and/or inside the heads of the villagers. Thus, "economic theory has rather little to offer" and "both the barriers to growth and the causes of growth seem to be largely internal rather than external" (Hagen, quoted in

---

[12] These data are presented in Table 2 of the conference version of the paper (www.iaae-agecon.org).

Stevens and Jabara, 1988, p. 94). This view is the opposite of Schultz's mentioned earlier, that inputs from outside of agriculture are key, and that, if profitable, they will be adopted. The noneconomic approach lost luster with the perceived failure of community development schemes, and it seems the last nail in its coffin was the Green Revolution. New varieties were adopted along with purchased inputs, apparently without need of cultural or psychological transformation in rural communities.

In order to test these ideas in equation (2) technological and cultural variables are required. The data available on either type of variable have neither the conceptual specificity nor the precise measurement required. There are data that provide plausible indicators of differences in technology between countries. Several studies, such as Evenson and Kislev (1975), use a country's research and/or extension expenditures as a cause of agricultural growth (with long lags). Two of their indicators measured as of the mid-1960s, public research expenditures and agricultural science publications (Evenson and Kislev, 1975, Appendix I), are used here. However, these data are not reported for 24 of the countries discussed in this paper. The CGIAR publishes later data for more countries, of which average public research expenditures in each country during 1976–1980 are used here. These data are used to estimate a research expenditure variable that covers 53 countries with information about spending in the 15 years preceding the 1980–2001 period over which equation (2) is estimated. Because identifying effects of new technology through these variables is far from assured, given the long lags found by most researchers between invention and implementation and the pervasive role of 'spill-ins' from other countries' research and from the CGIAR centers, indirect indicators of technology implementation are also used as X_0 variables. These are initial (1980) estimates of fertilizer per hectare and tractors per hectare. In addition, the rate of growth of cereal yields over the two decades preceding 1980 is used to indicate a preexisting willingness to innovate that may carry over to the 1980–2001 period.

On the cultural side, the most plausible variable about personal characteristics is the extent of illiteracy in a country. High illiteracy plausibly indicates a high prevalence of traditional attitudes that are barriers to growth. Illiteracy may also serve as a proxy for (lack of) schooling, and schooling is an indicator of investment in human capital and improved labor quality

that Schultz pioneered as important in agricultural development, and which has been widely accepted by economists as a source of economic growth.

A quite different labor-centered view of agriculture and economic growth stems from the observation of large numbers of poor and seemingly underemployed people in rural areas. The stark labor-surplus ideas of early dual economy models have evolved to more nuanced assessments that still retain the thought that the path to rural development must overcome in some way the insufficiency of remunerative employment where the ratio of workers to other resources is high. Tomich et al. (1995), for example, characterize the economies where development is needed as CARLs (countries with abundant rural labor). The difference made by labor abundance is tested here using rural population density (workers per hectare) as an explanatory variable.

Table 1 reports the results of estimates of several specifications of equation (2) for the set of countries included in this discussion. The results are typical of cross-country regressions in being suggestive but far from definitive in sorting out causes of growth. Regression 1, the simple convergence model, indicates significant divergence—in countries that started out with the highest agricultural GDP per worker in 1980, that variable grew the fastest between 1980 and 2001.[13] The best performers tend to be the OECD countries. Regression 2 explores whether that is the only reason for divergence by including regional dummy variables. It turns out that the OECD dummy is positive but not statistically significant. The dummy for sub-Saharan African countries is, however, significant and negative; agricultural GDP per worker in these countries on average grew at a rate of 1.6% per year slower than countries in Asia, Latin America, and the remaining group (transition economies and Mediterranean countries) that define the intercept. The estimates of β in regressions 2 and 3 are not significantly different from zero. These results are what Figure 2(a)–(d) suggest, and the regressions confirm the lack of convergence, even "conditional" convergence (appearing when other growth-conditioning variables are held constant).

The growth-conditioning variables included in regression 3 of Table 1 have jointly significant effects on the rate of agricultural GDP increase, and many have

---

[13] Here and later taking significance at the 10 percent level—requiring a 't' statistic of 1.7 or more in absolute value.

Table 1

Regressions explaining growth in agricultural GDP per worker, 71 countries, 1980–2001

Independent variable	Dependent variable: Percent growth in Ag. GDP per worker			
	1	2	3	4
Intercept	−0.303 (−3.34)	0.009 (0.50)	0.077 (2.08)	0.251 (3.20)
Ag. GDP per worker, 1980	0.007 (5.60)	0.002 (0.92)	−0.002 (−0.71)	−0.019 (−2.84)
Africa		−0.016 (−2.25)		
Asia		0.001 (0.13)		
Latin America		−0.006 (−0.81)		
OECD countries		0.011 (1.25)		
Fertilizer per ha., 1980			0.097 (3.31)	0.050 (1.01)
Tractors per ha., 1980			−0.073 (−1.66)	−0.213 (−3.62)
Growth in crops per worker, 1961–1980			0.261 (1.98)	0.599 (2.64)
Illiteracy rate of youth, 1980			0.0007 (0.54)	−0.0002 (−0.06)
Ag. research spending, percent of GDP, 1965–80			−0.0051 (−1.49)	−0.0025 (−0.50)
Std. dev. of inflation rate			0.0007 (1.09)	0.0024 (1.62)
Restraints on economic freedom (Heritage)			−0.008 (−1.21)	−0.028 (−2.86)
Absence of corruption (Transparency International)			0.003 (1.84)	0.002 (1.02)
Restrictions of civil liberties (Freedom House)			−0.006 (−1.61)	−0.030 (−3.29)
Rural population per ha., 1980			−0.014 (−1.34)	0.042 (2.06)
Trade in goods (% of GDP)				0.042 (2.43)
PSE				0.036 (2.96)
$\bar{R}^2$	0.309	0.426	0.532	0.714
Number of countries	71	71	49	27

the expected signs, but none of them emerges individually as a predominant determinant of agricultural growth.

Consider, for example, the illiteracy variable. This is the variable most directly related to human capital and also to ideas about cultural prerequisites to growth. The variable has an unexpected positive sign, indicating that countries with higher illiteracy grew faster. But this sign is not robust—it changes as other right-hand side variables are added or deleted—and the variable is not statistically significant. This lack of significance is not a complete surprise as it parallels the findings in Craig et al. (1997) and Mundlak (1999), who also estimated no significant effects of literacy on productivity or agricultural GDP per worker. However, Hayami and Ruttan (1985, Chapter 6) found education an important cause of productivity growth in agriculture (although their literacy variable alone was often insignificant). So did Antle (1985) and Fulginiti and Perrin (1993).[14] It could be argued that illiteracy data are imperfect both conceptually and practically

as measures of human capital or skill, but that still leaves no clear answer about the importance of schooling to growth. Based on empirical findings in the United States from states, counties, and individual farms (Gardner, 2002), it is possible that farmer education as a contributor to agricultural productivity has nothing like the importance that Schultz's ideas about human capital suggested and that early empirical work, notably Welch (1970), found.

The other variable related to labor supply is rural population density (persons per square kilometer) at the beginning of the period, 1980. If abundant labor is a hindrance to growth, this variable should have a negative sign. But the variable is insignificant in Table 1, and in all other specifications of the equation tried. Similar results were obtained for an alternative specification of this variable, namely agricultural workers per hectare of arable land.

Variables intended to indicate the initial presence of technological innovation had mixed performance. The 1980 level of fertilizer use per hectare has a positive sign and is statistically significant in regression 3, but tractors per hectare in 1980 are insignificant. Growth of crop output per worker in the preceding period,

---

[14] See Huffman (2001) for a thorough review of econometric studies.

1961–1980, has a significantly positive sign, suggesting that countries with a history of productivity improvement have the momentum that carries through to later growth. However, the research variable is insignificant. This result persists whether research spending is measured per dollar of agricultural GDP or as an absolute amount (the latter being appropriate if research is a pure public good within the country). The alternative technology variable from Evenson and Kislev (1975), research publications, has an even lower *t* statistic.

Of the four economic-political environment variables, only the absence of corruption as scored by Transparency International is significant. Its coefficient says that if a country's noncorruption index goes up by four positions on a scale of 1 to 7, that adds 1% annually to the country's rate of growth of agricultural GDP per worker.

Regression 4 adds two variables more explicitly related to a country's economic policies. The first is the value of international trade in goods as a fraction of national GDP. The second is an index of governmental support to agricultural commodity markets, the producer subsidy equivalent (PSE). PSEs have been calculated in a number of ways, and all of those ways have been subject to criticism. The 1985–1989 average PSE as estimated in U.S. Department of Agriculture (1994) is used here. This measure covers the largest number of countries using a consistent calculation method for the same time period. Even so, using this variable reduces the number of countries in the sample to 27, and increases the weight of OECD countries in the sample. The estimated value of goods in international trade in 1980 is also unavailable for many of the 49 countries used in regression 3.

The results are broadly the same as in regression 3, but now rural population density has a significantly positive sign, suggesting that more people per hectare is actually growth-increasing. Also, the two political liberty variables turn out to be significant and with the expected signs in regression 4: the *Heritage Foundation/Wall Street Journal* index of economic freedom has a negative sign as expected (the index is higher the greater the restraints), and it is statistically significant. The same is true of Freedom House's more general index of whether a country is "free" (index value 1), "partly free" (index value 2), or "unfree" (index value 3). However, neither the corruption index nor the economic instability measures (the standard deviation of the rate of inflation) are significant.

The variables added in regression 4, the importance of international trade and the PSE, are both significantly positive. A country can increase its growth of agricultural GDP per worker by trading more and by subsidizing its agriculture. (The national welfare consequences may, of course, be quite different, notably because the PSE boosts agriculture at the cost of taxpayer outlays.) Note that the coefficient means that the PSE would have to increase by 30% of the value of agricultural output in order to boost the rates of agricultural GDP growth by 0.1%.

3.2. Real household income growth

Growth in the agricultural sector is important insofar as it helps achieve growth in real standards of living. But analysis of the relationship between agricultural sector growth and average rural incomes or rural poverty involves several complications to an already complicated set of issues. Note first that the most easily demonstrable gains from productivity growth in agriculture are those of urban consumers of food products who benefit from lower prices. This contribution of agriculture is an important benefit in a whole-economy view but for rural incomes low prices mean less market returns. How can the contribution of agricultural sector growth to rural household incomes be identified?

Consider two views about the causes of real income growth in rural areas: first, agriculture as the engine of growth, with investment in agriculture generating real income growth in rural areas; and second, economy-wide demand for labor as the engine of growth in agriculture, with a growing real wage a sufficient condition for rural household income growth. The reason for drawing this contrast is that in recent work on economic growth in American agriculture (Gardner, 2002), there is more evidence that the second view captures the dominant forces behind the catch-up of farm to urban household incomes levels. Putting aside the Depression, from the "Golden Age of Agriculture" in 1897–1914 through the 1960s, U.S. agriculture was a technologically dynamic magnet for investment, with high and sustained rates of productivity growth after the mid-1930s. Yet the median income of farm households remained low relative to nonfarm incomes. Incomes of farm families rose above 60% of those of the nonfarm population only during 1910–20. The trend, if any, was negative until 1960. Therefore, a vigorous agricultural sector is not a sufficient condition for high incomes or for real income growth in the rural sector (relative to the urban sector).

Later history casts doubt on the necessity of agricultural sector growth for farm household income growth. Although U.S. agricultural productivity (multifactor productivity as estimated by USDA) has continued to grow at about the same rate of 1.5 to 2% annually for the whole 1948–1999 period, since 1980 investment in the sector has turned negative and real agricultural income per farm has declined. Yet this is the very period in which farm household incomes at last caught up to nonfarm household incomes, and indeed by the end of the twentieth century were well ahead (see Gardner, 2002, pp. 78, 84). Therefore, a vigorous agricultural sector is not necessary for high household incomes or income growth in the rural sector.[15]

Even if agricultural growth is neither necessary nor sufficient for household income growth, it could nonetheless be helpful. But in cross-sectional analyses of U.S. states and counties, no significant relationship between sectoral growth and rural household income growth was found, neither at the median-income level nor for relatively low-income groups nor for the incidence of farm poverty (Gardner, 2000). Instead, what matters is the linkage of farm factor markets, particularly the farm labor market, with the nonfarm economy.

Notwithstanding the preceding, it would be premature to dismiss agricultural sector growth as an engine of growth in developing countries, for reasons that have been continually emphasized in the literature on agriculture and development at least since Johnston and Mellor (1961). The most basic is macroeconomic, where agriculture's share of the labor force is large. An increase in real output per worker resulting from agricultural productivity growth increases labor productivity in the whole economy, and hence increases real income per capita. This is the point at which the oft-repeated statistics about a near majority of the developing world's labor force being rural becomes relevant to agriculture as an engine of growth. Similarly, if agriculture generates a large fraction of a country's consumption or export-earning goods, then improved

productivity in agriculture directly increases real GDP per capita substantially. If a public investment of $1 billion in a sector of the economy generates an increase in total factor productivity of 2% over the next 20 years in that sector, it is best, other things equal, to carry out the investment in the largest sector available. On those simple grounds, agriculture has a leg up in many economies.

Figure 3 shows the relationship between growth of agricultural GDP per worker and national GDP per capita for 52 developing countries during 1980–2001. The association is positive and significant. But what is the direction of causality? Investigation of lags in the sample of countries during 1961–2001 does not show agriculture as leading.

The prime experience in the developing world that impels consideration of agriculture as an engine of rural income growth is East Asia. This is where both agricultural GDP and overall GDP per capita have generated the most positive long-term record. The World Bank (2002) concludes that in these countries "agricultural development created a dynamism in rural areas, which, in later stages, was combined with rapid industrialization" (p. 47). The reasons for this conclusion are not spelled out, however. Both agricultural and overall GDP growth for South Korea, the outstanding example of rapid and sustained agricultural GDP growth, are shown in Figure 4. No strong message about causality is apparent. National GDP per capita grew faster than agricultural GDP per worker throughout the 1961–2001 period, the latter at the high rate of 4.7% annually and the former at the extraordinary rate of 6.1%, increasing from $1,350 per capita (in 1995 dollars) in 1961 to $13,500 in 2001. Agricultural GDP grew at a faster rate after 1980 than before, suggesting that national GDP led agricultural growth rather than the Bank's asserted causality.

Data are available on the growth rates of agricultural GDP per worker and national GDP per capita for 66 countries.[16] Within each of the regional developing country groupings (Africa, Asia, and Latin America), the countries that grew fastest in national GDP per capita also grew fastest in agricultural GDP per worker, with a few notable exceptions such as Brazil. Yet in the region where the fastest growth occurred, Asia,

---

[15] To put the point more concretely, if for example the mechanical cotton picker had never been invented, U.S. cotton laborers would not have appreciably higher incomes today, yet there would be a lot more of them (assuming the U.S. remained competitive in world markets).

[16] These are reported in Table 4 of the conference version of this paper (www.iaae-agecon.org).

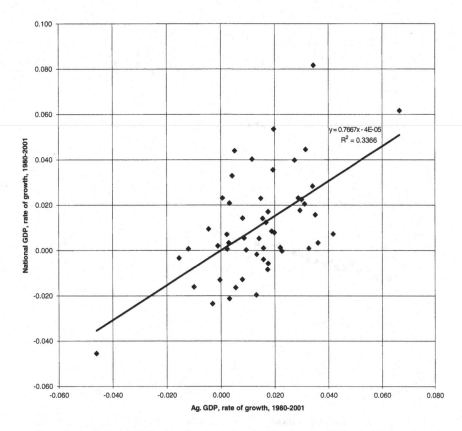

Figure 3. Agricultural GDP per worker and national GDP per capita, 52 developing countries.

agricultural growth lagged behind national growth; while in Africa and Latin America agricultural growth was higher.[17]

Regressions like those whose results are reported in Table 1 were also estimated with the growth of national GDP per capita as the dependent variable. These regressions (not shown in detail) gave results more nearly in accord with expectations. The estimated β coefficient is significantly negative, indicating that the lower-income countries in 1980 grew faster during 1980–2001. Factors causing faster growth in agricultural GDP have positive effects on national GDP growth. The political and economic institutional variables are significant, but the corruption index appears more closely related to GDP growth than to agricultural

growth. Because of its prominence in recent literature the hypothesis that countries with access to international waters (oceans or seas connected to them) grow faster than land-locked countries was also tested. Indeed, in this sample the "coastal" countries (75% of the sample) did grow significantly faster during 1980–2001, other things held constant. And, consistent with that finding, greater participation in international trade in the initial period (1980) is significantly related to faster GDP growth in 1980–2001.

However, these results say nothing about rural as compared to urban incomes, as these income differences are not in the data set. One study that carried out an econometric investigation of specifically rural incomes in a developing country context is Estudillo et al. (2001), who worked on wage rates of agricultural workers in the Philippines. Their findings parallel those cited above for the United States—the cause of growth in agricultural wage rates is growth of labor demand in the nonfarm economy. The implications of agricultural

[17] Note that these comparisons do not say anything about agriculture's share of national GDP, which turns on aggregate values. These growth rates are per person, so if the rural population is declining it is quite possible for agricultural GDP per person to rise while aggregate agricultural GDP, and its share in total GDP, decline.

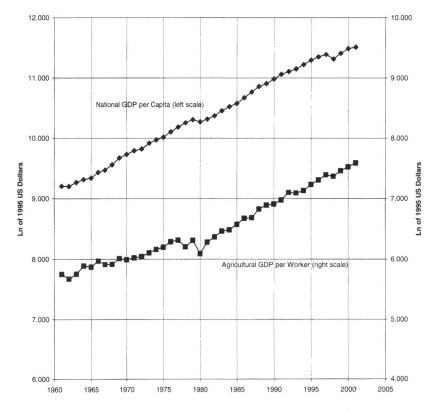

Figure 4. Agricultural GDP and national GDP: Korea.

GDP growth for rural poverty are even more complicated to analyze. The preponderance of evidence appears to support the conclusion that agricultural productivity growth is poverty reducing (see Hazell and Haddad, 2001; IFAD, 2001). But for the United States at least, real income growth in the nonfarm sector was found to be more fundamentally important in increasing low farm incomes than any specifically agricultural variable (Gardner, 2000). Timmer (2002) reports similarly but more nuanced findings for a sample of developing countries in terms of the linkages between nationwide per capita income and incomes in the lowest quintile, but he does not distinguish between agricultural and nonagricultural sources of national GDP.

4. Concluding discussion

The chief candidates for causes of growth in agricultural value-added (GDP) and rural household income growth are:

1. Macroeconomic and political stability
2. Property rights and incentives
3. Productivity-enhancing new technology
4. Access to competitive input and product markets
5. Real income growth in the nonagricultural economy

The first four of these have been discussed with reference to agricultural GDP growth, the fifth with reference to rural household income growth. Case studies have found all of these factors to be important, but compelling as the case may be, on the basis of observation and thought, for these causes being important, the cross-country empirical evidence on their role, in this paper and elsewhere, is mixed. The Green Revolution showed that some success can be achieved even without significant changes in (1), (2), (4), and (5); but transforming those gains into permanent increases in rural living standards has proven elusive.

What is scarcest is observations of sustained growth in developing countries. Where growth in rural household incomes has been achieved, all five factors are substantially present. In all these countries agriculture as a share of national GDP has fallen substantially, and only in a few agricultural exporting countries can agriculture really be said to be making a major contribution to national economic growth. Even in the fast-growing East Asian countries where agriculture has grown, this appears to be as much due to government subsidies as to growth dynamics generated by technological change or other productivity improvements within agriculture (Tsay, 2002).

In the context of overall real income growth in a country, it is important to distinguish between agricultural GDP in aggregate and agricultural GDP per worker. With appropriate institutions and policies GDP per worker may increase in all countries, but there is no expectation that the size of agriculture as a sector (aggregate agricultural GDP) would increase in all countries. Because of location-specific changes in technology or changes in relative factor prices because of different countries' factor endowments, some countries are expected to gain, but others lose, comparative advantage in agriculture under economic growth. So success in growth should not automatically be identified with increasing agricultural GDP (or output). On the other hand, increasing productivity or agricultural GDP per worker is generally an indicator of success, in the sense of providing the material basis for an improved standard of living for both rural and urban residents. That is why this paper focuses on GDP per worker rather than aggregate GDP or agricultural output.

Even when agricultural productivity grows, it is apparent that rural household incomes may not grow, as the earlier discussion of the United States indicated. It appears likely that a similar lesson will emerge from the East Asian countries where rural household incomes are growing: what is necessary is real average income growth in the economy as a whole, and that may be sufficient for rural income growth even if agriculture shrinks. In this context, some of the factors that did not show well in the cross-country regressions explaining agricultural GDP per worker may nonetheless be important causes of rural workers' income growth. It is well attested that education, to take the prime example, is valuable in increasing workers' earnings. But this value is not nearly so evident in farm production.

Although the arguments and evidence have hardly been touched upon here, it seems likely that the preceding conclusion applies also to rural poverty. To remedy rural poverty, what is most needed is improvement in the labor market generally more than, say, improved crop varieties. This is not to say that agricultural research, rural infrastructure investment, or the development of agricultural export sectors are not valuable or that their net effect on poverty is not in the right direction. The literature cited earlier suggests otherwise. Agricultural research and rural education and infrastructure development efforts have been highly profitable investments with a regularity that defies most commercial innovations. Agricultural economics is the discipline that can analyze the possibilities for these and other profitable investment opportunities in farm commodity and input markets. Yet it is becoming evident that rural income growth and poverty alleviation are not sub-fields of agricultural economics.[18]

In closing, I have to say that I am uneasy about the preceding conclusions. What they mostly rest upon is a failure to find sufficiently strong associations between variables representing hypothesized causal factors in agricultural GDP growth and differences across countries in actual growth rates of agricultural GDP per worker. Many of the variables are crude proxies for the real variables, and for these proxies measurement errors are likely. So the conclusions are even more than usually tentative. I am continuing these investigations together with Isabelle Tsakok and would be happy to eat what I have written here if further data and analysis change the story.

References

Alston, J., C. Chan-Kang, M. Marra, P. Pardey, and T. Wyatt, "A Meta-Analysis of Rates of Return to Agricultural R&D," Research Report 133 (International Food Policy Research Institute: Washington, DC, 2000).

[18] Evolution in how the agricultural sector is treated in the World Bank illustrates the problems that can arise. Traditionally an agricultural unit within the Bank made sense because of loans that had a central technical and business focus on agricultural production, so one needed technical experts in these matters. But as the Bank came to focus more on broader structural and policy matters in client countries, and on poverty reduction as a goal, to be useful an agricultural unit had to broaden greatly to become a rural development unit with expertise in all manner of socioeconomic issues. But when this is done, as exemplified in the Bank's recent rural strategy document (World Bank, 2002), it is not clear why the most appropriate policies and lending are not those centered in other parts of the Bank, such as education or health.

Antle, J., "Infrastructure and Aggregate Agricultural Productivity: International Evidence," *Economic Development and Cultural Change* 31 (1985), 609–619.

Barro, R., and X. Sala-i-Martin, *Economic Growth* (McGraw Hill: New York, 1995).

Clarke, S., *Regulation and the Revolution in United States Farm Productivity* (Cambridge University Press: New York, 1994).

Craig, B. J., P. G. Pardey, and J. Roseboom, "International Productivity Patterns: Accounting for Input Quality, Infrastructure, and Research," *American Journal of Agricultural Economics* 79 (1997), 1064–1076.

Danbom, D., *The Resisted Revolution: Urban America and the Industrialization of Agriculture* (Iowa State University Press: Ames, 1979).

Deaton, A., "Data and Econometric Tools for Development Analysis," in J. Behrman and T. N. Srinivasan, eds., *Handbook of Development Economics*, Volume IIIA (Elsevier Science: New York, 1995), pp. 1785–1882.

Eicher, C., "Institutions and the African Farmer," *Third Distinguished Economist Lecture* (CIMMYT: Mexico City, 1999).

Estudillo, J., A. Quisumbing, and K. Otsuka, "Income Distribution in Rice-Growing Villages during the Post–Green Revolution Period," *Agricultural Economics* 24 (2001), 71–84.

Evenson, R. E., "Economic Impacts of Agricultural Research and Extension," in B. Gardner and G. Rausser, eds., *Handbook of Agricultural Economics*, Volume IA (Elsevier Science: New York, 2001), pp. 573–628.

Evenson, R. E., and D. Gollin, "Assessing the Impact of the Green Revolution, 1960–2000," *Science* 300 (2003), 758–762.

Evenson, R. E., and Y. Kislev, *Agricultural Research and Productivity* (Yale University Press: New Haven, 1975).

Freedom House, "Freedom in the World Country Ratings 1972–73 to 2001–2002," available on-line at http://www.freedomhouse.org/ratings/, accessed 20 July 2003.

Frisvold, G., and K. Ingram, "Sources of Agricultural Productivity Growth and Stagnation in Sub-Saharan Africa," *Agricultural Economics* 13 (1995), 51–61.

Fulginiti, L. E., and R. K. Perrin, "Prices and Productivity in Agriculture," *Review of Economics and Statistics* 75 (1993), 471–482.

Gardner, B., "Economic Growth and Low Incomes in Agriculture," *American Journal of Agricultural Economics* 82 (2000), 1059–1074.

Gardner, B., *American Agriculture in the Twentieth Century: How It Flourished and What It Cost* (Harvard University Press: Cambridge, MA, 2002).

Griliches, Z., "Hybrid Corn: An Exploration in the Economics of Technological Change," *Econometrica* 25 (1957), 501–522.

Hagen, E. E., *On the Theory of Social Change—How Economic Growth Begins* (Dorsey: Homewood, IL, 1962).

Havens, A. E., and E. M. Rogers, "Adoption of Hybrid Corn: Profitability and the Interaction Effect," *Rural Sociology* 26 (1961), 409–414.

Hayami, Y., and V. Ruttan, *Agricultural Development: An International Perspective* (Johns Hopkins University Press: Baltimore, 1985).

Hazzell, P., and L. Haddad, "Agricultural Research and Poverty Reduction," Food, Agriculture and the Environment Discussion Paper (International Food Policy Research Institute: Washington, DC, 2001).

Huffman, W. E., "Human Capital: Education and Agriculture," in B. Gardner and G. Rausser, eds., *Handbook of Agricultural Economics*, Volume 1A (Elsevier Science: New York, 2001), pp. 333–381.

International Fund for Agricultural Development (IFAD), *Rural Poverty Report: The Challenge of Ending Rural Poverty* (Oxford University Press: Oxford, 2001).

Johnston, B. F., and J. Mellor, "The Role of Agriculture in Economic Development," *American Economic Review* 51 (1961), 566–593.

Lele, U., "Sources of Growth in East African Agriculture," *World Bank Economic Review* 3 (1989), 119–144.

Mundlak, Y., "The Dynamics of Agriculture," Elmhirst Memorial Lecture, Proceedings of the Twenty-third International Conference of Agricultural Economists, Aldershot, England (Ashgate Publishing Company, 1999), pp. 18–48.

Mundlak, Y., *Agriculture and Economic Growth* (Harvard University Press: Cambridge, MA, 2000).

Mundlak, Y., "Explaining Economic Growth," *American Journal of Agricultural Economics* 83, no. 5 (2001), 1154–1167.

Nerlove, M., "Growth Rate Convergence, Fact or Artifact? An Essay on Panel Data Econometrics," in J. Krishnakumar and E. Ronchetti, eds., *Panel Data Econometrics: Future Directions* (Elsevier Science: Amsterdam, 2000).

O'Driscoll, G. P., Jr., E. J. Feulner, and M. A. O'Grady, "Index of Economic Freedom," *The Heritage Foundation/Wall Street Journal*, available on-line at http://www.heritage.org/research/features/index/, accessed 18 June 2003.

Olson, M., *The Rise and Decline of Nations* (Yale University Press: New Haven, 1982).

Quah, D. T., "Empirics for Economic Growth and Convergence," *European Economic Review* 40 (1996), 1353–1376.

Pearson, S., F. Avillez, J. Bentley, T. Finan, R. Fox, T. Josling, M. Langworthy, E. Monke, and S. Tangermann, *Portuguese Agriculture in Transition* (Cornell University Press: Ithaca, NY, 1987).

Ryan, B., and N. Gross, "The Diffusion of Hybrid Corn in Two Iowa Communities," *Rural Sociology* 8 (1943), 15–24.

Schultz, T. W., *Transforming Traditional Agriculture* (Yale University Press: New Haven, 1964).

Schultz, T. W., *Economic Crises in World Agriculture* (University of Michigan Press: Ann Arbor, MI, 1965).

Schultz, T. W., *Distortions of Agricultural Incentives* (Indiana University Press: Bloomington, IN, 1978).

Simon, H., "Effects of Increased Productivity upon the Ratio of Urban to Rural Population," *Econometrica* 15 (1947), 31–42.

Singh, I., L. Squire, and J. Strauss, eds., *Agricultural Household Models: Extension, Applications, and Policy* (Johns Hopkins University Press: Baltimore, 1986).

Stevens, R. D., and C. L. Jabara, *Agricultural Development Principles* (Johns Hopkins University Press: Baltimore, 1988).

Timmer, C. P., "Agriculture and Economic Development," in B. Gardner and G. Rausser, eds., *Handbook of Agricultural Economics*, Volume 2A (Elsevier Science: Amsterdam, 2002), pp. 1487–1546.

Tomich, T. P., P. Kilby, and B. F. Johnston, *Transforming Agrarian Economies* (Cornell University Press: Ithaca, NY, 1995).

Transparency International, "Transparency International Corruption Perceptions Index," available on-line at http://www.transparency.org, accessed 20 July 2003.

Tsay, Y.-Y., "Economic Growth and Agriculture's Relative Decline," mimeo (2002).

U.S. Department of Agriculture, "Estimates of Producer and Consumer Subsidy Equivalents," Economic Research Service, Statistical Bulletin No. 913 (December 1994).

Welch, Finis, "Education in Production," *Journal of Political Economy* 78 (1970), 35–59.

World Bank, "Reaching the Rural Poor," Rural Sector Board (October 2002).

Plenary 1

Rural poverty dynamics: development policy implications

Christopher B. Barrett*

Abstract

This article explores the useful distinction between chronic and transitory poverty in understanding rural welfare dynamics, highlighting the possibility of poverty traps and their implications for "cargo net" policies to build up productive assets and "safety net" policies to protect such assets. We discuss the methodological difficulties in identifying and explaining either poverty traps or the critical thresholds that are their defining feature. A few empirical examples from sub-Saharan Africa illustrate the likely existence of poverty traps that help to explain chronic rural poverty.

JEL classification: O1, Q1

Keywords: economic growth; chronic poverty; cargo nets; safety nets; transitory poverty

1. Introduction

"Most of the people in the world are poor, so if we knew the economics of being poor we would know much of the economics that really matters. Most of the world's poor people earn their living from agriculture, so if we knew the economics of agriculture we would know much of the economics of being poor."

—T. W. Schultz (1980)

T. W. Schultz's words are no less true today than when he opened his 1979 Nobel Lecture with them almost a quarter of a century ago. Economics has nonetheless advanced significantly in its understanding of poverty since Schultz's seminal contributions. This paper summarizes a few key findings from a rich and growing body of research over the past 25 years about the nature of rural poverty and, especially, the development policy implications of relatively recent findings and ongoing work.

As will be explained in greater detail below, economists have begun to focus more precisely on the useful distinction between transitory and chronic poverty. Each has a different implication for poverty alleviation policy. Policymakers' and researchers' greatest concern revolves around chronic poverty, which seems to result from low initial endowments of productive assets, inability to generate high returns from the assets one owns, severe shocks that wipe out accumulated wealth, or some combination of these. Asset stocks appear central to the story of chronic poverty because returns on assets can be endogenously increasing for any of several reasons and financial market failures can impede the capacity of the poor to invest in productive assets to surmount thresholds at which the returns on assets are increasing. This has significant implications for both the design of research to identify such thresholds and for the targeting and emphasis of policies intended to address chronic poverty.

2. Rural poverty dynamics: what we know

Poverty is a complex, multifactorial concept reflecting a low level of well-being. Economists tend to use income or expenditure flows as a proxy for welfare and thus to use inherently arbitrary—albeit often rigorously constructed—poverty lines to define who is and is not poor. This approach is appropriately contested within the social sciences, and there has been considerable advance in the use of multidimensional poverty measures (Bourguignon and Chakravarty, 2002;

* Department of Applied Economics and Management, Cornell University, Ithaca, NY 14853-7801, USA.

Duclos et al., 2003). For the sake of simplicity, however, income is accepted as the dominant welfare measure by economists, and is used in this paper.

Although the measurement of poverty is an important technical concern, the focus here is not on where a poverty line is located, nor in precisely how many people fall below it, nor how far below it, at a given point in time. Although these are indisputably important issues, they are inherently static concerns. Rather, the focus here is on the dynamics of the measures of well-being, and only loosely in relation to a poverty line—who climbs above it, descends below it, or oscillates around it—because poverty dynamics is the more fundamental policy concern.

The reason for the primacy of poverty dynamics is that some but not all of the poor need help through policy. As the rest of this section explains more fully, policy research needs to distinguish between transitory and chronic poverty. One class of policies—safety nets—can effectively block the descent of the transitory poor into chronic poverty. Another class of interventions—termed cargo nets, for reasons explained below—can help the chronically poor find a way out of poverty. Picking the right policy to help a given poor subpopulation depends on an accurate understanding of rural poverty dynamics.

At this point, a brief digression into the simple mathematics of income dynamics may help frame the ensuing discussion with a little more precision. Taking the standard approach of using income as an (imperfect) measure of well-being, for any individual observational unit (a person, household, village, or nation), measured income, Y, is merely the sum of earned returns from productive assets, temporary income shocks, and measurement error:

$$Y = A'R + \varepsilon^T + \varepsilon^M, \tag{1}$$

where A is a vector of productive assets controlled by the household, R is a vector of returns on the assets in A, ε^T represents transitory exogenous income that is independent of asset productivity (e.g., lottery winnings, gifts),[1] and ε^M represents researcher measurement error. Returns are stochastic, thus

$$R = r + \varepsilon^R, \tag{2}$$

where r is the expected return and ε^R is an exogenous shock to physical productivity (e.g., due to rainfall or pests) or input or output prices. Assume all shocks (ε^M, ε^R, and ε^T) are mean zero, constant variance, and serially independent. This framework depicts income as a function of asset holdings, casting it in a familiar portfolio management framework. The mean and variance of income are thus simply

$$E[Y] = A'r, \tag{3}$$

$$V[Y] = A'V[\varepsilon^R]A + V[\varepsilon^T] + V[\varepsilon^M], \tag{4}$$

respectively. Expected income fundamentally depends on one's endowment of productive assets and the sorts of returns one can reap from those assets, thanks to production technologies and markets. Income variability results not only from stochastic returns to land, labor, financial savings, and other productive assets, but also from (due to volatility both in unearned transitory income and, in an econometrician's sample) measurement error.

Substituting (2) into (1) and then totally differentiating yields an expression for income change as a function of change in asset stocks, change in expected returns on assets, and various shocks:[2]

$$dY = dA'R + A'dr + A'd\varepsilon^R + d\varepsilon^T + d\varepsilon^M. \tag{5}$$

Of course, because the errors are all mean zero and serially independent, ex ante expected income change reduces to just

$$E[dY] = dA'r + A'dr. \tag{6}$$

This equation embodies the core of poverty reduction strategies over at least the past half century. In the initial term on the right-hand side of equation (6), dA reflects (dis)investment patterns, including involuntary asset shocks due to, for example, theft, natural disasters, injuries, or permanent illness. For many years,

[1] Remittances from migrant household members would fall under $A'R$, as it relates to an allocation of assets (labor power) to a particular activity and location.

[2] This simple partitioning is very similar to Dercon's (2000) innovative approach, but without the necessity of imposing the assumptions that there is a unique concave production technology, that markets are complete and competitive, or that households maximize profits (equivalently, that households' consumption and production decisions are separable). Those assumptions inherently rule out the poverty trap phenomena, which are discussed below.

antipoverty policy has focused on changing Y through dA, via land reform to transfer land to the poor, education and health programs to build the human capital of the poor, post-drought livestock restocking to reconstitute herds adversely impacted by climatic shocks, etc. Over the past fifteen to twenty years, increasing attention has been paid to the second term, emphasizing dr, the change in expected returns to productive assets. Policymakers and development scholars have expressed renewed concern about technological advances for smallholder farmers—most recently in the form of biotechnological and agroecological approaches to boosting yields—and about market-oriented sectoral and macroeconomic reforms intended to improve the output/input price ratios net of market access costs faced by the rural poor.

2.1. Transitory and chronic poverty

Recent research has underscored, however, that much poverty is transitory in nature.[3] Put differently, because the errors in equations (1) and (2) are mean zero, many realizations are necessarily negative, leading to lower-than-expected incomes that push people a bit below the poverty line for a relatively brief period of time, although their expected incomes lie above the poverty line. Moreover, temporarily low incomes are sometimes chosen by people as part of a long-term accumulation strategy, as almost any graduate student knows from personal experience. The incomes of the transitorily poor—whether temporary poverty is by chance or by choice—subsequently recover as new draws are made on income's stochastic elements (ε^R, ε^T) or as they begin to enjoy the payoff from voluntarily foregone income, often without any external assistance from charities or governments. While even transitory poverty is plainly undesirable—and safety nets to keep the transitorily poor from falling into chronic poverty are critically important—the obvious capacity of the transitorily poor to pull themselves up by their bootstraps means that policy interventions on their behalf are not always needed. Indeed, costly government interventions that risk

disturbing their self-sufficiency may sometimes be undesirable.

Care should be taken, however, not to jump to the erroneous conclusion that interventions on behalf of the poor are therefore unnecessary or undesirable. For one thing, exit rates from (and entry rates into) poverty tend to be overstated due to measurement error, which can inadvertently lead to overestimation of transitory poverty and a policy bias against intervention to assist the poor. The basic problem is that $d\varepsilon^T$, the change in transitory income, $d\varepsilon^R$, the change in returns on assets, and $d\varepsilon^M$, the change in measurement error, all necessarily lead to regression toward the mean in panel data. While the former two constitute true change in transitory income—due to interruptions in interhousehold transfers or to crop yield shocks, for example—they cannot be easily separated in data from changes in measurement error across periods due to questionnaire revisions, respondent fatigue or replacement, new field enumerators, etc. Because $d\varepsilon^T$ and $d\varepsilon^R$ are essentially impossible to identify separately from $d\varepsilon^M$, measurement error tends to inflate estimates of transitory poverty by creating artificial variability in incomes, leading to an upward bias in estimates of the share of the poor who are able to pull themselves out of poverty unassisted (Baulch and Hoddinott, 2000; Luttmer, 2002).[4]

2.2. Safety nets and cargo nets

The problem of getting estimates of transitory poverty rates correct matters because the chronically poor[5] cannot climb out of poverty without external assistance. Such assistance can come directly in the

---

[3] See in particular one of the original studies in this vein, by Grootaert and Kanbur (1995), the excellent recent volume by Baulch and Hoddinott (2000), and the various studies cited therein.

[4] In sampling, there are additional problems of inference associated with prospective bias due to nonrandom attrition from the sample over time due to respondent death, refusal to participate in later survey rounds, residence-based sampling after endogenous household division or union, and failure to trace migrant households. The evidence is quite mixed as to how significant a problem these phenomena pose. See the Spring 1998 issue of the *Journal of Human Resources*, Alderman et al. (2000), Falaris (2003), and Rosenzweig (2003) for details.

[5] The terms "chronically" and "persistently" poor are used interchangeably, even though some analysts try to distinguish between the two based on frequency of observations below a poverty line or mean income over a period relative to the poverty line.

form of transfers, or indirectly in the form of policy reforms that relax constraints on the choice sets faced by the chronically poor, enabling them to take advantage of previously inaccessible opportunities and to exit poverty of their own accord.

Interventions to combat chronic poverty can take one of two forms: preemptive and redemptive. The first, preemptive interventions, are *safety nets*, which aim to prevent the nonpoor and transitorily poor from falling into chronic poverty. Because people can become transitorily poor up to some threshold level and still recover on their own, often quickly, the role of safety nets is to keep them from crossing that threshold, from becoming chronically poor. Safety nets should restrict entry into the ranks of the chronically poor. In the preceding notation, safety nets truncate the lower tails of the distributions of ε^T and ε^R. Emergency feeding programs, crop or unemployment insurance, and disaster assistance are common examples of formal safety net interventions by governments and outside agencies. Social solidarity networks and systems of informal mutual insurance often provide safety nets internal to communities. The prospective partial displacement of the latter by the former should serve as a caution on the design of safety nets, however, so as to minimize the crowding out of informal safety nets.[6]

The second, redemptive form of poverty reduction intervention is meant to lift people or to help them climb out of poverty. These can be referred to as *cargo nets*. Safety nets catch people, keeping them from falling too far; then people step off the net and climb back up on their own. Cargo nets, by contrast, are used to help climbers surmount obstacles or even to lift objects, overcoming the structural forces (gravity, in the case of literal cargo nets) that would otherwise keep them down. In the notation used above, cargo nets shift A and r. Familiar examples of cargo net policies include land reform, targeted school feeding programs, targeted microfinance, or agricultural input subsidization projects, etc. Safety nets block pathways into chronic poverty for the nonpoor and transitorily poor. Well-designed and implemented cargo nets can set people onto pathways out of chronic poverty.

2.3. Identifying and explaining chronic poverty

Because different people need different types of assistance through policy or project interventions, researchers and policymakers must be able to distinguish between them. The descriptive task of distinguishing the chronically poor from the transitorily poor is a significant challenge. One can establish ex post whether people recovered after falling below a poverty line, provided one has sufficient time series data on the same individuals or households. But at the time when policymakers need to decide on prospective interventions, it can be difficult to predict ex ante from data who will recover and who will not, hence the attention paid over the past decade to identifying the correlates of "chronic" or "persistent" poverty.[7] Analysts can use past panel data to identify good predictors of future well-being in order to predict which of today's poor are likely to become nonpoor by some future date. If done accurately, such estimation can provide a basis for targeting interventions among the poor, enabling policymakers to distinguish between the nonpoor and the transitorily poor, for whom cargo nets—as distinct from safety nets—are unnecessary and possibly even unintentionally harmful, and the chronically poor who need direct assistance if they are to escape poverty.

The trick therefore lies in our ability to decompose poverty between those who are, to use Carter and May's (2001) terminology, "structurally" poor—that is, expected to remain chronically poor unless they receive assistance—and those who are "stochastically" poor, who one would expect to exit poverty of their own accord before long, i.e., the transitorily poor. This sort of decomposition has great potential as a tool for informing policy design because governments, donors, and operational agencies (e.g., NGOs, or multilateral agencies of the United Nations) faced with large numbers of structurally poor individuals or households confront a distinctly different challenge than those serving large numbers of stochastically poor persons. Once we know how to distinguish the transitorily or stochastically poor from the structurally or chronically poor using panel data econometric methods, the next challenge is to identify the mechanisms that lead to chronic

---

[6] See Cox and Jimenez (1992), Dercon and Krishnan (2003), and Albarran and Attanasio (2004) for empirical evidence on such crowding out effects.

[7] See, for example, Baulch and Hoddinott (2000), World Bank (2000), Carter and May (2001), or the papers in Hulme and Shepherd (2003).

poverty in order for interventions to treat causes rather than merely symptoms.

Some people are born into poverty and have difficulty escaping because (i) they do not enjoy the education, health, or nutrition required to accumulate crucial physical stature and cognitive capacity early in life (Loury, 1981; Strauss and Thomas, 1998; Basu, 1999), (ii) they do not inherit land or capital sufficient to add value to their human capital, or (iii) they cannot effectively employ the assets they own to generate income (Carter and May, 1999). There is no good empirical evidence on intergenerational earnings transmission in low-income rural settings. In the meantime, the evidence from higher income countries such as Finland and the United States suggests that even where governments offer relatively generous support for children's education, health and nutrition, and where financial markets are relatively accessible even to poor people, estimated elasticities of intergenerational earnings transmission are high, on the order of 0.6–0.8, and primarily attributable to credit constraints rather than to inherited ability.[8]

A variant of the "meager inheritance" explanation of chronic poverty looks at somewhat larger scales to explain chronic poverty on the basis of geography, both at the macro-scale of nation states and subcontinental regions (Bloom and Sachs, 1998; Gallup and Sachs, 1998) and at intra-national scale (Hentschel et al., 2000; Elbers et al., 2001; Jalan and Ravallion 2002; Ravallion and Datt, 2002). Natural resources such as soils, forests, water, and wildlife are a fundamental input to rural economies; health shocks due to climate-dependent infectious disease are a primary threat to livelihoods; local governance influences patterns of public goods provision, and the perishability and low value-to-bulk ratio of raw commodities makes market access crucial to profitability. Because of the coordination problems intrinsic to many natural resource

management and marketing decisions, a meso-level poverty trap can emerge where the collective endowment is weak and mechanisms to resolve coordination problems do not yet exist (Barrett and Swallow, 2003). Geography plainly matters for patterns of poverty and poverty dynamics.

Where some face poverty because of meager inheritance and a bad start to life, others start off more fortunate but fall into poverty because of an adverse shock or series of shocks. Natural disasters and civil strife are tragic not just because of the temporary displacement and deprivation they bring but, most of all, because they can wipe out in a moment what households have labored years to accumulate through disciplined savings and investment. Brief disturbances can have persistent effects (Hoddinott and Kinsey, 2001). These two effects are often mutually reinforcing, as those who start off with a bad lot are far more likely to suffer serious adverse shocks that knock them back down as they struggle to climb out of poverty (Dercon, 1998; Barrett and Carter, 2001). Easterly (2001, p. 197) reports that "between 1990 and 1998, poor countries accounted for 94 percent of the world's 568 major natural disasters and 97 percent of disaster-related deaths." Worldwide, the poor are several times more likely to suffer injury or illness than are the rich (Prasad et al., 1999).

3. Rural poverty dynamics: what we still need to learn

To date, explanations of chronic poverty have thus revolved around (i) individual, household, or community-level asset endowments (Dercon, 1998; Carter and May, 1999; Maluccio et al., 2000; Haddad and Ahmed, 2003), (ii) exogenous changes in returns to asset endowments (Gunning et al., 2000; Maluccio et al., 2000), or (iii) the impact of shocks and their persistence on welfare (Glewwe and Hall, 1998; Hoddinott and Kinsey, 2001; McPeak and Barrett, 2001; Elbers et al., 2002; Gertler and Gruber, 2002; Yamano and Jayne, 2002; Barrett et al., 2003; Dercon, 2004). The latter class of explanations, however, offers an important clue toward an emerging area of research that is of particular importance to understanding rural poverty dynamics.

Shocks can have persistent effects only in the presence of hysteresis that generates irreversibility or

[8] See Lucas and Pekkala (2003) for an especially interesting study using an extensive data set from Finland. They find low transmission from parents' earnings to those of their children, but high transmission rates from total family income to children's earnings, implying financial liquidity rather than intrinsic ability is the most likely cause. Recent research on the United States suggests that about 65 percent of fathers' earnings differentials relative to the broader population is transmitted to their children (Mazumder, 2001), with that transmission rate growing over the past couple of decades (Levine and Mazumder, 2002).

differential rates of recovery. These effects suggest important nonlinearities in the relationship between assets, stocks, and income growth; nonlinearities commonly associated with the concept of *poverty traps*. This is a burgeoning area of research in which many of us are presently engaged. Much remains to be learned about the empirics and theory of poverty traps.

3.1. Uncovering poverty traps and threshold effects

The pivotal feature of poverty traps is the existence of one or more critical thresholds of wealth that people have a difficult time crossing from below.[9] Consequently, poverty persists for a long time, often measured in generations. Above the threshold, endogenous growth processes carry people toward a high-productivity steady state, while below the threshold, people sink toward a low-productivity subsistence equilibrium.[10] These thresholds give rise to the important distinction between cargo nets and safety nets. The appropriate positioning of safety nets lies just above thresholds at which natural path dynamics break in different directions.

The idea of multiple equilibria in this general context has been around for at least seventy-five years, dating to Young (1928), Rosenstein-Rodan (1943), and Myrdal (1957), if not earlier. In recent years, however, economists have begun to formalize such concepts and to appreciate the central role of threshold effects that generate bifurcated welfare dynamics, with some people staying or climbing out of poverty and others mired in a long-term poverty trap. Without such thresholds, all poverty would be transitory with everyone converging toward a single-equilibrium income level, as posited by neoclassical economic growth theory (Solow, 1956). Overwhelming empirical evidence against such unconditional convergence has motivated considerable research over the past fifteen or so years on "new" theories and empirics of economic

growth.[11] To date, this work has focused heavily at the macroeconomic level of nation states, but the logic applies equally at meso and micro levels, where it may actually prove more useful for policy purposes (Barrett and Swallow, 2003).

The idea of multiple dynamic equilibria and its implication of threshold effects becomes especially salient because it gives rise to significant potential *endogenous* change in returns-to-asset endowments. There are at least three distinct mechanisms by which this can occur. First, risk avoidance behavior can cause endogenous selection of low-return portfolios that have relatively low variability in returns (Rosenzweig and Binswanger, 1993; Zimmerman and Carter, 2003). Second, credit market imperfections can constrain the feasible matching of variable input choices with quasi-fixed factors of production (i.e., assets), leading to a positive correlation between an agent's ex ante asset stock and the rate of return on those assets (Bardhan et al., 2000). Third, there can be locally increasing returns due to discrete choices of technologies or occupations (Banerjee and Newman, 1993).

By any of these three mechanisms, as asset stocks increase, expected returns on assets, r, increase, generating an added boost to income beyond that associated with adding to asset stock at a constant (much less diminishing) rate of return. This is a significant refinement of the dominant recent approach which, as previously described, has focused on inducing *exogenous* changes in returns to the productive assets of the poor due, for example, to market liberalization policies. Poverty traps depend fundamentally on endogenous change in returns on asset holdings, so that income is at least locally increasing in asset stocks. We have learned in recent years that returns can indeed prove endogenous, at the micro level of individuals or households, growing with one's asset stock, at the meso level of communities, due to interhousehold externalities and coordination, or at the macro level of countries, due to political economy effects.[12] The net effect of weakness in these processes are patterns of persistent poverty that replicate themselves across

---

[9] A crucial point, and one commonly misunderstood, is that these thresholds are not deterministic. Rather they reflect the point at which the expected path dynamics bifurcate, where $E[dA]$ or $E[dY]$—depending on whether one is working in asset or income space—switches signs. Intuitive examples of such thresholds include homelessness or permanent physical disability.

[10] There may be more than two stable dynamic equilibria. See Zimmerman and Carter (2003) for an example with three stable dynamic equilibria.

[11] Easterly (2001) offers an especially accessible, even entertaining treatment of the evolution of growth theory and the empirical evidence on economic growth.

[12] This distinction parallels that between internal and external economies of scale in the international trade literature.

multiple scales, termed as "fractal poverty traps" by Easterly (2001) and Barrett and Swallow (2003).

The possible endogeneity of returns to assets can be readily seen by returning briefly to the mathematics of income dynamics. To establish the effect of changing asset stocks on income, we implicitly differentiate equation (5) with respect to A and take expectations

$$E[dY/dA] = r + A'E[dr/dA]. \tag{7}$$

Equation (7) indicates that income growth can occur not just due to exogenous change in rates of return—dr from equation (6)—or to growth in the asset stock with a constant rate of return—the first term of equation (7)—but also due to induced growth in rates of return as people accumulate assets. This is a testable hypothesis. Because we observe neither r nor dr/dA, we need to estimate them:

$$dY/dA = \alpha + \beta'A + \psi, \tag{8}$$

where our estimate of α provides a best estimate of r and our estimate of β represents the $E[dr/dA]$ vector with what is likely to prove a heteroskedastic regression error, ψ. Rejection of the null hypothesis that $\beta = 0$ provides strong evidence in favor of the endogeneity of rates of return, with rejection in favor of the alternate hypothesis that at least one element of $\beta > 0$, and none negative, signaling locally increasing returns characteristic of threshold effects associated with poverty traps.[13] These returns can be to scale—of a single asset—or to scope, reflecting complementarity across assets, which endogenously boosts productivity and thus income. This can perhaps be seen most easily by specifying (9) with respect to each of the multiple asset stocks held by households or individuals in the relevant population:

$$dY/dA_i = \alpha_i + \sum_j \beta_{ij}A_j + \psi_i. \tag{9}$$

Simultaneous estimation of the system of equations represented by (9), each for a different asset, A_i, would establish the assets for which overall returns appear to be endogenously increasing, enabling development professionals to focus more precisely on the assets

that most matter to helping the poor to climb out of chronic poverty. Furthermore, comparison of the expected marginal returns to each asset could establish the relative expected income gains achievable from transfers of, or de novo investment in, each type of asset. Such estimates establish expected benefits, which can then be compared against cost estimates for different types of interventions so as to improve the likely yield from scarce funds invested in asset accumulation among the poor.

Equations (8) and (9) can be understood from a slightly different perspective as implying that r in equation (3) is endogenous, then estimating income levels, rather than changes in income, as a polynomial function of A. A second-order example would be the simple regression model

$$Y = A'r$$
$$= \theta + \sum_i \lambda_i A_i + \sum_i \sum_j \gamma_{ij}A_i A_j + \xi, \tag{10}$$

where convexity of $E[Y]$ in A signals endogenously increasing returns on assets consistent with the existence of a poverty trap, and concavity would indicate convergence.

Establishing the existence of endogenously increasing rates of return to assets is only one part of the research challenge. The more practical—and more difficult—task is to identify the thresholds at which welfare dynamics appear to bifurcate. These are the points where one can usefully distinguish between the transitorily poor who remain above the threshold and therefore should recover on their own, and the (perhaps newly) chronically poor who were born or have fallen below the threshold and whose path dynamics will carry them toward a meager, subsistence equilibrium in the absence of assistance.

Thresholds can sometimes be found via autoregressions of welfare measures such as assets, income, or expenditures on past values of the same measure. The methodological problem, however, is that the autoregressions have to allow for relatively high-order polynomial relations in order that one can feasibly find thresholds. Such thresholds can be tricky to identify, especially using parametric estimation methods because in theory one should find few observations around the unstable dynamic equilibria that define thresholds. This requires sufficient sample sizes, not only in cross-section but perhaps especially in the time domain, in

---

[13] Conversely, if at least one element of β is negative and none positive, this would imply convergence, based on decreasing returns to scale.

order to capture low-probability observations in the neighborhood of threshold points. Without sufficiently dense data or flexible estimators, inflection points will typically be washed out in global parametric estimation from just two or three observations per unit, more likely manifesting themselves more subtly as heteroskedastic and positively autocorrelated errors. If considerable within-sample heterogeneity in exogenous conditions causes the location of thresholds to vary considerably between households within the population under study, then uncovering them empirically becomes harder still. In sum, detection of thresholds associated with multiple dynamic equilibria and poverty traps can be extremely difficult, even if they exist. As U.S. Defense Secretary Donald Rumsfeld infamously asserted prior to the 2003 Iraqi war, the absence of evidence is not the same as evidence of absence. The methodological challenges of precisely identifying poverty traps and threshold effects remain formidable.

Nonparametric methods can be very effective in locating thresholds, as demonstrated in Figure 1, taken from Lybbert et al. (2004). This graphic depicts the nonparametric autoregression of the natural logarithm of household herd size one and ten years ahead on current values using 17 years' herd history data on 55 pastoralist households from the Borana Plateau of southern Ethiopia. The solid, 45-degree line represents dynamic equilibrium, points where current and expected future herd sizes are the same. The S-shaped asset dynamics reveal a threshold household herd size—reflecting an unstable dynamic equilibrium—of approximately 12–15 cattle. Below that level, pastoralists effectively become sedentarized because the lactating herd is too small to split so as to support both migrating herders and a nonmigratory base camp of women, children, the elderly and the infirm.

Below the threshold herd size, livestock holdings tend to collapse toward an equilibrium of about one head of cattle because of optimal portfolio rebalancing—manifest as net sales of livestock—and frequent agroclimatic shocks to which they cannot respond through migration. Above the threshold, the herd can be split, enabling migratory extensive grazing of the dry herd (and a few lactating animals used to feed trekking herders) in response to spatio-temporal variability in forage and water availability, thereby achieving a higher dynamic equilibrium herd size of 50–75 head. Because the density of observations just below the threshold is low, second- and third-order polynomial parametric regressions did not initially uncover this relationship; hence the value of nonparametric methods for empirical inquiry into poverty traps.

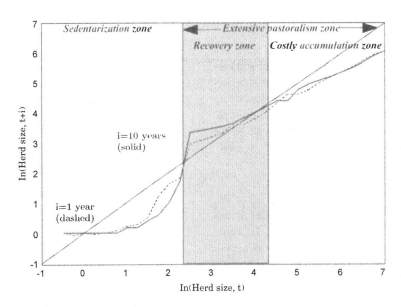

Nadaraya-Watson estimates using Epanechnikov kernel with bandwidth (h=1.5)

Figure 1. Nonparametric estimates of expected herd size transitions in southern Ethiopia, from Lybbert et al. (2004).

Note as well the crucial difference from the conventional empirical approach of looking for differences in growth rates across quantiles of a wealth or income distribution. Prospective differences in accumulation dynamics differ relative to the thresholds that define the boundary of a poverty trap, not relative to the seams between distribution quantiles. Consequently, unless the quantile divides just happen to correspond with those thresholds—which they almost surely will not—the quantile-based approach will generally miss threshold effects associated with poverty traps.

Qualitative research methods more familiar to the other social sciences can prove especially helpful in uncovering the thresholds that underpin chronic poverty (Hulme and Shepherd, 2003). Precisely because there should be few observations in the vicinity of unstable dynamic equilibria, the task of identifying thresholds can often defy statistical methods based on observational data. Yet the poor can often identify in open-ended conversations what it takes to be able to shift to a different production technology (e.g., from sedentarized cattle husbandry to extensive pastoralism in the preceding example), a different livelihood strategy (e.g., from petty trade to wholesaling), or to migrate to a place offering brighter prospects. If asked, the poor can often pinpoint the asset(s) responsible for endogenous returns. Economists are slowly warming to the integration of qualitative data collection with our more familiar quantitative methods, begetting a promising union for policy-oriented poverty research (Kanbur, 2003).

An indirect signal of threshold effects can sometimes be found in distributional data. Because poverty traps give rise to birfurcated welfare dynamics, people who initially start out close to one another can follow sharply divergent trajectories. In the presence of threshold effects, therefore, the tendency over time will be for people to cluster around a small number of stable equilibria that serve as local basins of convergence, with discernible troughs between these points. This will be apparent in cross-sectional income (or expenditure or other welfare measures) distribution data as multiple modes around dynamic equilibria, leading to what Quah (1996) has termed "twin-peakedness," which might be more generally thought of as "multi-peakedness."

We see two examples of twin-peakedness in Figures 2 and 3, which present nonparametric density

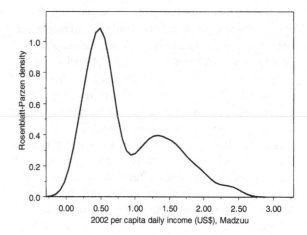

Figure 2. Bimodal income in western Kenya.

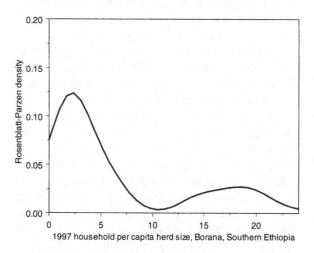

Figure 3. Bimodal cattle wealth in southern Ethiopia.

estimates of two different welfare distributions. Figure 2 plots the 1989 per capita daily incomes of a sample of households in Madzuu, a village in Kenya's western highlands where good soils, abundant rainfall, and moderate access to urban markets (such as Kisumu) create some, albeit limited, opportunities for upward economic mobility. Poverty rates nonetheless remain very high, with 61% below a $0.50/day per capita poverty line in both 1989 and 2002 and only 9% above it in both periods.[14] Figure 3 displays the density of 1997 per capita herd sizes among pastoralists on the Borana Plateau of southern Ethiopia,

---

[14] The daily per capita income figures are in inflation-adjusted 2002 U.S. dollar terms.

a region of relatively favorable agroecological potential for pastoralists, therefore similarly offering some
chance for upward economic mobility among a subpopulation typically far poorer than national averages.
In each case, there is a dominant mode at a low level,
around $0.50/day per capita income in Madzuu and
1–2 cattle per person in Borana, and a secondary mode
at a much more desirable level, about $1.30/day per
person in Madzuu and 15–20 animals per capita in
Borana. These sorts of bimodal distributions suggest
the existence of threshold effects that lead to birfurcated welfare dynamics, with some people heading toward a low-level stable equilibrium and others toward
a higher one.

Of course, in many very poor communities, unimodal distributions exist, not because thresholds are
not present, but more likely because too few people
cross them to create sufficient density at higher equilibria to find these effects in the data.[15] Such places reflect
geographic poverty traps (Jalan and Ravallion, 2002)
where the underlying agroecological conditions, market access, or sociopolitical stability—or some combination of these—are such that there exist few pathways
out of poverty in the absence of significant external
interventions.

Figure 4 exhibits one such example, from
Madagascar's poorest province, Fianarantsoa. The
graphic shows distinctly unimodal distributions of
daily per capita income distribution (again in constant
2002 U.S. dollars) of households surveyed in 1996 and
again in 2002, with a mode of only $0.20–0.25/day per
person. In communities as desperately poor as these,
income appears unimodal because virtually everyone
is caught in a geographic poverty trap. The distinct left-
ward compression of the income distribution in 2002,
relative to 1996, reflects the effect of a sharp covariate shock, the eight-month national crisis that befell
Madagascar following the violently disputed presidential elections of December 2001.

To summarize, longstanding hypotheses about multiple equilibria are receiving renewed attention in the
empirical literature on development microeconomics.

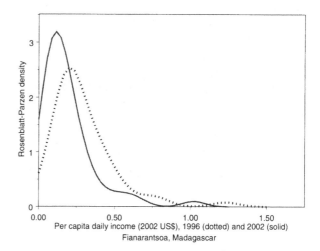

Figure 4. Intertemporal shifts in unimodal income distributions.

Highly suggestive evidence is emerging that indeed
Myrdal, Rosenstein-Rodan, and Young may have been
correct about the existence of distinct accumulation
trajectories, one or more of which are associated with
chronic poverty. If confirmed through further (qualitative, quantitative, or mixed-method) empirical research
and explained adequately with one or more theories of
poverty traps applicable to the contexts from whence
the data originate, these findings would have significant implications for development policy. In particular, empirical corroboration of the existence of poverty
traps would signal the necessity of renewed activism
by donors and governments to address the insufficiency
of asset holdings among the chronically poor.

3.2. Explaining poverty traps

There are multiple pathways out of rural poverty, so
one needs to beware of presenting too simplistic or mechanical a description. For some, the optimal pathway
is through agricultural intensification and commercialization. For others, it lies in migration to an urban area.
For others, the right strategy involves gradual transition
out of agriculture and into rural nonfarm activities.[16]
Some will use a combination of these strategies. The
key is not the particular path to be followed, which may

---

[15] In relatively wealthy communities in which few people face
serious obstacles to wealth accumulation, welfare distributions similarly appear unimodal, because relatively little of the population
falls below a threshold associated with a poverty trap. Because the
focus of this paper is poverty, we do not discuss this case any further.

---

[16] The role of the rural nonfarm economy in facilitating escape
from poverty has been widely undervalued in agricultural and development economics. A range of studies in recent years have uncov-
ered a positive relationship between nonfarm income and household

vary markedly across space, time, and even among individuals in the same location and moment. Rather, the key is the existence of *some* pathway out of poverty, a strategy in which current optimal choices predictably lead to the accumulation of sufficient productive assets so that the household can reasonably expect to earn an investible surplus above and beyond immediate consumption needs, enabling continued accumulation and steady growth in all or most welfare measures. A poverty trap exists when a household's optimal strategy does not lead to such accumulation, when the feasible choice set essentially precludes accumulation.

Why might this be? A range of sophisticated theoretical models have emerged over the past twenty years to explain the phenomenon of poverty traps.[17] This literature will not be reviewed here. Instead, the focus rests on two key features that underpin the logic of poverty traps in virtually every published model: endogenously increasing returns and financial market failures.

If returns on assets increase in wealth, this (somewhat tautologically) implies increasing returns to scale due to some mechanism, often modeled as resulting from externalities or societal-level coordination problems such as agglomeration economies. The key becomes understanding why any low-level dynamic equilibrium would exist in the presence of increasing returns. The pivotal feature seems to be discreteness. If occupational or technology choice is discrete—if people cannot combine different jobs or production technologies at arbitrarily fine scales, gradually shifting from lower-return technologies to the higher return one—and there exist nontrivial fixed or sunk costs to making the shift, an entry barrier emerges that will segregate a population into those who can clear the poverty trap threshold and those who cannot.

One implication is the potential importance of "transition technologies," options that are inferior to the highest return, state-of-the-art technologies, but

superior to the lowest return options presently chosen and—most importantly—accessible to those presently choosing lowest return strategies. Like the concepts of multiple equilibria and poverty traps, the idea of intermediate technologies has been around for a long time. Formal theorizing is, however, new and may shed light on its importance. Ideally, intermediate technologies build in their own demise by inducing people onto an improved accumulation trajectory that, in time, leads them to shift from the intermediate technology to a still-better option, which may actually predate the intermediate technology but which was previously inaccessible without the intermediate step. Moser and Barrett (2003) find some evidence of "technology adoption ladders" of this sort, with Malagasy peasants proving more likely to adopt a high-yielding rice cultivation practice if they first adopt off-season cropping of tubers in their rice fields, mainly because off-season cropping helps resolve seasonal cash liquidity constraints to the adoption of the new practice.

This particular example also underscores the centrality of financial market failures to the logic of poverty traps. If people could borrow freely, then everyone of like latent ability could make optimal investments for their circumstances and there would be no significant variation in interpersonal welfare within sites in equilibrium. Conditional convergence would be to steady states that vary only by inherited ability. Plainly, that is a fantasy world. In the real world, limits on credit and insurance access confront the chronically poor with binding constraints that limit their ability and willingness to invest today in order to reap higher steady-state income in the future. Asset accumulation failures are the predictable result (Dercon, 1998; Carter and Zimmerman, 2000; Gunning et al., 2000; Carter and May, 2001; Mude et al., 2004).

The problem is even more pernicious than mere accumulation failures. Productivity suffers too because, when people do not have access to credit or insurance so as to enable them to move consumption across periods, they inevitably find alternative markets through which they can get costly quasi-credit. For example, farmers will sell crops at low prices immediately after harvest, fully expecting to buy back the same crop months later at a considerably higher price. Given an immediate need for cash for any of a host of reasons, but lacking access to credit or cash savings, farmers commonly "borrow" through product markets. This

---

welfare indicators, in particular, that greater nonfarm income diversification causes more rapid growth in earnings and consumption (Barrett et al., 2001b). In places where the ranks of landless or near-landless poor are swelling rapidly, the rural nonfarm economy will become essential to poverty reduction strategies (Jayne et al., 2003).

[17] An incomplete listing of key papers would include Loury (1981), Romer (1986), Lucas (1988), Azariadis and Drazen (1990), Krugman (1991), Banerjee and Newman (1993), Galor and Zeira (1993), Mookherjee and Ray (2002), and Zimmerman and Carter (2003).

appears to economists as significant allocative inefficiency, although it can be an optimal strategy for a severely credit-constrained household. Other farmers will use labor markets for similar purposes, working for cash wages during planting season when a bit more time spent on their own farm would enable them to employ a cultivation method yielding significantly higher yields, and thus greater future marginal revenue product of labor. The premium on cash today from low wages can be more than sufficient to compensate for foregone productivity even a few months later (Moser and Barrett, 2003). This appears as technical inefficiency, choosing to operate within the feasible production frontier, even though it can be an optimal choice for the farmer.

So what can be done about financial market failures that beget poverty traps? The literature on rural finance and microfinance is vast and central to research on poverty. The good news is that much excellent recent and ongoing research exists on how to resolve financial market failures so as to empower the rural poor to conserve their scarce capital in the face of adverse shocks and to accumulate additional productive capital.[18]

Discreteness and financial constraints can jointly generate significant poverty traps. Faced with entry barriers to more remunerative livelihood strategies, or production technologies offering higher yields and lacking the liquid assets or borrowing capacity to meet those minimum entry requirements, the poor must commonly choose demonstrably lower return activities or portfolios or inferior technologies (Dercon and Krishnan, 1996; Bardhan et al., 2000; Barrett et al., 2001a, 2001b; Moser and Barrett, 2003; Zimmerman and Carter, 2003; Barrett et al., forthcoming). Exactly where the relevant threshold points associated with these entry barriers lie depends on exogenous biophysical and market conditions and the fixed or sunk costs inherent in accessing the more remunerative option. In places with good market access and favorable agroecological endowments, we hypothesize that poverty traps are less acute, trapping fewer people. Some of these factors are endogenous at the level of communities or nation states, as in the case of cooperatives that can permit smallholder producers to enjoy better output/input price ratios due to larger-scale transactions, or national-level institutions that ensure property rights, contract enforcement, and reasonably equal opportunities to all residents (North, 1990; Acemoglu et al., 2001, 2002; Barrett and Swallow, 2003). Others are exogenously determined by geography.

4. Development policy implications

Perhaps the most fundamental lesson of the past quarter century's research on rural poverty is the need to distinguish transitory from chronic poverty. Because the transitorily poor need no direct assistance in order to recover from and exit poverty, the necessary activism of donors and government in combating poverty depends inversely on the extent to which poverty is transitory. One must be attentive to the inherent upward bias in estimates of transitory poverty caused by measurement error with the caution not to interpret all poverty as demanding costly—and potentially injurious—external intervention. A central task for researchers is to help policymakers strike this balance effectively through careful empirical research.

The fundamental distinction between transitory and chronic poverty arises from the existence of threshold effects associated with multiple dynamic equilibria and poverty traps. Threshold points are likely to prove heterogeneous, varying with geography and perhaps individual and community attributes (e.g., gender, age, density of social solidarity networks) and they will certainly be endogenous to policies that change the incentives to switch livelihood strategies. This complicates the analytical task facing the research community. The existence of thresholds nonetheless makes it necessary to establish a fundamental distinction between safety nets set above the threshold to keep people from becoming chronically poor in the wake of adverse shocks, and cargo nets intended to facilitate the exit from poverty of the chronically poor. This also implies a central role for effective targeting in order that the appropriate policies are applied to the right subpopulations.

There are many different methods for targeting interventions.[19] Three in particular merit comment: geographic, indicator, and self-targeting. Geographic

---

[18] An incomplete list of especially exciting work in this area includes Zeller et al. (1997), Morduch (1999), and DeJanvry et al. (2003).

[19] Barrett (2002) reviews and assesses the strengths and weaknesses of alternative targeting modalities.

targeting is the perhaps the least expensive means of targeting and can be highly appropriate in areas of nearly universal chronic poverty, as in much of the drylands of Ethiopia, Kenya, and Madagascar, where more than 80% of the population falls below low national poverty lines. Geographic targeting can similarly be appropriate for short-term, safety net interventions such as food aid distribution in the wake of natural disasters in order that short-term disruptions to incomes and food availability do not cause long-term injury for affected populations.

However, because variation in incomes tends to be at least as much within regions (and even within villages) as between them (Jayne et al., 2003), geographic targeting alone will necessarily miss many, if not most of the poor. Beyond areas of intense, widespread poverty, donors, NGOs, and governments need to identify thresholds measurable in readily observable units (e.g., landholdings, herd size, educational attainment) and to target for assistance the chronic poor who fall below those thresholds, hence the importance of indicator targeting. It must be borne in mind, however, that indicator targeting only works well at combating chronic poverty if the indicators used are strongly and causally associated with lower measures of wellbeing. Identifying appropriate indicators often requires the sort of empirical analysis described earlier, or else stylized associations such as gender, ethnicity, or age will typically be used, often with little or no correlation with actual need, resulting in ineffective assistance (Clay et al., 1999).

Self-targeting mechanisms can be especially useful for safety nets. These instruments take advantage of the character of the transfer—e.g., a low-wage work requirement associated with public employment schemes, inferior subsidized foods, or significant queuing for food, clothing, or cash—to try to induce the nonpoor to self-select out of the beneficiary pool. When set up as standing policies that kick in automatically in response to income and other shocks that imperil vulnerable populations—which may include seasonal cycles of shortage that can preclude investment by smallholders (Barrett et al., 2001a)—then self-targeting programs such as food-for-work or other public employment schemes can be valuable tools for providing safety nets in response to quickly developing emergencies. Significant experience in South Asia, Southern Africa, and Argentina, in particular,

has demonstrated the potential of this approach (Ravallion, 1991; von Braun, 1995). Without an indicator targeting entry hurdle, self-targeting transfer schemes are often ineffective, however, in addressing chronic poverty, especially where land and credit markets both fail, causing considerable inter-household variation in marginal returns to labor, or when agencies try to accomplish multiple goals with self-targeting transfers (Barrett et al., 2004).

An important corollary of targeting is the need for triage in transfer programs. The literature is surprisingly silent on the value of directing certain transfers away from not only the nonpoor but also away from a subpopulation of the poor who are unlikely to benefit significantly from the transfers. Consider, for example, the implications of Figure 1 for herd restocking projects. Such transfers provide an excellent safety net intervention for those hitting the threshold at which they might become involuntarily sedentarized, enabling them to get back out onto the open range as viable pastoralists. But providing one or two cattle to a herder who has just lost his entire herd is unlikely to enable resumption of extensive pastoralism. Rather, he is likely to lose one of the animals in short order as he settles into a new, lower, sedentarized equilibrium; he may benefit more from skills training to improve his prospects in the labor market (McPeak and Barrett, 2001; Lybbert et al., 2004). Policymakers need to think through carefully when triage might be necessary in safety net programs and which assets will be most helpful to which poor people. Researchers need to help identify appropriate triage points and rules of thumb on different means of assistance. Ethical considerations may make assistance imperative, but the form of the assistance needs to pay attention to likely effectiveness, hence the need for triage with respect to form-specific transfers.

Targeting concerns revolve not just around whom to assist, where, or when, but equally how and with what. The "how" and "what" questions of targeting receive too little attention from researchers and policymakers but are of particular importance in addressing chronic poverty. The reason is straightforward: in order to enable the chronically poor to begin accumulating productive assets, one must know what factors currently most limit their choices. Is the problem chiefly due to an insufficiently productive asset stock, implying a need for improved technologies to boost yields or better market access to improve the terms of trade for the goods and services sold by the chronically poor? Decades of government interference in rural markets and global market distortions due to wealthy countries' domestic farm subsidies often play a significant role here. Or is the problem more an insufficient stock of productive assets, and, if so, of which type? Land, implying a possible rationale for progressive land reform? Human capital, implying a rationale for greater public investment in education, health and nutrition, perhaps especially for young children? Or is the need chiefly for deeper and broader access to financial services so as to free more households to undertake additional investment appropriate to their particular circumstances and talents?

These are the familiar pillars of decades of rural development strategies. There is little new to offer other than the simple observation that each case is different. Simple, blanket prescriptions rarely work. Effective policies to combat chronic poverty depend on careful, empirical policy research customized to local conditions. The research community has an obligation to develop tools and information that can provide policymakers with accurate and timely information on the who, what, where, when, and how targeting questions that are the essence of poverty reduction strategies.

Acknowledgments

This article has benefited greatly from conversations, collaborations, and/or comments from Larry Blume, Doug Brown, Michael Carter, Paul Cichello, Alain DeJanvry, Bruce Gardner, Ravi Kanbur, Dan Maxwell, John McPeak, Bart Minten, Chris Moser, Andrew Mude, Festus Murithi, Ben Okumu, Willis Oluoch-Kosura, Frank Place, Tom Reardon, David Sahn, David Stifel, Brent Swallow, and Erik Thorbecke, and from data collection and analysis assistance from Solomon Desta, Travis Lybbert, Paswel Phiri, Jean Claude Randrianarisoa, Jhon Rasambainarivo, and Justine Wangila, although none of them bear any responsibility for any remaining errors of mine.

This work has been made possible, in part, by support from the United States Agency for International Development (USAID) Grant No. LAG-A-00-96-90016-00 through the BASIS CRSP, and the Strategies and Analyses for Growth and Access (SAGA) cooperative agreement through the Africa Bureau, USAID, undergrant HFM-A-00-01-00132-00. The views expressed here are mine and do not represent any official agency.

References

Acemoglu, D., S. Johnson, and J. A. Robinson, "Colonial Origins of Comparative Development: An Empirical Investigation," *American Economic Review* 91, no. 5 (2001), 1369–1401.

Acemoglu, D., S. Johnson, and J. A. Robinson, "Reversal of Fortune: Geography and Institutions in the Making of the Modern World Income Distribution," *Quarterly Journal of Economics* 117, no. 4 (2002), 1231–1294.

Albarran, P., and O. Attanasio, "Do Public Transfers Crowd Out Private Transfers? Evidence from a Randomized Experiment in Mexico," in S. Dercon, ed., *Insurance Against Poverty* (Oxford University Press: Oxford, UK, 2004).

Alderman, H., J. R. Behrman, H.-P. Kohler, J. A. Maluccio, and S. C. Watkins, "Attrition in Longitudinal Household Survey Data: Some Tests for Three Developing-Country Samples," World Bank Policy Research Working paper 2447 (2000).

Azariadis, C., and A. Drazen, "Threshold Externalities and Economic Development," *Quarterly Journal of Economics* 105, no. 3 (1990), 501–526.

Banerjee, A. V., and A. F. Newman, "Occupational Choice and the Process of Development," *Journal of Political Economy* 101, no. 2 (1993), 274–298.

Bardhan, P., S. Bowles, and H. Gintis, "Wealth Inequality, Wealth Constraints and Economic Performance," in A. B. Atkinson, and F. Bourguignon, eds., *Handbook of Income Distribution*, volume 1 (Elsevier Science: Amsterdam, 2000).

Barrett, C. B., "Food Security and Food Assistance Programs," in B. Gardner, and G. Rausser, eds., *Handbook of Agricultural Economics*, volume 2B (Elsevier Science: Amsterdam, 2002).

Barrett, C. B., M. Bezuneh, and A. A. Aboud, "Income Diversification, Poverty Traps and Policy Shocks in Côte D'Ivoire and Kenya," *Food Policy* 26, no. 4 (2001a), 367–384.

Barrett, C. B., M. Bezuneh, D. C. Clay, and T. Reardon, "Heterogeneous Constraints, Incentives and Income Diversification Strategies in Rural Africa," *Quarterly Journal of International Agriculture* (forthcoming).

Barrett, C. B., and M. R. Carter, "Can't Get Ahead for Falling Behind: New Directions for Development Policy to Escape Poverty and Relief Traps," *Choices* 16, no. 4 (2001), 35–38.

Barrett, C. B., S. Holden, and D. C. Clay, "Can Food-for-Work Programs Reduce Vulnerability?" in S. Dercon, ed., *Insurance Against Poverty* (Oxford University Press, Oxford, UK, 2004).

Barrett, C. B., T. Reardon, and P. Webb, eds., "Nonfarm Income Diversification and Household Livelihood Strategies in Rural Africa: Concepts, Dynamics and Policy Implications," *Food Policy* 26, no. 4 (2001b), 315–331.

Barrett, C. B., S. M. Sherlund, and A. A. Adesina, "Macroeconomic Shocks, Human Capital and Productive Efficiency: Evidence from West African Farmers," Cornell University Working paper (2003).

Barrett, C. B., and B. M. Swallow, "Fractal Poverty Traps," Cornell University Working paper (2003).

Basu, K., "Child Labor: Cause, Consequence, and Cure, with Remarks on International Labor Standards," *Journal of Economic Literature* 37, no. 2 (1999), 1083–1119.

Baulch, B., and J. Hoddinott, eds., *Economic Mobility and Poverty Dynamics in Developing Countries* (Frank Cass: London, 2000).

Bloom, D. E., and J. D. Sachs, "Geography, Demography and Economic Growth in Africa," *Brookings Papers on Economic Activity* no. 2 (1998), 207–273.

Bourguignon, F., and S. R. Chakravarty, "The Measurement of Multidimensional Poverty," DELTA Working paper (2002).

Carter, M. R., and J. May, "Poverty, Livelihood and Class in Rural South Africa," *World Development* 27, no. 1 (1999), 1–20.

Carter, M. R., and J. May, "One Kind of Freedom: The Dynamics of Poverty in Post-Apartheid South Africa," *World Development* 29, no. 12 (2001), 1987–2006.

Carter, M. R., and F. Zimmerman, "The Dynamic Costs and Persistence of Asset Inequality in an Agrarian Economy," *Journal of Development Economics* 63, no. 2 (2000), 265–302.

Clay, D. C., D. Molla, and D. Habtewold, "Food Aid Targeting in Ethiopia: A Study of Who Needs It and Who Gets It," *Food Policy* 24, no. 3 (1999), 391–409.

Cox, D., and E. Jimenez, "Social Security and Private Transfers in Developing Countries: The Case of Peru," *World Bank Economic Review* 6, no. 1 (1992), 155–169.

DeJanvry, A., E. Sadoulet, C. McIntosh, B. Wydick, M. Valdivia, A. Trigueros, G. Gordillo, and D. Karlan, "Up the Lending Ladder: Extending Financial Services for the Rural Poor Through Credit-Reporting Bureaus," University of Wisconsin-Madison BASIS Brief no. 16 (2003).

Dercon, S., "Wealth, Risk and Activity Choice: Cattle in Western Tanzania," *Journal of Development Economics* 55, no. 1 (1998), 1–42.

Dercon, S., "The Impact of Economic Reforms on Households in Rural Ethiopia," University of Oxford Centre for the Study of African Economies working paper (2000).

Dercon, S., ed., *Insurance Against Poverty* (Oxford University Press: Oxford, UK, 2004).

Dercon, S., and P. Krishnan, "Income Portfolios in Rural Ethiopia and Tanzania: Choices and Constraints," *Journal of Development Studies* 32, no. 6 (1996), 850–875.

Dercon, S., and P. Krishnan, "Risk Sharing and Public Transfers," *Economic Journal* 113 (2003), C86–C94.

Duclos, J.-Y., D. Sahn, and S. D. Younger, "Robust Multidimensional Poverty Comparisons," Université Laval Working paper (2003).

Easterly, W., *The Elusive Quest for Growth* (MIT Press: Cambridge, MA, 2001).

Elbers, C., J. W. Gunning, and B. Kinsey, "Convergence, Shocks and Poverty: Micro Evidence on Growth under Uncertainty," Tinbergen Institute Working paper (2002).

Elbers, C., J. O. Lanjouw, and P. Lanjouw, "Welfare in Villages and Towns: Micro-Level Estimation of Poverty and Inequality," Vrije Universiteit Working paper (2001).

Falaris, E. M., "The Effect of Survey Attrition in Longitudinal Surveys: Evidence from Peru, Côte d'Ivoire and Vietnam," *Journal of Development Economics* 70, no. 1 (2003), 133–157.

Gallup, J. L., and J. D. Sachs, "Geography and Economic Growth," in B. Pleskovic, and J. E. Stiglitz, eds., *Proceedings of the Annual World Bank Conference on Development Economics* (World Bank: Washington, DC, 1998).

Galor, O., and J. Zeira, "Income Distribution and Macroeconomics," *Review of Economic Studies* 60, no. 1 (1993), 35–52.

Gertler, P., and J. Gruber, "Insuring Consumption Against Illness," *American Economic Review* 1 (2002), 51–70.

Glewwe, P., and G. Hall, "Are Some Groups More Vulnerable to Macroeconomic Shocks Than Others? Hypotheses Tests Based on Panel Data from Peru," *Journal of Development Economics* 56, no. 1 (1998), 181–206.

Grootaert, C., and R. Kanbur, "The Lucky Few amidst Economic Decline: Distributional Change in Côte d'Ivoire as Seen through Panel Data Sets, 1985–88," *Journal of Development Studies* 31, no. 4 (1995), 603–619.

Gunning, J. W., J. Hoddinott, B. Kinsey, and T. Owens, "Revisiting Forever Gained: Income Dynamics in the Resettlement Areas of Zimbabwe, 1983–96," in B. Baulch, and J. Hoddinott, eds., *Economic Mobility and Poverty Dynamics in Developing Countries* (Frank Cass: London, 2000).

Haddad, L., and A. Ahmed, "Chronic and Transitory Poverty: Evidence from Egypt, 1997–99," *World Development* 31, no. 1 (2003), 71–85.

Hentschel, J., J. O. Lanjouw, P. Lanjouw, and J. Poggi, "Combining Census and Survey Data to Study Spatial Dimensions of Poverty: A Case Study of Ecuador," *World Bank Economic Review* 14, no. 1 (2000), 147–166.

Hoddinott, J., and B. Kinsey, "Child Growth in the Time of Drought," *Oxford Bulletin of Economics and Statistics* 63, no. 4 (2001), 409–436.

Hulme, D., and A. Shepherd, eds., "Chronic Poverty and Development Policy," *World Development*, 31, no. 3 (2003), 399–665.

Jalan, J., and M. Ravallion, "Geographic Poverty Traps? A Micro Model of Consumption Growth in Rural China," *Journal of Applied Econometrics* 17, no. 4 (2002), 329–346.

Jayne, T. S., T. Yamano, M. T. Weber, D. Tschirley, R. Benfica, A. Chapoto, and B. Zulu, "Smallholder Income and Land Distribution in Africa: Implications for Poverty Reduction Strategies," *Food Policy* 28, no. 2 (2003), 253–275.

Kanbur, R., ed., *Q-Squared: Combining Qualitative and Quantitative Methods of Poverty Appraisal* (Permanent Black: Delhi, 2003).

Krugman, P., "Increasing Returns and Economic Geography," *Journal of Political Economy* 99, no. 3 (1991), 483–499.

Levine, D. I., and B. Mazumder, "Choosing the Right Parents: Changes in the Intergenerational Transmission of Inequality between 1980 and the Early 1990s," Federal Reserve Bank of Chicago working paper WP 2002–08 (2002).

Loury, G. C., "Intergenerational Transfers and the Distribution of Earnings," *Econometrica* 49, no. 4 (1981), 843–867.

Lucas, R., "On the Mechanics of Economic Development," *Journal of Monetary Economics* 22, no. 1 (1988), 2–42.

Lucas, R. E. B., and S. Pekkala, "Abilities, Budgets and Age: Inter-Generational Economic Mobility in Finland," Boston University Institute for Economic Development Discussion Paper 130 (2003).

Luttmer, E. F. P., "Measuring Economic Mobility and Inequality: Disentangling Real Events from Noisy Data," University of Chicago working paper (2002).

Lybbert, T. J., C. B. Barrett, S. Desta, and D. L. Coppock, "Stochastic Wealth Dynamics and Risk Management among a Poor Population," *Economic Journal* 4, no. 498 (2004), 750–777.

Maluccio, J., L. Haddad, and J. May, "Social Capital and Household Welfare in South Africa, 1993–98," in B. Baulch, and J. Hoddinott, eds., *Economic Mobility and Poverty Dynamics in Developing Countries* (Frank Cass: London, 2000).

Mazumder, B., "Earnings Mobility in the US: A New Look at Intergenerational Inequality," Federal Reserve Bank of Chicago working paper WP 2001–18 (2001).

McPeak, J. G., and C. B. Barrett, "Differential Risk Exposure and Stochastic Poverty Traps among East African Pastoralists," *American Journal of Agricultural Economics* 83, no. 3 (2001), 674–679.

Mookherjee, D., and D. Ray, "Contractual Structure and Wealth Accumulation," *American Economic Review* 92, no. 4 (2002), 818–849.

Morduch, J., "The Microfinance Promise," *Journal of Economic Literature* 37, no. 4 (1999), 1569–1614.

Moser, C. M., and C. B. Barrett, "The Complex Dynamics of Smallholder Technology Adoption: The Case of SRI in Madagascar," Cornell University working paper (2003).

Mude, A. G., C. B. Barrett, J. G. McPeak, and C. Doss, "Educational Investments in a Dual Economy," Cornell University working paper (October 2004).

Myrdal, G., *Economic Theory and Underdeveloped Regions* (G. Duckworth & Co.: London, 1957).

North, D. C., *Institutions, Institutional Change and Economic Performance* (Cambridge University Press: Cambridge, 1990).

Prasad, K., P. Belli, and M. Das Gupta, "Links between Poverty, Exclusion and Health," World Bank background paper for the World Development Report 2000/2001 (1999).

Quah, D., "Twin Peaks: Growth and Convergence in Models of Distribution Dynamics," *Economic Journal* 106, no. 437 (1996), 1045–1055.

Ravallion, M., "Reaching the Rural Poor through Public Employment: Arguments, Lessons, and Evidence from South Asia," *World Bank Research Observer* 6, no. 1 (1991), 153–176.

Ravallion, M., and G. Datt, "Why Has Economic Growth Been More Pro-Poor in Some States of India Than Others?" *Journal of Development Economics* 68, no. 2 (2002), 381–400.

Romer, P., "Increasing Returns and Long Run Growth," *Journal of Political Economy* 94, no. 4 (1986), 1002–1037.

Rosenstein-Rodan, P., "Problems of Industrialization of Eastern and South-Eastern Europe," *Economic Journal* 53 (1943), 202–211.

Rosenzweig, M. R., "Payoffs from Panels in Low-Income Countries: Economic Development and Economic Mobility," *American Economic Review* 93, no. 2 (2003), 112–117.

Rosenzweig, M. R., and H. Binswanger, "Wealth, Weather Risk and the Composition and Profitability of Agricultural Investments," *Economic Journal* 103, no. 416 (1993), 56–78.

Schultz, T. W., "Nobel Lecture: The Economics of Being Poor," *Journal of Political Economy* 88, no. 4 (1980), 639–651.

Solow, R., "A Contribution to the Theory of Economic Growth," *Quarterly Journal of Economics* 70, no. 1 (1956), 65–94.

Strauss, J., and D. Thomas, "Health, Nutrition, and Economic Development," *Journal of Economic Literature* 36, no. 2 (1998), 766–817.

von Braun, J., ed., *Employment for Poverty Reduction and Food Security* (International Food Policy Research Institute: Washington, DC, 1995).

World Bank, *World Development Report 2000/1* (Oxford University Press: Oxford, 2000).

Yamano, T., and T. S. Jayne, "Measuring the Impacts of Prime-Age Adult Death on Rural Households in Kenya," Tegemeo Institute of Agricultural Policy and Development working paper 5 (2002).

Young, A. A., "Increasing Returns and Economic Progress," *Economic Journal* 38 (1928), 527–542.

Zeller, M., G. Schrieder, J. von Braun, and F. Heidhues, *Rural Finance for Food Security for the Poor: Implications for Research and Policy* (International Food Policy Research Institute: Washington, DC, 1997).

Zimmerman, F. J., and M. R. Carter, "Asset Smoothing, Consumption Smoothing and the Reproduction of Inequality under Risk and Subsistence Constraints," *Journal of Development Economics* 71 (2003), 233–260.

Six characters (and a few more) in search of an author: how to rescue rural development before it's too late

Simon Maxwell*

Abstract

Rural development has not received the priority and attention warranted by the present and future concentration of poverty in rural areas. Is this perhaps because rural development agencies present conflicting narratives? A framework is presented within which to answer that question, and is then applied to the recent policy statements of the European Union (EU), Food and Agriculture Organization (FAO), International Fund for Agricultural Deveopment (IFAD), and the World Bank. Each policy statement is compelling in its own way, but the strategies are not consistent. A narrative is needed, which recognizes the rapid pace of change in rural areas, acknowledges the need for diversification out of agriculture, builds market institutions for growth, and works effectively within the current international consensus on poverty reduction, emphasising opportunity, empowerment, and security.

JEL classification: Q15

Keywords: poverty; agriculture; rural development; policy; aid

1. Introduction

In Pirandello's play, "Six characters in search of an author," the director is mustering his actors for a rehearsal, when six "characters" burst into the theater, and demand that he assemble a play from their experience. The director is intrigued. He calls for scenery, props, and paper. He instructs his actors to observe. The characters lay out their lives. But the project is not so easy. The characters argue. They have different perspectives, different stories. There is no coherence. At the end of the play, the director loses patience. In almost its last line, he shouts out in exasperation: "You can all go to Hell, every last one of you" (Pirandello, 1995, p. 65).

In this little parable, we can identify the "director" as the international development community, especially governments and donors. The "characters" are the rural poor, or perhaps the agencies that speak on their behalf, each with its own priorities and institutional interests. And the director's last line is the response of the development community to requests for more attention and more resources for rural development. The sentiment is not expressed quite so brutally, to be sure, but is no less real: funding for rural development has fallen by two thirds in ten years (IFAD, 2001, p. 41); in the latest round of country strategy papers prepared by the EU, committing 7.4 billion euros, agriculture and rural development account for only 7.8% of the total;[1] in general, "Poverty Reduction Strategy Papers," the new vehicle for development policy at country level, are weak when it comes to rural development (Cord, 2002). Meanwhile, 75% of the poor are to be found in rural areas, and this is likely to continue (IFAD, 2001, p. 15).

Should we blame the director for this state of affairs, or should we blame the characters? In Pirandello's case, we might well argue that a competent impresario should happily engage with a tapestry of competing narratives. In the real world of rural

* *Overseas Development Institute, 111 Westminster Bridge Road, London SE1 7JD, UK.*

[1] Social infrastructure accounts for a further 5–6% and food aid for about 2% (Philip Mikos, pers. comm.).

development, however, the characters cannot so easily be forgiven. Is there not a risk that they have damaged their case by pursuing conflicting paths to the reduction of rural poverty?

That is the charge we need to explore. Diversity is often desirable, we know that, not least in rural development (Chambers, 1993). But debate and diversity may do us no favors when they present policymakers and funders with conflicting messages.

There is a reason, and it can be found in the literature on how policy is formed. Policy making is not a simple, logical, linear process. As Clay and Schaffer (1986, p. 192) remind us, "the whole life of policy is a chaos of purposes and accidents." In simplifying the process, an important part is played by "policy narratives" (Roe, 1991). Never mind that narratives are sometimes wrong and often contested (Leach and Mearns, 1996; Keeley and Scoones, 2003). Roe is surely right when he observes that

> Rural development is a genuinely uncertain activity, and one of the principal ways practitioners, bureaucrats and policy makers articulate and make sense of this uncertainty is to tell stories or scenarios that simplify the ambiguity (1991, p. 288)

There have certainly been powerful narratives about rural development in the past. Ellis and Biggs (2001, p. 441), for example, observe that

in retrospect, it is evident that one major body of thought, albeit with plenty of side-excursions and add-ons, has dominated the landscape of rural development thinking throughout the last half-century. This is the "agricultural growth based on small-farm efficiency" paradigm.

Ellis and Biggs (2001) go on to review other approaches, including the process approach most strongly associated with Rondinelli (1983), and a variety of versions of the livelihoods framework (Carney, 1998; Scoones, 1998). Other authors have couched the changing narratives in the language of the Washington and post-Washington consensus (Maxwell and Heber-Percy, 2001; Kydd and Dorward, 2001) or have been focused on the need to build effective markets (Omamo, 2003). More recently, there has been a focus on developing recommendations for high and low potential areas or similar classifications (e.g., Farrington et al., 2003 on "weakly integrated areas"). For example, Richards et al. (2003) present a framework based on two axes, one capturing the degree of "connectedness" or market access, and the other the volume and value of livelihood assets (Figure 1). There has been a general focus in the new poverty agenda on social services (Maxwell, 2003). This has been criticized for neglecting productive sectors (Belshaw, 2002; Williamson and Canagarajah, 2003), though Farrington et al. (2003)

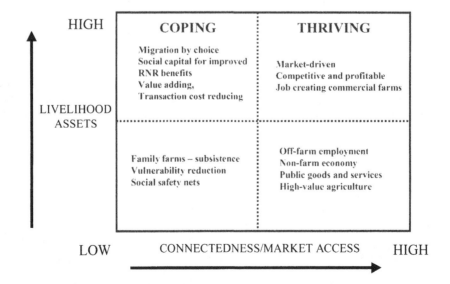

Figure 1. Conceptual poverty × location matrix. *Source*: Richards et al., 2003.

explore how productive and social expenditures might be linked.

A key question remains. Do we, in this decade, have a "narrative" about rural development that will be sufficiently convincing to reverse government and donor neglect? The international agencies have certainly grappled with that question. There has been a flurry of policy reviews and statements in recent years. But are they consistent? Are they compelling? Are they right? That is what we need to find out.

2. What to look for

The first step needs to be some way of filleting the different agency perspectives. Caroline Ashley and I took on this job in the special issue of *Development Policy Review* that we edited in 2001 on "Rethinking Rural Development." That review (a) provided some stylized facts about the past, present, and future of rural poverty, (b) identified the narratives and sticking points in the rural development debate, and (c) provided an initial sketch of a new narrative on rural development, with five principles and ten more specific conclusions (Ashley and Maxwell, 2001).

The stylized facts and the sticking points need not cause undue delay, but a brief summary is needed. First, "trends and discontinuities in the character of rural areas generate a rural development *problematique* sharply different from that of the past" (Ashley and Maxwell, 2001, p. 397). Rural areas are highly heterogeneous, particularly as between high potential and low potential, well connected and weakly connected, and peri-urban or remote; but across the world, the character of rural space is being changed by demographic transitions, the diversification of livelihoods, the spread of market relations, technical change, and, importantly, the gradually shrinking contribution of the agricultural sector to national GDP, export earnings, and tax revenue.

On the sticking points, there are six key issues: (i) can agriculture be the engine of rural growth?; (ii) can small farms survive?; (iii) can the rural nonfarm economy take up the slack?; (iv) the challenge of new thinking on poverty (especially with regard to social protection issues and the distribution of assets, income and power); (v) governance (especially the transition from a discourse about participation to a wider debate

about decentralization and "political deepening"); and (vi) implementation issues, in retrospect a rather overburdened category, covering rural development planning, but also conflict.

The five principles and ten more specific points are listed below. Note that it is not necessary to agree that these are right in order to accept that they may provide a way of discriminating between the competing narratives on offer. Nevertheless, at the limit, this list does challenge the conventional wisdom on rural development, especially the "agricultural growth based on small farm efficiency" paradigm that has been so influential since the 1960s. Significant doubts are also cast on the speed and sequencing of liberalization.

The five principles were that a successful rural development strategy should:

(i) recognize the great diversity of rural situations;
(ii) respond to past and future changes in rural areas;
(iii) be consistent with wider poverty reduction policy;
(iv) reflect wider moves to democratic decentralization; and
(v) make the case for the productive sectors in rural development, as a strategy both to maximize growth and to reduce poverty.

And the ten specific points were that a strategy should:

(i) offer different options for peri-urban, rural, and remote locations;
(ii) favor livelihood-strengthening diversification options for multi-occupational and multi-locational households;
(iii) accept the force of the post-Washington consensus—that market institutions need to be in place before liberalization, and that states have a key role to play, for example, in supplying (national and global) public goods;
(iv) explicitly take on inequality, in assets and incomes, with targets, timetables, and concrete measures;
(v) propose measures to counter the anti-South bias of technical change, recognizing the need for public support to research;
(vi) demonstrate that agricultural strategies will be consistent with natural resource protection, including water management;

(vii) recognize the importance of investment in infrastructure and human capital;
(viii) respond to the "obligation" to protect the poor, with new social protection measures, including in conflict areas, and for HIV/AIDS;
 (ix) propose pragmatic steps toward greater de-concentration and devolution; and
 (x) identify the place for agriculture and rural development in Poverty Reduction Strategy Programmes (PRSPs) and sector programs.

We recognized in 2001 that this "tick-list" needed prioritization in particular cases, and that choices had to be made. In particular, there is a risk that strategies say everything and mean nothing. A coherent narrative needs to trace a line through these various options.

3. Agency narratives on rural development

Most aid agencies, and indeed most governments, have a rural development policy—and many have reviewed their policies within the past five years. Reviewing the policies of a dozen different national and international agencies in 2001, Farrington and Lomax (2001) were able to claim a substantial degree of convergence on key issues, like the priority given to poverty reduction, the importance of environmental sustainability, and the shift to sector approaches. On the other hand, there was less uniformity on the role of government in general, the role of sector ministries in particular, and the management of decentralization. Farrington and Lomax also highlighted a debate about the value and future of projects as a vehicle for delivering rural development. Four issues were identified as "emerging challenges," where donor policies were "diverse" (Farrington and Lomax, 2001, p. 536). These were (a) diversification, (b) decentralization, (c) globalization, and (d) institutional strengthening.

The degree of consensus can be explored in more detail by looking at the policies of four "market leaders" in the international arena, and at the key policy documents they have produced. The four are:

The European Union	Fighting Rural Poverty: European Community policy and approach to rural development and sustainable natural resources management in developing countries (July 2002)
FAO	Anti-Hunger Programme: Reducing hunger through sustainable agricultural and rural development and wider access to food (July 2002a)[2]
IFAD	Rural Poverty Report 2001: The Challenge of Ending Rural Poverty (2001)
The World Bank	Reaching the Rural Poor: A Renewed Strategy for Rural Development (October 2002)

The documents listed are not precisely equivalent in scope or level of detail; but each provides an insight into the thinking of the agency.

3.1. The European Union

The latest policy of the European Union is contained in a "Communication" from the European Commission, entitled "Fighting Rural Poverty" and dated July, 2002. The paper lays out the rationale for a rural focus, discusses the nature of rural poverty and changing approaches to rural development, and then lays out a policy and strategy. The policy sets out six objectives:

 (i) Promote broad-based rural economic growth;
 (ii) Ensure more equitable access to productive assets, markets and services;
(iii) Support human and social development;
(iv) Ensure sustainable natural resources management;
 (v) Reduce vulnerability to risks; and
(vi) Address the social and political exclusion of the rural poor.

The more detailed strategy builds on these points. It describes supporting actions (like sound macroeconomic management and trade liberalization), explores the six areas listed above in more detail, and picks up cross-cutting issues like gender equality. It has a section on country programming, *inter alia* discussing the role of PRSPs and Sector-Wide Approaches (SWAPs). And finally, it has an important section on policy coherence and complementarity, dealing with issues like

---

[2] Other relevant documents are "Rome Declaration on World Food Food Summit Plan of Action" (November 1996); and "International Alliance Against Hunger: the Declaration of the World Food Summit Five Years Later" (August 2002).

reform of the Common Agricultural Policy and the Common Fisheries Policy.

3.2. FAO

FAO organized the World Food Summit in 1996, and followed this up with "WFS—Five Years Later" in June 2002. The WFS itself provided a framework, structured around six "Commitments" (Box 1). The "Five Years Later" meeting updated the analysis and reviewed funding requirements. The key issues reviewed were conflicts and natural disasters; freshwater resources; the evolution of technology; globalization, public goods, and trade; food safety; and the right to food.

The Anti-Hunger Programme draws these together. It identifies five priorities for action in food, agriculture, and rural development. These are:

(i) Improve agricultural productivity and enhance livelihoods and food security in poor rural communities;

(ii) Develop and conserve natural resources;

(iii) Expand rural infrastructure (including capacity for food safety, plant and animal health) and broaden market access;

(iv) Strengthen capacity for knowledge generation and dissemination (research, extension, education, and communication); and

(v) Ensure access to food for the most needy through safety nets and other direct assistance.

The Programme document sets out a policy framework for action in these areas, at international and domestic levels.

3.3. IFAD

IFAD published its major report, "The Challenge of Ending Rural Poverty," in 2001. This was not a formal "policy," but a useful guide to IFAD's overall approach. The four substantive chapters deal respectively with (a) assets, (b) technology and natural resources, (c) markets, and (d) institutions. Underlying

---

Box 1: The seven commitments of the World Food Summit

1. We will ensure an enabling political, social, and economic environment, designed to create the best conditions for the eradication of poverty and for durable peace, based on full and equal participation of women and men, which is the most conducive to achieving sustainable food security for all.

2. We will implement policies aimed at eradicating poverty and inequality and improving physical and economic access by all, at all times, to sufficient, nutritionally adequate and safe food, and its effective utilization.

3. We will pursue participatory and sustainable food, agriculture, fisheries, forestry, and rural development policies and practices in high and low potential areas, which are essential to adequate and reliable food supplies at the household, national, regional and global levels, and combat pests, drought, and desertification, considering the multifunctional character of agriculture.

4. We will strive to ensure that food, agricultural trade, and overall trade policies are conducive to fostering food security for all through a fair and market-oriented world trade system

5. We will endeavor to prevent and be prepared for natural disasters and man-made emergencies and to meet transitory and emergency food requirements in ways that encourage recovery, rehabilitation, development, and a capacity to satisfy future needs.

6. We will promote optimal allocation and use of public and private investments to foster human resources, sustainable food, agriculture, fisheries and forestry systems, and rural development, in high and low potential areas.

7. We will implement, monitor, and follow up this Plan of Action at all levels in cooperation with the international community.

the analysis on these topics are four overriding points (pp. 3ff):

(i) The critical role of food staples in the livelihoods of the rural poor;
(ii) The requirement for better allocation and distribution of water;
(iii) The importance, if poverty targets are to be met, of redistributing assets, institutions, technologies, and markets; and
(iv) The need for special attention to women, ethnic minorities, hill people, and residents of semi-arid areas.

3.4. The World Bank

A new World Bank strategy, "Reaching the Rural Poor," was approved by the Board in October 2002. The strategy replaces its 1997 predecessor, "From Vision to Action," which was itself intended to be an inclusive rural development strategy.

The new strategy is described as having five key features and five strategic objectives. The five key features are:

(i) Focusing on the poor;
(ii) Fostering broad-based growth;
(iii) Addressing the entire rural space;
(iv) Forging alliances of all stakeholders; and
(v) Addressing the impact of global development on client countries.

The chapters of the strategy relate to the objectives, which are:

(i) Fostering an enabling environment for broad-based rural growth;
(ii) Enhancing agricultural productivity and competitiveness;
(iii) Encouraging nonfarm rural economic growth;
(iv) Improving social well-being, managing risk, and reducing vulnerability; and
(v) Enhancing sustainable management of natural resources.

It is worth noting that, like the EU strategy, the World Bank paper has a good deal to say about trade protectionism, especially in Organization for Economic Cooperation and Development (OECD) countries.

4. Is there a consistent and compelling narrative?

We need to be realistic in exploring the consistency and power of the rural development narratives contained in these documents, and for three reasons. First, policy documents are only paper statements and may or may not reflect what actually happens on the ground: policy is, after all, what policy does. Second, such documents are always compromise statements, reflecting political and bureaucratic interests inside and outside the agency. And, third, all the agencies have their own mandates and histories: it would be surprising, for example, if FAO were to lead on the importance of primary education or health, rather than agriculture. But, of course, the last of these points is the point, partly: the "message" on rural development depends to some extent on the messenger. That is why it is useful to compare and contrast.

This task is carried out in Table 1, using as an organizing framework the ten specific headings listed by Ashley and Maxwell (2001) and reproduced above. A degree of interpretation and simplification is obviously required, but the main lessons are clear.

First, it is evident from the texts that all these strategies are strongly driven by the agriculture and natural resource departments of the respective agencies. These are documents that cover topics like education and health, but are not primarily driven by those issues, nor, most likely, strongly owned by the social infrastructure departments of the agencies, which cover both rural and urban areas. A strong institutional commitment to integrated rural development, which was a feature of development discourse in the 1970s, does not leap out of these pages.

Second, and partly as a consequence, the documents are all stronger on agriculture and natural resource management than on other issues. Thus, all have important things to say about rural infrastructure, technology, the provision of public goods, environmental protection (instrumentally in support of farming, if for no other reason), and the decentralization of rural services. However, there are straws in the wind in all the documents about diversification out of agriculture, for both households and districts.

Third, there is a strong poverty focus throughout, though a somewhat variable commitment to equality. Better access to asset and services are major themes, along with safety nets. There is little discussion of

Table 1
Four rural development strategies compared

Dimension	EU	FAO	IFAD	World Bank
1. Offer different options for peri-urban, rural, and remote locations	The diversity of rural areas is recognized, though evidence is mostly provided on the diversity of regional problems (e.g., inadequate infrastructure in SSA, inequality in Latin America).	The WFS documentation was particularly strong in distinguishing between high- and low-potential areas, and seeking a balance in development effort between them. The emphasis in the Anti-Hunger Programme is more generically on "poor rural communities" (and on Africa). The costs of poor connectedness are stressed, however.	Mountains and semi-arid areas are identified as requiring special attention, but there is little systematic treatment of the problems of different areas. The classification of the poor (Table 2.1) distinguishes rain-fed farmers, smallholder farmers, pastoralists, artisanal fishermen, landless, indigenous people, female-headed households, and displaced people—i.e., not principally by location. However, IFAD has worked intensively in marginal environments.	There is emphasis throughout on heterogeneity and on different farming types (e.g., commercial, small family farms, subsistence etc.)
2. Favor livelihood-strengthening diversification options for multi-occupational and multi-locational households	One of the six core principles is to promote broad-based rural economic growth. The main focus is on raising the productivity of the natural resources sectors, but the growth of the nonfarm sector (and associated infrastructure) is mentioned.	There is reference to rural development in Commitment 3 of the WFS, but the emphasis is clearly on agriculture and sector support services— particularly food, and particularly in the fyl (five years later) declaration and the Anti-Hunger Programme. The latter has one paragraph (p. 73) on the nonfarm rural economy, in a section on the domestic policy environment.	The report clearly recognizes that "most poor rural households diversify their sources of income" (p. 22). However, the report has a strong focus on agriculture, and particularly on food staples.	Agriculture is clearly identified as the leading sector and the primary engine of economic growth, but with emphasis on links to the wider rural economy, overall food chains, and diversification into high-value crops. A chapter is also devoted to the nonfarm economy, with a strong emphasis on supporting rural entrepreneurship.

(Continued)

Table 1
(Continued)

Dimension	EU	FAO	IFAD	World Bank
3. Accept the force of the post-Washington consensus—that market institutions need to be in place before liberalization, and that states have a key role to play, for example, in supplying (national and global) public goods	This point is clearly made. The paper discusses the importance of trade liberalization and the removal of price distortions, but also says that "liberalisation . . . must be carefully managed and sequenced . . . and must be accompanied by actions to create the conditions for equitable and environmentally sustainable market-led development" (p. 9).	No explicit discussion of this item. The emphasis in the Anti-Hunger Programme is on "stable and predictable macroeconomic policies," with no discussion of sequencing. However, the Programme is strong on state investment in infrastructure, services, and safety nets.	The report has a chapter on markets, emphasizing the benefits of market access and liberalization, but also the constraints, especially for the poor. The report emphasizes high transport and transaction costs, lack of collective organizations, discrimination, and "cultural and social distance." It has little to say about sequencing or the role of government.	There is a strong emphasis on liberalization and on "completing" reforms, e.g., removing the remnants of marketing boards, and removing other obstacles to the effective operation of markets (such as fertilizer subsidies). However, the report also recognizes (somewhat in passing) that governments need to ensure that parastatal institutions are replaced by satisfactory arrangements, that trader entry is not constrained and that newly liberalized markets function adequately.
4. Explicitly take on inequality, in assets and incomes, with targets, timetables, and concrete measures	More equitable access is another of the six key principles. The main focus is on land, rural finance, and economic and social infrastructure.	Little discussion of inequality, but a strong focus on poverty reduction under Commitment two of the WFS, and to safety nets in the Anti-Hunger Programme. There is an emphasis on poverty and hunger throughout.	This is a major theme, particularly in the chapter on assets. The report pays particular attention to land, water, and livestock, but also deals with housing, health, nutrition, and education.	There is little explicit discussion of redistribution, apart from a brief mention of land reform. However, there is a strong focus throughout on poverty reduction.
5. Propose measures to counter the anti-South bias of technical change, recognizing the need for public support to research	There is a strong commitment to supporting agricultural research and extension, including with respect to global public goods.	Technical change is identified as a priority in Commitment three of WFS, and "the evolution of technology" is one of the six new challenges picked up in fyi. Global public goods (e.g., genetic diversity) is another challenge identified. The Anti-Hunger Programme also emphasizes technology, especially for the poor.	This is a major theme, especially in the chapter on technology and natural resources. There is a careful analysis of the technology requirements of the poor, with many examples—in crops, pest control, land management, and water.	Agricultural growth will 'increasingly be knowledge-based, especially in high-potential areas. Priorities are new public–private partnerships, biotechnology, and sustainable pest control.

6. Demonstrate that agricultural strategies will be consistent with natural resource protection, including water management	Promoting sustainable natural resources management is one of the six key principles. There is a particular focus on community-based institutions.	Conserving natural resources is one of the six priorities of the Anti-Hunger Programme, especially with respect to water, genetic resources, fisheries, and forests. The fyl papers identify freshwater resources as one of six key issues, especially the conflict between "water for agriculture and rural development" and "water for nature."	Improved natural resource management is largely treated as an instrumental input to poverty reduction, rather than a good in its own right—but is a recurrent theme. Water issues are prominent throughout.	There is a short chapter on enhancing the sustainable management of natural resources, noting the importance of land degradation, water management, forests, fisheries, and global warming.
7. Recognize the importance of investment in infrastructure and human capital	Investment in human capital is one of the six principles: "major investments are required in order to improve the coverage, quality, and affordability of health and education services in rural areas."	The fyl papers contain an analysis of investment required in agriculture, particularly for research, extension, and public infrastructure and services. There is mention under Commitment two to health and education. The Anti-Hunger Programme includes rural infrastructure as one of the six priorities. Education is dealt with mainly in an extension context.	Better transport infrastructure is seen as high priority. Education and health are discussed in the context of asset redistribution, but are not major themes. However, the chapter on institutions deals extensively with strengthening groups, e.g., for managing common property resources or for micro-finance.	Adequate infrastructure is identified as a *sine qua non*, but in practice little is discussed.
8. Respond to the "obligation" to protect the poor, with new social protection measures, including in conflict areas, and for HIV/AIDS	Managing risk and providing safety nets is one of the six principles.	There is a short section in the fyl papers on "transitional assistance to the food insecure." The right to food is a recurring theme, and is strongly emphasized in the fyl papers. The Anti-Hunger Programme cites safety nets as one of six priorities, with a cost estimate of $5.2 billion (20% of the total investment package proposed).	There is little in the report on social protection.	A chapter is devoted to social well-being, risk, and vulnerability, focusing especially on nutrition and health, HIV/AIDS, education, and food security.

(Continued)

Table 1
(Continued)

Dimension	EU	FAO	IFAD	World Bank
9. Propose pragmatic steps toward greater de-concentration and devolution	Building more effective, accountable, and decentralized institutions is one of the six key principles. The paper covers decentralization and the reform of public sector institutions, among other topics.	No significant discussion of this item, though fiscal and administrative decentralization are mentioned in the Anti-Hunger Programme.	Decentralization is a theme of the chapter on institutions, for example, with respect to natural resources.	Better governance is a recurrent theme, including administrative and fiscal decentralization. Participation and social inclusion are discussed in the chapter on social well-being.
10. Identify the place for agriculture and rural development in PRSPs and sector programs	There is a strong section on country programming, including a discussion of PRSPs and sector-wide approaches, and of public expenditure reform. A methodology is proposed for country-level rural development strategy work.	Commitment seven of the WFS notes the importance of national plans.	There is little discussion of the modalities of aid, except for a review of partnership possibilities at the end of the report.	There is strong support for national rural development strategies, and an extended discussion of how rural development priorities can be incorporated successfully into PRSPs and other planning processes.

Note: The ten dimensions are taken from Ashley and Maxwell (2001). Additional sources include EU (2002), FAO (2002a), IFAD (2001) and World Bank (2002).

any problems that might arise from the declining competitiveness of small farms, nor of issues to do with taxation.

Fourth, there are some rather surprising weak spots in all the documents. None really grapples with urbanization and the resulting transformation of supply chains, for example, through the growing role of supermarkets (Reardon and Berdegue, 2002). Similarly, with the sole exception of the EU, none really grapples in detail with the debate about the sequencing of and limits to liberalization, the so-called post-Washington Consensus on food, agriculture, and rural development.

Fifth, there are then some interesting differences between the agencies. The EU policy, and perhaps that of the World Bank, is the most complete—but it is worth pointing out that these are also the most recent, and from the two agencies in this group with the broadest remit. The EU covers all the main bases, and contains a well-thought-through strategy on how to incorporate rural issues in the national planning processes. The FAO is strong on agriculture, as might be expected, but is also notable for its attention to safety nets and the right to food. It is unfortunate that the high-potential/low-potential framework, strongly present in the WFS papers, has been diluted in later presentations. The IFAD is notable for its focus on food staples on small farms, and for the institutional environment needed to support this. The World Bank is pretty strongly directed to agricultural growth and to market-friendly approaches.

Each of these is quite compelling in its own way, and certainly has been in the past. But it would be brave to argue that the four strategies are fully consistent with each other. It is hard to be compelling as a community when there are different narratives on the table. A brutal caricature would be as follows:

FAO: an updated version of agriculture-led development from the 1960s.

IFAD: an updated version of the small-farm Green Revolution from the late 1960s and early 1970s.

EU: an updated version of integrated rural development from the mid- and late 1970s.

World Bank: an updated version of market-led growth strategies from the 1980s.

5. Conclusion

Whether or not the different narratives are consistent and compelling, there remains the question about whether they are right. The answer to that question is bound to be location-specific, but five general points came out of our earlier review, and are worth repeating here.

First, it is extraordinarily important to understand the rapid pace of change in rural areas throughout the world. In our earlier work, we tried to look ahead. We (Ashley and Maxwell, 2001, pp. 400–401) suggested that:

- Rural populations will begin to stabilize, possibly with a lower dependency ratio initially as birth rates fall, but then a higher one, as migration (and AIDS) remove young adults.
- The connectedness of rural areas will improve, with more roads and other infrastructure (including telecommunications).
- Human capabilities will improve, with better education and health.
- The great majority of rural people will be functionally landless, either without land altogether, or with only a small homestead plot.
- Most rural income in most places will be nonagricultural in origin (though with linkages to agriculture in many cases).
- Most farms will be predominantly commercial, i.e., buying most inputs and selling most of their output.
- Farms (other than part-time subsistence or homestead plots) will be larger than at present, and become larger.
- For those farms able to engage in the commercial economy, input and output marketing systems will be integrated, industrialized, and sophisticated.
- As a result of all the above, disparities between rural areas will increase.
- Agriculture's contribution to GDP will be no more than 10%.
- Agriculture will contribute no more than 10% to exports (perhaps more in Latin America and sub-Saharan Africa).
- Agriculture will become a net recipient of government revenue.

Second, the evidence does not suggest that agriculture, especially small-farm agriculture, is a particularly easy business to be in. This is despite the strong production and consumption linkages that can be found when small-farm growth does materialize. When agricultural growth does happen, it may be in niche or high-value products for urban markets—and is likely to favor larger farms with better access to technology, information, and management skills.

Third, all the evidence suggests that poor people themselves finesse the small farm problem by diversifying in sector and space. This is true even in India, where agriculture remains the main source of rural livelihoods but where migration, for example, is immensely important (Deshingkar and Start, 2003). An effective rural development policy will need to put diversification out of agriculture at the heart of its interventions.

Fourth, growth remains central to rural poverty reduction, but there is much more work to do on the implications for rural development of the post-Washington consensus. It is one thing to make general statements about the need for careful sequencing of liberalization. It is quite another to argue, as some do, that the parastatals abolished with such difficulty in the 1980s should now be resurrected, or that trade should be controlled so as to stabilize prices. Would it not be more reasonable to concentrate attention on the measures necessary to create missing markets and reduce transactions costs?

Finally, those strategies that focus on how to influence the content of PRSPs are surely right. The dominant development consensus in the world is poverty reduction, driven by the Millennium Development Goals, inspired by the three-pronged poverty reduction framework of the 2000/2001 World Development Report, and supported through sector-wide approaches and new forms of budget support. Rural development is possible within that context, provided there is a strong narrative in which growth, empowerment, and security are linked.

Acknowledgments

Thanks are due to John Farrington, Rob Tripp, and Steve Wiggins for comments on an earlier draft. The responsibility for the content is entirely mine.

References

Ashley, C., and S. Maxwell, "Rethinking Rural Development," *Development Policy Review* 19, no. 4 (2001), December 2001.

Belshaw, D., "Presidential Address—Strategising Poverty Reduction in Sub-Saharan Africa: The Role of Small-Scale Agriculture," *Journal of Agricultural Economics* 53, no. 2 (2002), 161–193.

Carney, D., ed., *Sustainable Rural Livelihoods: What Contribution Can We Make?* (Department for International Development: London, 1998).

Chambers, R., *Challenging the Professions: Frontiers for Rural Development* (IT Publications: London, 1993).

Clay, E. J., and B. B. Schaffer, eds., *Room for Manoeuvre: An Explanation of Public Policy in Agriculture and Rural Development* (Heinemann: London, 1986).

Cord, L., "Rural Poverty, in World Bank (2002)," *A Sourcebook for Poverty Reduction Strategies, Volume 2: Macroeconomic and Sectoral Approaches* (World Bank: Washington DC, 2002).

Cord, L., et al, "The Coverage of Rural Issues in PRSPs: A Review of Preliminary Experiences," Draft working paper (PREM and WBI World Bank, 2002).

Deshingkar, P., and D. Start, "Seasonal Migration for Livelihoods in India: Coping, Accumulation and Exclusion," Working Paper 220 (ODI: London, 2003).

Ellis, F., and S. Biggs, "Evolving Themes in Rural Development 1950s–2000s," *Development Policy Review*, 19, no. 4 (ODI: London, 2001).

European Commission, "Fighting Rural Poverty: European Community Policy and Approach to Rural Development and Sustainable Natural Resources Management in Developing Countries," Communication from the Commission, COM (2002) 429 final, Brussels, 25.7.2002 (2002).

Farrington, J., N. C. Saxena, T. Barton, and R. Nayak, "Post Offices, Pension and Computers: New Opportunities for Combining Growth and Social Protection in Weakly-Integrated Rural Areas?" *Natural Resource Perspectives 87* (ODI: London, 2003).

Farrington, J., and J. Lomax, "Rural Development and the 'New Architecture of Aid: Convergence and Constraints'," *Development Policy Review* 19, no. 4 (2001).

FAO, *Rome Declaration on World Food Security and World Food Summit Plan of Action* (Rome: 1996), pp. 13–17.

FAO, *Anti-Hunger Programme: Reducing Hunger Through Agricultural and Rural Development and Wider Access to Food* (Rome: 2002a).

FAO, *International Alliance Against Hunger: Declaration of the World Summit: Five Years Later* (Rome: 2002b), pp. 10–13.

IFAD, *Rural Poverty Report 2001: The Challenge of Ending Rural Poverty* (Oxford University Press: Oxford, 2001), for IFAD.

Keeley, J., and I. Scoones, *Understanding Environmental Policy Processes: Cases from Africa* (Earthscan: London, 2003).

Kydd, J., and A. Dorward, "The Washington Consensus on Poor Country Agriculture: Analysis, Prescription and Institutional Gaps," *Development Policy Review* 19, no. 4 (2001), 467–478.

Leach, M., and R. Mearns, eds., *The Lie of the Land: Challenging Received Wisdom on the African Environment* (James Currey: Oxford, 1996).

Maxwell, S., and R. Heber-Percy, "New Trends in Development Thinking and Implications for Agriculture," in K. G. Stamoulis, ed., *Food, Agriculture and Rural Development: Current and Emerging Issues for Economic Analysis and Policy Research* (FAO: Rome, 2001).

Maxwell, S., "Heaven or Hubris: Reflections on the New 'New Poverty' Agenda," *Development Policy Review* 21, no. 1 (2003).

Omamo, S. W., "Policy Research on African Agriculture: Trends, Gaps, and Challenges," Research Report 21 (2003). ISNAR.

Pirandello, L., *Six Characters in Search of an Author and Other Plays* (Penguin: London, 1995).

Reardon, T., and J. A. Berdegué, "The Rapid Rise of Supermarkets in Latin America: Challenges and Opportunities for Development," *Development Policy Review* 20, no. 4 (2002), 371–388.

Richards, M., S. Maxwell, J. Wadsworth, E. Baumeister, I. Colindres, M. Laforge, M. Lopéz, H. Noé Pino, P. Savma, and I. Walker, "Options for Rural Poverty Reduction in Central America," ODI Briefing paper (2003), London.

Roe, E., "Development Narratives, or Making the Best of Blueprint Development," *World Development* 19 (1991), 287–300.

Rondinelli, D. A., *Development Projects as Policy Experiments: An Adaptive Approach to Development Administration* (Methuen: London, 1983).

Scoones, I., "Sustainable Rural Livelihoods: A Framework for Analysis," Working Paper 72 (IDS, University of Sussex, 1998).

Williamson, T., and S. Canagarajah, "Is There a Place for Virtual Poverty Funds in Pro-Poor Public Spending Reform? Lessons from Uganda's PAF," *Development Policy Review* 21, no. 4 (2003), 449–480.

World Bank, "Rural Development: From Vision to Action. A Sector Strategy," *Environmentally and Socially Sustainable Development Studies and Monographs Series 12* (World Bank: Washington, DC, 1997).

World Bank, *Reaching the Rural Poor: Strategy and Business Plan* (World Bank: Washington, DC, 2002).

Achieving success in rural development: toward implementation of an integral approach

Alain de Janvry and Elisabeth Sadoulet*

Abstract

Starting from an analysis of the reasons for failure of previous "integrated" approaches to rural development, we propose here a conceptual framework for an "integral" approach better in tune with current opportunities for and constraints on rural poverty reduction. The approach is centered on the process whereby well-being is determined, linking household asset positions, the context where the assets are used, pathways from poverty through specific livelihood strategies, and well-being outcomes. This in turn identifies entry points for policies and programs to improve the well-being outcomes that consist in interventions to: increase access to assets, improve context, provide social protection, and achieve social incorporation of the rural poor.

JEL classification: R200

Keywords: rural poverty; rural development

1. Why is it necessary to give increased attention to rural development to meet the Millennium Development Goals?

Rural development has been badly neglected by international development agencies, and this requires urgent redress if there is any chance that the Millennium Development Goals (MDGs) will be met. Neglect is best appreciated when contrasting the 25% share of the World Bank's lending portfolio going to rural development to the 75% share of world poverty that is rural (World Bank, 2002). Not only is the number of rural poor worldwide three times larger than the number of urban poor (Figure 1), but the incidence (Figure 2) and the depth of poverty are also much higher in rural areas. Consequently, the first MDG of halving world extreme poverty and hunger by 2015 cannot be reached without a focus on rural poverty reduction. Rural areas also systematically lag behind urban areas in meeting the other MDGs that concern education, the status of women, child mortality, maternal health, the incidence of endemic diseases, and

environmental stress. To meet the MDGs, rural poverty and lags in rural social development need to be attacked directly, for which rural development must be placed much higher on international development agencies' priority lists.

Why has rural development been so badly neglected when it is common knowledge that rural poverty dominates world poverty, and when combating poverty has become the top priority for most international development agencies? The answer is in part because past efforts at rural poverty reduction have had little success. Aggregate income growth, the main instrument for poverty reduction, has proven to be a weaker force for reducing rural rather than urban poverty (e.g., Figure 2 shows that the income elasticity of the incidence of poverty is -0.07 in urban areas and -0.05 in rural areas). For Latin America (Figures 3 and 4), these elasticities are -0.09 and -0.06, respectively. The ratchet effect of economic booms and busts is consequently larger for urban than for rural areas (see Figure 5 for Mexico), with poverty in the former rising more steeply during recessions and falling more sharply during growth spells.

Hence, if growth is to be more effective in helping reduce rural poverty, the qualitative nature of growth

* *Department of Agriculture and Resource Economics, University of California at Berkeley, CA, USA.*

Alain de Janvry and Elisabeth Sadoulet

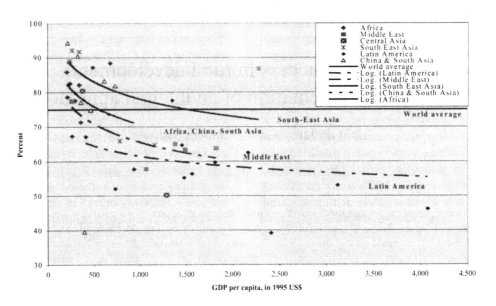

Figure 1. Percentage of total poverty that is rural, by regions.

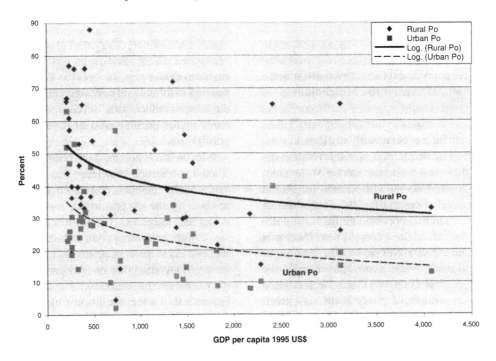

Figure 2. World rural and urban poverty.

needs to be modified. This issue is discussed in this paper, with consideration for the regional dimension of
growth, i.e., ensuring that it reaches the areas where
the rural poor are located. In addition, the decline

observed in the share of total poverty that is rural as
income per capita rises (Figure 1 for the world and
Figures 3 and 4 for Latin America) has not been due
to success in raising rural incomes faster than urban

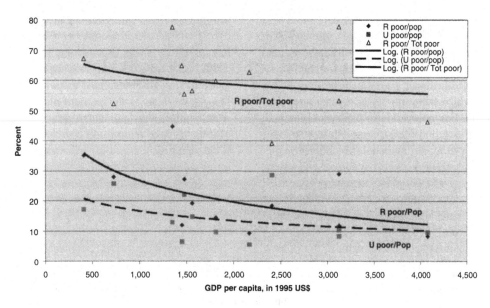

Figure 3. Latin America (number of rural and urban poor as shares of total population). Countries: Bzl, Col, CR, DR, Ec, El S, Guat, Nica, Pan, Para, Peru.

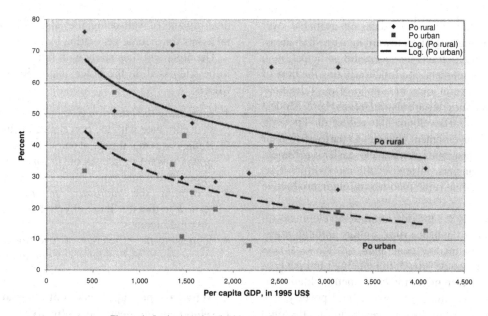

Figure 4. Latin America (incidence of rural and urban poverty).

incomes, as much as to migration to urban areas. This is because migration to the metropolitan areas of ill-prepared rural migrants has generally only contributed to displacing poverty to the urban sector. This role of migration can be seen in contrasting the more rapid fall in the number of rural than urban poor (Figure 2) while the incidence of poverty falls by much less in the rural than in the urban areas. This is indeed an important reason why the number of urban poor is rising faster than the number of rural poor (see Figure 6

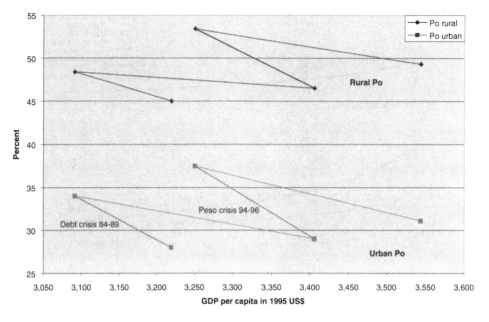

Figure 5. Mexico: Poverty (Po) and the ratchet effect of economic crises.

for Central America where data are available over time). Reducing migration toward metropolitan areas, preparing rural migrants for successful participation in regional labor markets, and functionalizing the flow of remittances to rural areas in support of rural development are all issues that are also addressed here. Modest success with rural development to reduce rural poverty has induced development agencies to partly shift their anti-poverty programs toward the easier instrument of welfare transfers, instead of the more difficult attempts at raising rural incomes through productive activities.

It is precisely (1) the drift in rural development toward seeking to enhance rural welfare through transfers instead of income generation, (2) the weakness of aggregate income growth in reducing rural poverty, and (3) reliance on migration to the metropolitan areas as the main instrument to contain rural poverty, that need to be corrected. Income generation by the rural poor themselves needs to be placed squarely back on the poverty reduction agenda to achieve the MDGs. In addition, social expenditures in education, health, and the status of women need to be both increased in rural areas to reduce the gap with urban areas, and made more efficient to be competitive in qualifying for aid budgets. Not only are these expenditures important

determinants of well-being, they also create fundamental assets for income generation by the poor.

The argument thus starts with four issues that need to be addressed, in the order of priority, to reach the MDGs:

1. *Increasing the attention* given to rural development in accordance with the prevalence of rural poverty.
2. Redressing the thrust of rural development programs toward emphasis on *income generation* as opposed to transfers.
3. Amending the *qualitative nature of economic growth* so that it is more effective in helping generate rural incomes.
4. Increasing the efficiency of investments in *social development and social protection* in rural areas.

2. Why have past approaches at integrated rural development not been more effective?

Rural development flourished in the 1970s and 1980s, in part under the leadership of the World Bank and USAID, following the approach of what has been referred to as "Integrated Rural Development." This approach mobilized the public sector to deliver services to the rural poor, trying to integrate the many

(Costa Rica, Guatemala, Honduras, Panama)

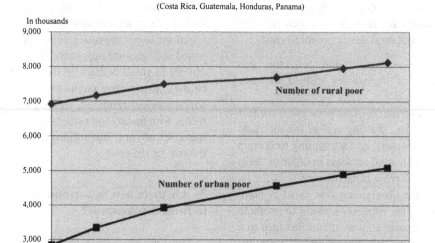

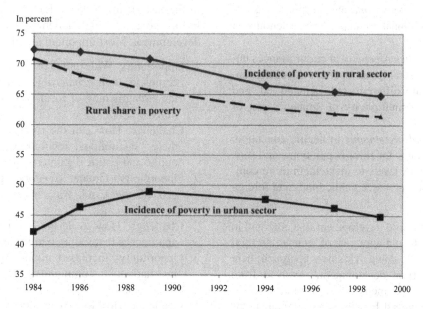

Figure 6. Poverty in Central America (Costa Rica, Guatemala, Honduras, Panama).

public services required for poverty reduction through the role of an implementation agency. It principally focused on agriculture, seeking to extend to small farmers the benefits of the Green Revolution.

This approach met with limited success, as the results have generally not been sustainable beyond the end of state support. The approach itself was made dysfunctional by a decline in the role of the state and in the size of public budgets following structural adjustment. Difficulties with the approach led to a drift in attacking rural poverty toward welfare transfers instead of income generation and social development.

Important lessons were, however, learned from the experience with integrated rural development, including the need to reorient rural development toward:

- Relying more on *individual and collective* initiatives of the poor themselves instead of state tutelage;
- Achieving *competitiveness* of the poor in the context of market forces, i.e., seeking access for the rural poor to dynamic markets, and attempting to correct the market failures that affect them in order to "help the poor play by the rules of the market";
- Capitalizing on the observation that *rural is more than agriculture*. This suggests seeking to promote multisectoral approaches, rural–urban linkages in a territorial perspective, and pluriactivity whereby rural households are engaged in a multiplicity of farm and nonfarm activities;
- Placing rural development efforts in the context of supportive *international and macro* policies, as opposed to the traditional urban policy bias denounced by Lipton (1977);
- Recognizing the *heterogeneity* of circumstances under which the rural poor operate, requiring approaches that are differentiated, demand-led, and that allow for a multiplicity of strategies out of poverty;
- Increasing *social investments* in health, education, and women's status for rural populations, and raising their *efficiency* levels to make them more competitive in attracting scarce public funds and foreign aid budgets.

The approach to rural development that emerged in the 1990s turned the old approach of "integrated rural development" upside down. This new approach, here termed "integral rural development," evolved from practice, through experimentation with rural development projects organized by pioneering organizations such as IFAD, the Inter-American Foundation, and innovative NGOs. It was recently adopted and perfected by the World Bank (2002), the Inter-American Development Bank (2004), the Inter-American Institute for Cooperation in Agriculture (2003), and INRA/CIRAD (2001). This approach is still largely experimental and incompletely defined. It is a process as opposed to a blueprint, characterized by pragmatic adaptation to local conditions, yet resting on several fundamental principles that contrast it to the "integrated" approach, most

particularly in emphasizing decentralization, participation and collective action, devolution of managerial functions to communities, following a territorial as opposed to a sectoral approach, pursuing the advantages offered to small holders by the "new agriculture" (see below), introducing payments for environmental and social services rendered, seeking coordination mechanisms with macro and sectoral policy, and reconstructing a set of rural institutions following de-scaling of the role of the state.

3. Are there new opportunities for success in rural development?

There are several reasons why a new approach to rural development, going from integrated to integral, has a chance to succeed in helping meet the MDGs. Opportunities come from major changes that have occurred in the last 10 to 15 years. Each of the opportunities in turn poses specific challenges for success in rural development. Most important are the following five:

1. Opportunity: Widespread progress, even if far from complete, with *democracy, decentralization*, and the strengthening of local civil society *organizations*.
   Challenge: How can the rural poor capitalize on these "institutional revolutions" to secure economic and political gains for themselves?
2. Opportunity: Greater *freedoms*, gained with the end of the Cold War, in experimenting with new approaches.
   Challenge: How to design experiments with new approaches that will maximize learning?
3. Opportunity: Increased importance given by the international community to *environmental* problems.
   Challenge: How to use the concern for the environment in support of rural development, in particular in order to secure transfers of funds toward rural areas?
4. Opportunity: Modest increases in *foreign aid budgets*.
   Challenge: How to make a convincing case in directing these additional resources toward rural development?
5. Opportunity: *Steep learning curve* with elements of an "integral" approach to rural development.

Challenge: How to use what has been learned in localized success stories for design and scaling up?

4. What have we learned from experience with new approaches to rural development?

So, while greater opportunities exist, what is "new" about how rural development can be approached that justifies putting it high on the agenda of international development agencies? To answer this question, a conceptual framework is proposed in Figure 7 that must address the following four issues:

(i) Explain the determinants of rural well-being.
(ii) Identify the entry points (policies and programs) for rural development interventions that can improve well-being.
(iii) Identify the processes through which pro-poor rural policies and programs are determined.
(iv) Identify instruments for greater efficiency in implementing these policies and programs.

4.1. The process through which rural well-being is determined

If the levels of rural well-being are to be raised, the process through which it is determined needs to be understood. This, in turn, will help to identify entry points that rural development interventions can use to alter outcomes toward higher levels of well-being. This process can be conceptualized as shown in Figure 7.

4.1.1. Rural poverty must be understood in terms of the behavior of the actors involved: the rural poor and their organizations

Rural households are endowed with assets that establish their capabilities. In turn, the levels of well-being they will achieve with these assets depends on the opportunities and constraints offered by the context in which they operate. Behavior in using assets may be individual, or it may be collective. It is always influenced by the behavior of others, both poor and nonpoor. The purpose of rural development will be to change the capabilities (assets) and the opportunities and constraints (context) that determine the well-being outcome. The cornerstone for the formulation of any approach to rural development is thus an understanding of the actors and how they define their livelihood strategies individually and collectively. This is particularly important if rural development is to reach the poorest of the poor with their own particular idiosyncratic behaviors. This includes female-headed households, pastoralists in marginal zones, forest dwellers, artisanal fishermen, and indigenous people. Their livelihood strategies are all the more complex because they have to deal with lack of opportunities, marginal environments, market failures specific to them, and precarious conditions leaving little margin for error.

4.1.2. Well-being, the objective of rural development, is multidimensional

An important dimension of well-being is income and the consumption expenditures that it affords. Insufficient income (income below a poverty line) implies poverty. Uncertain income implies vulnerability, often with irreversible consequences, such as the fire sale of productive assets, or taking children out of school in response to an income shock, or hunger spells that affect subsequent physical and mental development. Furthermore, unequal incomes undermine participation and willingness to contribute to social undertakings. Yet there are other dimensions to well-being that are complementary to income, and are more of a public goods nature. This includes human development (health, education, nutrition), sustainability in the use of natural resources, and social status and rights. In this article, the term "poverty" is reserved for the income deficit in well-being, and reducing it should be a primary objective of rural development. Improving overall well-being, however, implies balance in raising the different components that define it. Successful rural development thus requires an integral perspective on the different dimensions of well-being, seeking, in particular, interventions that will achieve joint gains in more than one dimension (e.g., productivity, sustainability, and health by means of technological innovations in agriculture).

4.1.3. Household assets are highly heterogeneous and yet complementary

Following the livelihoods approach, a household's asset portfolio determines its capability to generate income. These assets are multidimensional, implying both complementarities among assets and a

Alain de Janvry and Elisabeth Sadoulet

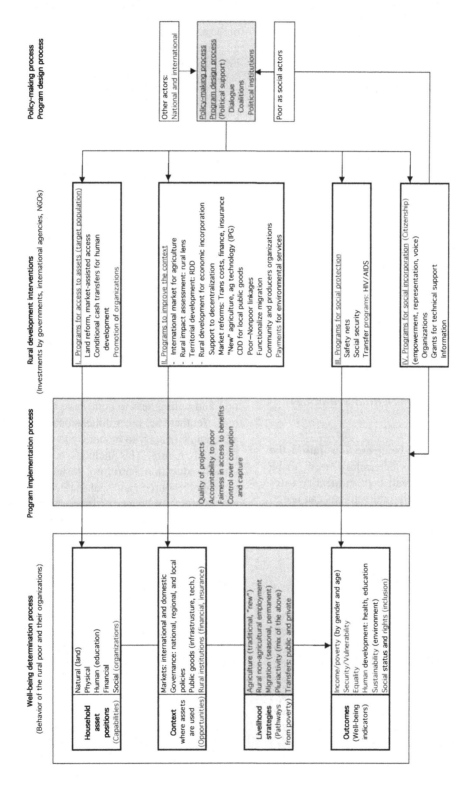

Figure 7. Conceptual framework for integral rural development.

multiplicity of entry points for rural development programs aimed at increasing asset positions. Assets include natural capital (land, water, trees), physical capital (tools and equipment), human capital (education and skills), financial capital (liquid assets such as livestock and bank deposits), and social capital (relations of trust with others and membership in organizations). Asset positions need to be assessed by gender and age, as household endowments may be unequally accessible for members, women and youth in particular. Heterogeneity of asset positions implies differential abilities to respond to opportunities across households and individuals. Hence, any specific policy reform will have a widely differentiated incidence across categories of rural poor. A key feature here is that poverty is due to lack of access to assets, most particularly land and education. Getting these households out of poverty thus requires either increasing their access to land or diversifying their activities off-farm, for which education is key. It is indeed a distinguishing feature of the rural poor that their sources of income tend to be highly diversified, combining agriculture with off-farm activities.

4.1.4. The income generation value of assets depends on the quality of the context where they are used

Good asset endowments in an unfavorable context will not help households escape poverty. Access to land, for instance, is only effective in getting households out of poverty if their competitiveness can be secured, which depends on the nature of the markets, institutions, public goods, and policies where they operate. Important aspects of context that determine the well-being value of asset use include:

- The international price level for tradable agricultural commodities, affected in particular by OECD farm policies and trade protection between developing countries;
- The national policy environment, both macro (growth, real exchange rate) and sectoral (trade policies, subsidies);
- The nature and quality of governance at the national and local levels;
- Public goods such as agricultural research and extension services, and infrastructure;
- The performance of markets, particularly in terms of competitiveness, transactions costs, and constraints

on access for the poor, including markets for the assets they hold;
- Rural institutions such as financial services and community organizations for service and mutual insurance.

This multidimensionality of context implies, again, complementarities and a multiplicity of entry points for rural development programs aimed at increasing the productivity of the assets used by the poor.

4.1.5. There are several pathways from poverty

Finding a way out of rural poverty can be achieved by traveling along several alternative pathways. Most frequent has been *migration* of the rural poor to the urban environment. If these migrants are not prepared or able to compete on the urban labor market (where education is key), this may only result in displacing poverty from rural to urban environments. *Agriculture* offers a second pathway. What is important here is not farming of traditional staple crops on small expanses of land that will never generate enough income to erase poverty, but opportunities offered to smallholders by the "new agriculture." It consists of producing high value-added crops, seeking remunerative market niches, pursuing quality and food safety, relying on certification and labeling of products, seeking contracts with supermarkets, agroindustries, and agroexporters, adding value by moving up the marketing chain, etc. A third pathway from poverty is through *off-farm activities*, including employment in small and medium enterprises, self-employment in microenterprises (e.g., cultural industries), and remittances from household members.

Combining agriculture (in particular, the production of staple crops on small plots of land to secure home consumption) and off-farm activities leads to the pervasiveness of *pluriactivity*. This is particularly important in the least well-endowed regions where the "new agriculture" has low potential. Pluriactivity offers many advantages to participating households: while individual household members tend to be specialized in one activity, the household as a totality is a diversified enterprise, engaged in a portfolio of activities that reflect each member's comparative advantage while together providing risk diversification. Finally, *transfers* also offer a pathway from poverty, especially for those unable to work or exposed to uninsured shocks. This multiplicity of pathways from poverty also means that rural development interventions provide opportunities out of poverty, and pursue diverse and multiple strategies. This is why a territorial approach to rural development that goes beyond a sectoral focus on agriculture and, instead, seeks success along all possible pathways from poverty has greater potential.

4.2. Entry points for investing in integral rural development

How can this approach help define entry points for investments in rural development? The conceptual framework in Figure 7 identifies four entry points. Each of these, in turn, suggests an array of investments that can help reduce rural poverty. The key is to maintain an appropriate balance between investment types in accordance with particular situations. In general, however, the suggestion is to increase investments in programs that promote income generation and social development, while reducing programs that seek to address poverty through cash transfers to individuals with the capacity to work.

4.2.1. Entry Point I: Programs to increase access to assets

Among the multiplicity of assets that sustain livelihood strategies, only three that are particularly important for rural poverty are discussed here: land, education, and social capital.

4.2.1.1. Land. Programs to increase access to land for the rural poor, and improve the security of access to land (through granting formal titles or recognizing informal rights to individuals or to communities), are both fundamental and controversial. Lessons learned from recent experiences indicate that there are many alternative strategies that can be followed to promote access to land in support of rural poverty reduction (Deininger, 2003). For instance, many countries have introduced constitutional constraints on land use, requiring minimum thresholds of land productivity or imposing ceilings on land ownership. Farms that do not satisfy these requirements are subject to expropriation. This is, for example, the legitimacy advocated by the MST (Landless Movement) in Brazil in invading large underused estates. Other countries such as Bolivia have tried to revisit the legality of properties with dubious land titles, seeking in this fashion to reclaim

large expanses of land that could be distributed to the poor. Remaining public lands (Venezuela) or quotas of lands improved by public irrigation projects (Dominican Republic) can also be used for land settlements. Finally, several countries have pioneered subsidies to land market transactions, favoring access to land for the poor. Large programs have been put in place in Brazil, Colombia, South Africa, and at a smaller scale in many countries through land funds, sometimes managed by NGOs. These programs are still at early stages and in much need of experimentation and evaluation. However difficult to achieve, historical experience shows that access to land is, relatively, the easy part in attempting to reduce rural poverty through land reform, while securing the sustainable competitiveness of beneficiaries is by far the harder task. This is why so many hard-fought land reforms have been effective in promoting access to land, but not in reducing poverty.

4.2.1.2. Rural education. Poverty is inherited if poor parents do not educate their children and do not provide them with good health and nutrition. To break this vicious cycle of inheritance of poverty, many countries have introduced conditional cash transfer programs, where poor parents (usually mothers) receive payments in exchange for sending their children to school and to health visits (Morley and Coady, 2003). Programs of this type are in place in Mexico (Progresa), Brazil (Bolsa Escola), Honduras (PRAF), Nicaragua (RPS), Argentina, Costa Rica, Jamaica, Colombia, and Turkey. These programs have been successful in enhancing school achievements, but they are expensive, largely because they are targeted at the poor without further selection criteria. This is inefficient because many of the poor can afford to send their children to school. As a consequence, it is better to target transfers to maximize educational gains, choosing as beneficiaries children most likely to start going to school (when they were likely not to go) or to stay at school (when they were likely to drop out) thanks to the transfers. This is important to make human development programs more efficient, one of the fundamental requirements in attracting larger foreign aid budgets.

4.2.1.3. Social capital. Organizations support collective action whose purpose is to service members in a context of market failures, the development of joint enterprises for income generation, acquiring market

power, gaining political representation, accessing information, and obtaining training. Programs in support of the development and improved performance of organizations are the cornerstone of an integral approach. A major challenge is to transform organizations that were created by the state or by donors for the appropriation of rent into organizations for the generation of value. Another challenge is to balance the advantages offered by the traditional community for trust, solidarity, and reciprocity with the requirements of effective organizations in democratic leadership, professionalism, and capacity to absorb external influences. Assuming that traditional communities can alone sustain devolution programs and provide the basis for effective producers' organizations has all too often been proven wrong (Abraham and Platteau, 2002). At the same time, traditional community organizations, if they do not stifle the emergence of producer organizations, can help the latter be more successful if they are effective in the provision of mutual insurance, public goods, and the management of common pool resources.

4.2.2. Entry Point II: Programs to improve the quality of the context where assets are used

This is, of course, a huge work agenda, but it has to be prioritized in the purposeful perspective of enhancing the well-being generation capacity of the assets held by the rural poor. Four aspects of the context that are most important as components of an integral approach to rural development are mentioned here.

4.2.2.1. International market for agriculture. Even from the perspective of integral rural development that takes a territorial as opposed to a sectoral approach, agriculture remains the main source of income for the poor in most rural regions, either directly as farmers or indirectly as workers or entrepreneurs in activities linked to agriculture. Yet, the profitability of agriculture worldwide is being undermined by trade restrictions on the markets of OECD countries and by the huge subsidies given to their farmers. Agricultural trade is also restricted by protection on developing country markets. Clearly, rural development and rural poverty reduction cannot succeed under these conditions. These price distortions should be eliminated, and the Doha negotiations should be conducted from the perspective of making agricultural policies in the

industrialized countries consistent with the MDGs. In the transition toward a new world order for agriculture, developing countries' peasantries need to be given protection and subsidies so that market distortions do not result in their premature elimination, clearly for the wrong reasons and at enormous human and efficiency costs.

4.2.2.2. Rural impact assessment. Rural development initiatives and agricultural policies are often contradicted by other policies affecting the exchange rate, interest rates, the pricing of industrial goods, and effective demand for rural goods. It is thus important to secure consistency between macro policy and rural development. A useful concept for this is the Canadian practice of "rural lens," whereby any national policy initiative must be scrutinized from the point of view of its potential impact on rural areas and the well-being of rural people. This is important to elevate concerns with rural development in the national policy agenda. Many costly investments in rural development have simply been wasted because they were contradicted by policy initiatives in other sectors of economic activity.

4.2.2.3. Territorial development. It is widely agreed that economic growth is essential for poverty reduction, yet aggregate economic growth has had a modest impact on rural poverty other than through migration to the metropolitan areas. This is because growth all too often does not create new opportunities in the regions where the rural poor are located. Endowing the poor with assets is of no consequence to rural poverty if they are not located in a context where growth offers them new employment and investment opportunities. This is the purpose of regional (territorial) development. While extensive in the 1990s, decentralization has almost exclusively been pursued at the municipal level, a unit of decision making appropriate for the delivery of public goods and the targeting of social expenditures, but not for the promotion of income-generating strategies. For this, larger economic regions are needed. The dynamics of growth in their rural hinterlands is importantly based on the intensification of rural–urban linkages centered on secondary cities. Indeed, in many countries a reconsideration of decentralization in support of economic projects is observed, seeking to either define regions administratively or to promote the formation of associations of municipalities for the implementation of specific economic projects in the corresponding territory (e.g., a watershed development project that cuts across municipalities, or the promotion of regional specialty products in localized agricultural systems). To be successful in their economic initiatives, territories need to endow themselves with institutions for three functions:

- Institutions for *consultation and coordination*: regional development councils or roundtables that include all segments of society in the region, i.e., public, corporatist, and nonprofit.
- Institutions for *planning*: bureaus for the identification and strategic creation of comparative advantages, the planning of public investments, etc.
- Institutions for *promotion*: regional chambers of commerce, etc., that function, in part, to develop the region's corporate image, promote culture, define quality products, seek certification and labeling, and advertise regional specialties.

It is only if these regional employment and investment opportunities exist that rural development can be effective in assisting the poor to benefit from these sources of income. This more decentralized regional growth pattern may have a cost when compared to centralized growth in the metropolitan areas, yet it may be the opportunity cost that has to be incurred to increase the poverty reduction capacity per point of economic growth. The quality and quantity of growth thus needs to be reconsidered for maximum aggregate poverty reduction, including rural poverty.

4.2.2.4. Rural development for economic incorporation. Once regional development creates investment and employment opportunities, poverty reduction requires rural development as a set of interventions to *help the rural poor seize at least part of these opportunities*. There are some important principles for success in this, derived from the practice of integral rural development programs.

The first is that many *markets fail* the poor, preventing them from deriving full benefits from the assets they control. The most notorious are:

- High transactions costs on product and factor markets associated with distance, poor infrastructure, imperfect information, and lack of market power;

- Lack of access to credit due to collateral requirements by commercial lenders and lack of public credit histories in gaining access to loans from microfinance institutions;
- Extensive noninsured risks implying the need to engage in costly risk management at the opportunity cost of reduced expected incomes; and
- Missing markets for environmental services delivered by rural people, including watershed management, biodiversity conservation, and carbon capture.

Overcoming these market failures requires institutional innovations such as service cooperatives, microfinance institutions and credit bureaus, mutual and regional insurance schemes, and payments for environmental services.

The second is to use the potential offered by the *new agriculture* to derive higher value added from limited land resources, one of the fundamental defining characteristic of rural poverty. Some of the main principles of the "new agriculture" are the following:

- Seek to identify *dynamic market opportunities* as the driving force in choosing investment priorities: export markets, market niches, demands derived from linkages with dynamic local activities (e.g., tourism).
- Apply *frontier technology* to peasant farming systems. Because most of these technologies are global or regional public goods, they need to be delivered by international organizations such as the CGIAR and its associated national partners.
- Focus on crops and animals with *high value added* per hectare and capture additional value added in moving up in the product chain toward transformation and marketing.
- Seek *higher prices* through quality, food safety, certification (organic, fair trade), and labeling.
- Sell *environmental services* through contracts (watershed management for water and soil erosion, biodiversity, carbon capture) or the competitive sale of environmental goods and services (ecotourism).
- Seek production *contracts* with supermarkets, agro-industry, and agro-exporters.

The third is to use *demand-led approaches*, where the rural poor, their organizations, and their communities are offered "choice and voice" in identifying and managing the investments in public goods and services best meeting their perceived needs and opportunities. This is the Community-Driven Development (CDD) approach pioneered by IFAD and widely implemented by the World Bank. Finding effective ways of enabling communities to perform these functions while avoiding capture of benefits by local elites, scaling up successful programs, and accelerating disbursements remain important challenges that require experimentation. To support the emergence of new income-generating opportunities, demand-led approaches must be extended to the regional level, in Region-Driven Development (RDD) programs.

The fourth important principle is to seek *maximum linkages between the poor and the nonpoor*. While the targeting of welfare transfers should try to maximize exclusion of the nonpoor, the targeting of income-generating programs should maximize linkages to the nonpoor, so the poor benefit from the nonpoor's superior capacity to identify new markets, take risks, negotiate contracts, explore new technological alternatives, and exercise influence. The Petrolina development in Brazil (Damiani, 2002) is an excellent example, where the nonpoor opened distant markets and negotiated contracts with distributors for the fruits and vegetables they and the poor produce. Poor/nonpoor associations can be obtained through joint ventures, subcontracts, and more simply, employment opportunities in the local businesses of the nonpoor, likely the safest and easiest option out of poverty for those with limited entrepreneurial skills and social capital. To avoid exploitation by the nonpoor, organizations of the poor are important to help them develop countervailing power and appropriate a fair share of the value created through these associations.

The fifth important principle is to functionalize *migration* in support of rural development. Rural–urban migration is a necessary aspect of the "agricultural transformation," since the population engaged in agriculture needs to decline as GDP per capita rises, a process that has been important in holding in check rural relative to urban poverty. This powerful force needs to be functionalized for rural development in the following four dimensions:

- Assist rural areas release *prepared individuals* for urban employment, so that migration does not displace poverty to the urban centers, but to the contrary

is an integral component of an aggregate poverty reduction strategy;

- Help migrants find employment in their *regions of origin* through regional development, as opposed to leaving them with no options but to migrate to metropolitan areas or abroad;
- Channel *remittances* toward local investment, e.g., through the capitalization of local banks and lending locally on the basis of reputation;
- Make migrants *agents of local change* by promoting linkages between migrants and their communities of origin through clubs, the flow of ideas for new initiatives, and subcontracts with local entrepreneurs.

4.2.3. Entry Point III: Transfer programs for social protection

Many individuals are unable to generate income due to age, disabilities, and disease, or are the victims of uninsured income shocks. Only transfers can help them escape poverty. *Safety nets* are consequently essential, and are in much need in developing countries. Largely missing, for example, are unemployment insurance schemes while developing economies are typically highly unstable, exposing individuals who have little ability to withstand risk, to bear all the costs of absorbing income shocks. Helping the rural poor reduce the need for risk management can be an important source of higher expected incomes in their economic activities. Programs to provide insurance include guaranteed employment schemes and safety nets put into place ex ante relative to the incidence of risk.

Social security schemes for rural areas, which are largely missing in most of the developing world, are not only important sources for welfare gains for the elderly, but they can create major efficiency gains. This is because they allow reduction of private welfare transfers from young to old, which compete with productive investments, and they allow intergenerational transfers of land at a younger age, serving to combat the generalized phenomenon of aging among rural entrepreneurs.

Finally, *transfer programs* are essential for categories at risk, such as orphans from deceased parents due to the HIV/AIDS pandemic. What is wrong, however, is to use cash transfer programs as a way of reducing poverty among populations perfectly fit to work. Yet, this has increasingly been done in an attempt to reduce rural poverty because transfers are easier to organize than income generation programs. Transfer programs need to be expanded, but not as a substitute for income generation programs for those who can work, the "old way" of getting out of poverty. International organizations must not yield to facile approaches. To return to the "old way," they need to have on board people with the technical skills to help respond effectively to local demands for sustainable income-generating projects, the essence of integral rural development.

4.2.4. Entry Point IV: Programs to promote the social incorporation of the poor

In the end, the only sustainable way of reducing rural poverty is to make the rural poor be their own agents of change. This requires social incorporation, the process of acquiring full citizenship rights and the ability to be represented, to voice demands, and to bargain for better deals, both in the market and in the political arena. Promoting social incorporation as the cornerstone of an anti-poverty strategy is probably the most important difference between integrated and integral rural development. For the rural poor, who have historically been the most voiceless and the least represented segment of civil society, this is a sharp departure from their past status. This requires the promotion of *organizations* representing the rural poor, owned by them, and servicing their needs. Rural development programs can be specifically designed to seek the emergence and to enhance the effectiveness of these organizations, both at the local level and in high-level organizations. Innovative projects of this type have been pursued by IFAD, and more recently, by the World Bank (e.g., CLCOP [Cadre Local de Concertation des Organizations de Producteurs/Local Committee for Coordination of Farmers' Organizations] in Senegal and CPCE/OP [Comités Provinciaux de Concertation et d'Echanges des Organizations Paysannes/Provincial Committees for Coordination and Exchange between Farmers' Organizations] in Burkina Faso). Yet, much is still to be learned about the approach. What is needed is to experiment with different set-ups, engage in stakeholder participatory appraisal, and seek strategies for scaling up and accelerating disbursements, while maintaining accountability and efficiency. Lead organizations that are recognized as legitimate representatives by their constituent organizations are needed for effective access

and representation. Organizing for social incorporation is one of the most important aspects to be explored in an integral approach to rural development, the true frontier of the field.

In the end, using rural development successfully for poverty reduction is a political undertaking. This is because policies and programs are the outcome of a political and administrative process in which a multiplicity of social actors (Figure 7) participate. If these policies and programs are to effectively serve in poverty reduction, it is a sine qua non condition that the poor be fully represented in the process where they are made and designed. This requires rural development interventions in support of social incorporation, the fourth entry point discussed here. This is, however, not enough. Translating social incorporation into influence over the definition and implementation of pro-poor policies and programs requires a set of institutional innovations that will *inform* the poor, support *dialogue*, and enable the exercise of *influence*. They must allow their participation in local and national democratic processes, including the elaboration of country-level poverty reduction strategies.

Finally, social incorporation of the rural poor has a fundamental role to play in program *implementation*, in order to allow them to monitor and enforce accountability, achieve fairness in access to benefits, and gain control over corruption and capture (Figure 7). Social incorporation of the rural poor is thus the major challenge for an integral approach to rural development. Success or failure in meeting this challenge will determine the effectiveness of rural development in reducing poverty. Much experimentation and learning remains to be done, requiring special funding if it is to happen. In such complex undertakings as integral rural development, projects cannot follow blueprints and outcomes are rarely as initially expected. The key to success is the ability for participants to learn from successes and failures and to internalize lessons learned into adjusting projects for greater efficiency.

5. What are the broad principles of an approach to integral rural development?

Integral rural development offers a significant opportunity to reduce the enormous and stubborn mass of rural poverty. Yet, it will not happen easily. It is a complex undertaking, vastly underexplored, and where much has yet to be learned. In addition, the enormous heterogeneity of conditions under which the rural poor operate requires creativity and flexibility in project design and implementation, and this has to be done locally through decentralization and participation. The approach is not one where blueprints can be found, but one that requires systematic learning-by-doing and learning-from-others through horizontal exchanges and comparative analysis. It is consequently an approach that will demand time to mature and sustained support by donors, not quick entry–exits in search of rapid results. To put this approach to rural poverty reduction in motion toward the MDGs, there are several recommendations to development agencies that derive from the analysis in this paper. The following five priorities are put forward as leading issues requiring attention:

1. *Create capabilities*
   Objective: Make the rural poor into able agents of change.
   Instruments: Investments in social development (health, education, nutrition, family planning), improved access to assets including social capital under the form of organizations for the creation of value, provide greater social protection.
2. *Create citizenship*
   Objective: Assist social incorporation of the rural poor.
   Instruments: Promote and improve the effectiveness of local and peak organizations for representation of the rural poor, decentralization for greater access and accountability, and devolution for local control.
3. *Create opportunities*
   Objective: Increase opportunities for the rural poor to generate income in their regions of origin.
   Instruments: Pursue a territorial approach to rural development, opportunities offered by the "new agriculture," increase the profitability of agricultural commodities (Doha negotiations on agricultural protectionism and subsidies, and support for peasant agriculture in the transition to a new agricultural order), functionalize migration to local development, organize demand-led programs for public goods (choice), introduce payments for

environmental services, introduce new rural institutions in particular for finance and insurance, and promote poor–non-poor linkages.
4. *Create political support*
   Objective: Elevate rural development in the political agenda.
   Instruments: Introduce a "rural lens" approach to rural impact assessment, promote political reforms for participation of rural people.
5. *Create knowledge*
   Objective: Experiment with, and learn about integral rural development.
   Instruments: Analyze current experiences, experiment with alternative approaches, do real-time impact analysis, do impact analysis of demand-driven and decentralized projects, and engage in participatory learning. This learning process needs to be funded largely as public goods, as gains from lessons learned are widely shared. Otherwise, continued under-investment in learning at the project level will remain the norm, frustrating a transition into integral rural development.

References

Abraham, A., and J.-P. Platteau, "Participatory Development in the Presence of Endogenous Community Imperfections," *Journal of Development Studies* 39, no. 2 (2002): 104–136.

Damiani, O., "Diversification of Agriculture and Poverty Reduction: Effects on Small Farmers and Rural Wage Workers of the Introduction of Non-Traditional High-Value Crops in Northeast Brazil," Ph.D. dissertation, MIT (2002).

Deininger, K., *Land Policies for Growth and Poverty Reduction* (The World Bank: Washington, DC, 2003)

INRA/CIRAD, *Systèmes Agroalimentaires Localisés: Terroirs, Savoir-Faire, Innovations* (INRA: Paris, 2001).

Inter-American Development Bank, "Rural Poverty Reduction in Latin America and the Caribbean" (Washington, DC, 2004).

Inter-American Institute for Cooperation in Agriculture, "Poverty and the New Rurality" (San Jose, Costa Rica, 2003).

Lipton, M., *Why Poor People Stay Poor: A Study of Urban Bias in World Development* (Harvard University Press: Cambridge, MA, 1977).

Morley, S., and D. Coady, *From Social Assistance to Social Development: Targeted Education Subsidies in Developing Countries* (International Food Policy Research Institute: Washington, DC, 2003.

World Bank, "Reaching the Rural Poor: A Renewed Strategy for Rural Development" (International Bank for Reconstruction and Development: Washington, DC, 2002).

Plenary 2

Is there a future for small farms?

Peter B. R. Hazell*

Abstract

Small farms are seriously challenged today in ways that make their future precarious. Marketing chains are changing and becoming more integrated and more demanding of quality and food safety. This is creating new opportunities for farmers who can compete and link to these markets, but threatens to leave many others behind. In developing countries, small farmers also face unfair competition from rich country farmers in many of their export and domestic markets. The viability of many is further undermined by the continuing shrinkage of their average farm size. And the spread of HIV/AIDS is further eroding the number of productive farm family workers, and leaving many children as orphans with limited knowledge about how to farm. Left to themselves, these forces will curtail opportunities for small farms, overly favor large farms, and lead to a premature and rapid exit of many small farms, adding to already serious problems of rural poverty and urban ghettos. If small farmers are to have a viable future, then there is a need for a concerted effort by governments, NGOs, and the private sector to create a more enabling economic environment for their development. Appropriate interventions could unleash significant benefits in the form of pro-poor agricultural growth in many developing countries and more than pay for themselves in terms of their economic and social return. But they do not seem very likely at the moment and current trends are moving in the opposite direction.

JEL classification: Q12, Q13, Q17, Q18

Keywords: small farms; market liberalization; agricultural trade; farm size transformation

1. Introduction

Small farms still dominate the agricultural sector in much of the developing world and they are still significant players in the rural life of many rich countries. As part of the economic transformation process, rising labor costs drive most small farms out of business, and only part-time farmers and a few small specialized producers of higher-value products survive. Historically, this process has usually taken several generations to unfold, but the process may prove much faster in the future. New driver variables are quickening the pace, including the miniaturization of small farms under continuing rural population growth in poorer countries, the trade-distorting agricultural policies of most Organization for Economic Cooperation and Development (OECD) countries, a shift toward increasingly integrated and consumer-driven markets as part of market liberalization and globalization, and the demographic

impact of the spread of HIV/AIDS. The viability of small family farms is threatened today in all kinds of countries in historically unprecedented ways. Yet there are good reasons why policy makers should want to keep small farms around, and this will require deliberate policies to provide them with viable development pathways in an increasingly hostile world. This paper reviews the problem and discusses appropriate policy interventions.

2. The lure of small family farms

Why should we care about the future of small family farms? What is it that makes them important in the policy debate in rich and poor countries alike? Why does almost every rich country distort its agricultural markets and spend large amounts of public funds supposedly[1] to support its small farms, and why do many

---

* *Development Strategy and Governance Division, International Food Policy Research Institute, Washington, DC, USA.*

[1] Although OECD countries cite small farms as the intended beneficiaries of their agricultural policies, in reality large farms seem to capture most of the benefits (World Bank, 2003).

developing countries attempt land reforms or constrain land market transactions in order to create and retain more small farms?

For poorer countries, the attraction lies in their economic efficiency relative to larger farms, and the fact that they can create large amounts of productive employment, reduce rural poverty and food insecurity, support a more vibrant rural nonfarm economy (including rural towns), and help to contain rural–urban migration.

The efficiency of smaller farms is demonstrated by an impressive body of empirical studies showing an inverse relationship between farm size and land productivity (see Heltberg, 1998 for a recent review). Moreover, small farms often achieve their higher productivity with lower capital intensities than large farms. These are important efficiency advantages in countries where land and capital are scarce relative to labor.

The greater efficiency of small farms stems from their greater abundance of family labor per hectare farmed. Family workers are typically more motivated than hired workers and provide higher quality and self-supervising labor. They also tend to think in terms of whole jobs or livelihoods rather than hours worked, and are less driven by wage rates at the margin than hired workers. Small farms exploit labor using technologies that increase yields (hence land productivity) and they use labor-intensive methods rather than capital-intensive machines. As a result, their labor productivity is typically lower than that of large farms. This is a strength in labor-surplus economies, but it becomes a weakness for the long-term viability of small farms as countries get richer and labor becomes more expensive.

In poor, labor-abundant economies, not only are small farms more efficient, but because they also account for large shares of the rural and total poor, small farm development can be win-win for growth and poverty reduction. Asia's green revolution, for example, demonstrates how agricultural growth that reaches large numbers of small farms can transform rural economies and raise enormous numbers of people out of poverty (Rosegrant and Hazell, 2000). Recent studies also show that a more egalitarian distribution of land not only leads to higher economic growth but also helps ensure that the growth that is achieved is more beneficial to the poor (e.g. Deininger and Squire, 1998; Ravallion and Datt, 2002). Small farms also contribute to greater food security, particularly in subsistence

agriculture and in backward areas where locally produced foods avoid the high transport and marketing costs associated with many purchased foods.

Small-farm households also have more favorable expenditure patterns for promoting growth of the local nonfarm economy, including rural towns. They spend higher shares of incremental income on rural non-tradables than large farms (Mellor, 1976; Hazell and Roell, 1983), thereby creating additional demand for the many labor-intensive goods and services that are produced in local villages and towns. These demand-driven growth linkages provide greater income-earning opportunities for small farms and landless workers among others.

For rich countries, the potential efficiency gains of small farms are much less important and may not even exist, except for some specialty and labor-intensive products like horticulture. Small farms are attractive because they are key to maintaining a vibrant rural economy. They are important consumers of the services and products of rural towns, they help to maintain critical levels of rural population density needed to sustain key rural services and institutions, and they also have an important electoral voice. Small family farms are also still perceived as an attractive, wholesome, and stable way of life, perhaps because there are still many urban people around who grew up on farms or in rural areas.

But not all see small family farms as desirable. Some have seen them as technologically backward, a form of colonial exploitation, and even a form of self-exploitation (e.g. Karl Marx). A naive belief that large-scale mechanized farming necessarily means greater efficiency and productivity has led some policy makers to seek to consolidate holdings, often through compulsory means or land seizures. These range from large state farms in some post–Independence African countries, large settler farms in colonies or new territories, to cooperatives and state collectives in communist regimes.

3. The farm size transition

Having accepted that small farms have several attractive features, just how small should a small farm be and how many should a country have?

There are no easy answers to these questions. Size depends on the ability to create viable household livelihoods, and this varies enormously with the type of farming that is possible at any location, and the possibilities of combining farm with nonfarm sources of income. A "viable" small cereal farm, for example, might vary from just a couple of hectares in parts of Asia or Africa to 100 times as large in parts of Europe to 1,000 times as large in North America. But it can be much smaller if cereal farming is combined with a nonfarm source of income, as with many small, part-time rice farmers in Japan.

An important driver of the size distribution of farms is the stage of economic development of a country. Gross Domestic Product (GDP) per capita correlates closely with the relative costs of land and labor. In the early stages of development, small farms typically account for the lion's share of the farming population. Because farm labor is abundant but land is expensive, small farming is economically efficient. As per capita income rises, economies diversify and workers leave agriculture, rural wages go up, and capital becomes relatively cheaper. It then becomes more efficient to have progressively larger farms. There is a natural economic transition to larger farms over the development process, but one that depends critically on the rate of rural–urban migration, and hence on growth in the nonagricultural sector.

Small farms survive longer into the transformation process if they can adapt to the changing economic environment. Key adjustments include buying or renting additional land, diversifying into higher value production activities (e.g., fruits and vegetables, and niche markets like organics), and expanding into nonfarm sources of income or employment. Fortunately, opportunities to diversify into a broader range of farm and nonfarm activities also grow as countries become richer. This is because the demand for more diverse and higher-value foods increases with per capita incomes and urbanization, and the nonfarm economy grows more quickly than agriculture.

Few countries handle this farm size transition well. Many countries have successfully developed their economies, but farm consolidation and rural–urban migration have lagged behind economic growth, leaving a situation with too many small farms whose incomes fall below the national average. This leads to pressures for government support, and hence the kind of farm policies found in many OECD countries. In much of Southeast Asia the number of small farms is still increasing despite rapid growth in per capita GDP (Rosegrant and Hazell, 2000). Unless these farms can successfully diversify into nonfarm sources of income, it is likely they too will be headed toward protectionist policies (indeed this is already happening in South Korea and China).

Other countries attempt the transition too soon. An early concentration of land among large farms can occur through colonization (e.g., South Africa, Zimbabwe, and many Latin American countries), creation of large state farms (e.g., Syria and Tanzania), or collectivization of agriculture (USSR, Eastern Europe, and China). Many of these interventions have been costly failures, and have led to lost opportunities for more efficient growth and employment creation in agriculture, and have contributed to impoverishment of neglected small farms and excessive rural–urban migration in relation to available jobs (Eastwood et al., 2003). This has contributed to the kinds of dualistic patterns of development with high rural poverty found in many Latin American countries.

4. New challenges

Government policies have an important bearing on the timing and success of the farm size transition, especially policies and investments that affect the rate of rural–urban migration. The complexity of the transition problem is also changing in some important ways, raising concerns about the future viability of small farms in all types of countries. Four driving forces in particular deserve mention.

First, in many of the poorer countries, continuing rural population growth on a fixed land base is creating a situation where the subdivision of small farms has or is approaching the point where many farms may now be too small to be efficient or to survive. In Ethiopia, for example, large numbers of small farms are now too small to provide a subsistence living even in years with reasonable rainfall, and nonfarm income and employment opportunities are far too limited to provide an adequate compensating source of livelihood. As a result, many farmers have little choice but to practice unsustainable farming methods, and this is undermining current and future land productivity. It is hard to see how the inverse relationship between farm size and land productivity can be sustained under these conditions, or how livelihoods can be sustained without a growing dependence on food aid and other welfare transfers.[2] Some land consolidation seems essential for reversing these problems, but it cannot be undertaken on the required scale until viable pathways for rural-urban migration can be found. This, in turn, requires much faster rates of nonagricultural growth than many poor countries are achieving. In Africa, there is also new evidence to suggest that not only are small farms becoming smaller, but land is also becoming more concentrated among larger farms (Jayne et al., 2003). This implies a worrying expansion in the number of landless and near-landless workers in rural Africa, and a possible rapid exodus from the countryside despite insufficient urban employment opportunities.

Second, marketing chains are changing dramatically in all types of countries with trade liberalization and globalization. The small farmer is increasingly being asked to compete in markets that are much more demanding in terms of quality and food safety, and which are much more concentrated and integrated. Supermarkets, for example, are playing a much more dominant role in controlling access to retail markets (Reardon et al., 2003). As small farms struggle to diversify into higher-value products, they must increasingly meet the requirements of these demanding markets. These changes offer new opportunities to small farmers who can successfully access and compete in these transformed markets, but they are also a serious threat to those who cannot.

Third, the protectionist agricultural polices of many rich countries are reaching new heights in creating unfair competition for small farmers in developing countries. Farmers in developing countries not only have limited access to agricultural markets in richer countries, but they also face unfair competition in their own domestic markets from subsidized imports. The size of these distortions is immense. In 2000, the producer subsidy equivalent of these policies in the OECD countries was US$330 billion; worth about eight times the value of all official development assistance to developing countries in that year (World Bank, 2003). These policies are particularly damaging to small farmers in poor countries because they limit their opportunities to produce more of the products in which they have comparative advantage. This is not just a matter of

farmers in developing countries being squeezed out of export markets for tropical crops like cotton, sugar, and tobacco, but they are even pressured in their own domestic and regional markets for staple foods like cereals and livestock products.

Fourth, HIV/AIDS is taking a severe and increasing toll among small farms in many developing countries, reducing the number of able adult workers and leaving many children as orphans with limited knowledge about how to farm. Many small farms will eventually disappear as a result of HIV/AIDS, but only after a difficult transition problem during which local communities must find ways to cope with the human tragedies involved.

These driving forces are particularly challenging for Africa and South Asia, where small farms dominate the landscape and account for the lion's share of the agricultural sector output (Narayanan and Gulati, 2003). If agricultural growth is to play a key role in reducing rural poverty in these countries, then developing viable strategies for small farms is probably one of the most fundamental problems that policy makers will need to resolve.

5. Policy interventions

What kinds of policies are needed to help ensure that small farmers have a viable future? In rich countries, where small farm households account for just a tiny share of the population, the policy tool kit can include targeted subsidies, though these will eventually have to be delinked from production if the Doha round of the World Trade Organization (WTO) trade negotiations succeeds. Given the preponderance of small farms in most developing countries, widespread subsidies are not a viable financial option. Rather, small farms must find viable development pathways that enable them to play a key role in national economic growth and poverty reduction. This requires public policies and investments that create an enabling environment for small farm development. Some of the more important interventions appropriate for developing countries are discussed below.

5.1. Organizing small farmers for marketing

Small farms have always been at a disadvantage in the market place. They only trade in small volumes,

---

[2] Carter and Wiebe (1990) have provided evidence from Kenya showing that profits per hectare decline when farms get too small.

often have variable and substandard quality products to sell, lack market information, and have few links with buyers in the marketing chain. These inefficiencies can all too easily offset the efficiency advantages of small farms as producers. The problem has been exacerbated by market liberalization and globalization. Not only has the state been removed from providing many direct marketing and service functions to small farms, leaving a vacuum that the private sector has yet to fill in many countries (Kherallah et al., 2002), but small farmers must now also compete in ever more integrated and consumer-driven markets where quality and price are everything (Narayanan and Gulati, 2003). Small farmers will need to organize themselves to overcome these problems and to exploit the new opportunities that these market changes offer; otherwise they risk losing market access.

The private sector is emerging as a key player in linking larger-scale commercial farmers with markets (e.g., contract farming and supermarkets), but they have less interest and ability to deal with small-scale farmers on an individual basis. Voluntary producer organizations of various types will have important roles to play in filling this void and in linking small farmers to food processors, manufacturers, traders, supermarkets, and other food outlets (Kindness and Gordon, 2002). Such organizations can help serve businesses by providing an efficient conduit to reach small-scale producers, and help improve the quality and timeliness of small farmers' production and their access to agricultural research and extension, input supplies, and agricultural credit.

Unlike former state cooperatives that are widely discredited because of their poor performance and high cost, key design principles are organizations that are voluntary, economically viable, self-sustaining, self-governed, transparent, and responsive to their members. Supporting these kinds of organizations will require government and donor support, engaging with businesses and civil society groups. Producer-based organizations will need help in developing business and management skills, establishing information systems and connections to domestic and global markets, creating good governance practices, and creating the infrastructure to connect small farmers to finance and input supply systems.

Public policy can help ensure improved market access for small farmers by putting in place institutions to deliver finance, reduce risks, build social capital of producers and traders, transmit market information, grade and certify goods, and enforce contracts (Gabre-Madhin, 2001). Infrastructure investments are also crucial; the farmers least likely to benefit from globalizing markets are those who are more distant from roads and markets (Narayanan and Gulati, 2003).

5.2. Agricultural research and extension

Small farmers need improved technologies appropriate to their needs if they are to survive in today's marketplace. This typically means utilizing more labor-intensive technologies than large farms, though as small farms get smaller and/or labor becomes relatively more expensive, it becomes increasingly important to develop technologies that increase total factor productivity. Farmers not only need to produce more output and income per unit of land, but also to do this in ways that increase their labor productivity. Otherwise, they will simply be working harder to try and achieve, perhaps unsuccessfully, the same level of per capita income.

Smallholder farms also need to diversify into higher-value products to maintain their incomes, given diminishing land/labor ratios. Such diversification is already happening in many countries, especially in Asia and Latin America. The opportunities for income-enhancing diversification are much more constrained in countries with low and stagnant per capita incomes, as in much of Africa. In these cases, attention needs to be given to developing cash crops for export. Agricultural research for higher-value crops and livestock, and for post-harvest handling, is under-funded in many developing countries.

Publicly-funded research and extension still has a crucial role to play in meeting the technology needs of small farms. Private agricultural research and seed firms are less attracted to the problems of small farms because of the higher transactions costs incurred and lower volumes of business. Producer organizations can help bridge this gap.

Women now manage many small farms and research and extension systems need to cater to their specific needs. Targeted research is also needed for farm households impacted by HIV/AIDS. They typically need technologies for producing foods that use relatively little labor, but without the expense of mechanization (e.g., low tillage methods and choice of crops that require less labor).

To meet these challenges, there is need for a more client-oriented, problem-solving approach in public agricultural research systems. This approach will often translate into a need for more on-farm research and for more participatory approaches in which farmers have a greater say in selecting research priorities and in evaluating research outputs. Not all of the technological challenges facing small farms will be solved by more on-farm and participatory work. Modern science, including biotechnology conducted in a strict laboratory environment, may be critical, for example, in raising yield ceilings or for improving drought tolerance. However, even biotechnology will be more effective if it addresses priorities set on the basis of a client-oriented, problem-solving approach that draws many of its insights from interaction with farmers.

5.3. Agricultural credit for small farms

In many developing countries, the financial sector reforms undertaken as part of the structural adjustment programs have left a vacuum in the supply of seasonal credit for small farms. Private banks are servicing the needs of large commercial farms, and micro-finance institutions have mushroomed to cater to the financial needs of the poor. The seasonal nature of farm credit needs, and the highly covariate nature of most agricultural production and marketing risks, undermine the viability of borrowing groups for farm credit purposes. With the demise of publicly funded agricultural development banks, most small farmers now have to rely on self- or family financing, using livestock and other assets, as well as remittances from family members in nonfarm employment. Improving small farmers' ability to save and invest requires the development of an entire rural financial infrastructure in which farmers can access a full range of financial services, including credit and deposit banking at competitive interest rates. Although a return to the inefficient and highly subsidized agricultural development banks is not to be recommended, there is a clear need for some form of public intervention to help fill this void.

5.4. Risk management aids

Small farmers face a range of weather, disease, pest- and market-related risks that can discourage them from

investing in major land improvements and from adopting more profitable technologies and crop and livestock activities. Agricultural research can help reduce risk, for example, by improving drought or pest resistance in crops and helping develop better ways to conserve soil moisture. Investments in irrigation and watershed development can reduce drought exposure and control flooding while also increasing productivity.

Additionally, governments can help farmers cope with catastrophic weather events like drought by providing safety net programs, and by facilitating the development of credit and insurance arrangements that provide cash in times of need. Such interventions need to be designed to assist farmers to better manage risk and improve their productivity and incomes, but without creating incentives that lead to inappropriate land uses and environmental degradation. The experience with crop insurance has had mixed results. While it has sometimes helped farmers protect their incomes and food security and repay debt in drought years, the heavy subsidies that are invariably included have led to negative impacts on the way resources are managed (e.g., by encouraging farmers to grow crops in areas for which they are not suitable) (Hazell et al., 1986; Hazell, 1992). Better alternatives for catastrophic risk management are area-based rainfall insurance sold in small denominations so as to be affordable to small farmers and the development of more accurate and accessible drought forecasting information (Skees et al., 1999; Arndt et al., 2000). This kind of insurance could be sold by the private sector without the need for heavy subsidies.

Commodity futures markets also offer new possibilities for providing forward price contracts to small farms. Rather than expecting farmers to trade in these markets on their own account, market intermediaries, such as large traders, processors, or exporters, might be induced to offer farmers forward price contracts, and then to hedge the assumed price risk on their own account in the futures market. For this to happen, government must establish mechanisms for ensuring that contracts are enforced.

5.5. Tenure security and improved access to land

Farmers need assured long-term access to land if they are to pursue sustainable farming practices and to make long-term investments in improving the productivity of their resources. Many of the indigenous land tenure systems that prevail in the developing world already provide reasonable tenure security to those who have access to land, and they also seem to evolve to accommodate changing needs (e.g., greater privatization of rights) as population and commercialization pressures increase (Otsuka and Place, 2002). In these cases, the appropriate role for government is to seek ways of strengthening existing systems rather than imposing new systems. Legal registration of land by community groups and simple measures for recording land transactions and resolving disputes can often increase security by reducing land disputes between and within communities. By contrast, registration of individual plots will only be worthwhile in areas of high population density, where land has a high value, where formal lending institutions are also well developed, and land is already effectively privatized.

Many resources are owned and managed as common property (e.g., grazing areas, woodlands, water, and wetlands). There are usually good reasons for this; it can be a cost effective way of preventing intruders from using the resource, of maintaining flexible responses to drought, and of ensuring equitable access for all members of the community. But if these resources are to remain in common ownership and avoid being privatized or overused, then governments need to recognize local rights and capacities to manage these assets. Often, governments have undermined indigenous institutions by nationalizing important common property resources such as rangeland and forests, while being unable in practice to manage them effectively. As a result, many common property resources have degenerated into open access areas. There is now increased acceptance that the most successful institutions for managing common properties are likely to be local organizations, run by the resource users themselves. Government policy needs to support local management by such user groups, while at the same time ensuring that poor people are adequately represented in their management (Knox et al., 2002; Otsuka and Place, 2002).

Small farms also need access to efficient land markets for sale and rental purposes. Efficient land markets require an enabling legal environment, both in terms of legal recognition of the right to sell, rent, or mortgage land, and also effective means of enforcing contracts. Land markets facilitate land consolidation as farms get too small or as some farmers seek to migrate to urban areas. They can also provide a way of facilitating land redistribution in countries where land is excessively concentrated. For example, South Africa has been redistributing land to the poor by making use of the "willing buyer, willing seller" principle for voluntary land transfers (van Zyl et al., 1996). However, the effective application of this market-assisted approach requires well-developed mortgage financing, strict control of land prices to reduce speculation, and a range of complementary support services (credit, training, extension, and marketing). Better options are government-sponsored land redistribution (though politically difficult), more effective land rental markets (Mearns 1999), and organization of the poor to obtain greater access to common property resources and their management.

5.6. Nonfarm opportunities and migration

Rural nonfarm income and migration remittances are important components of the livelihood strategies of small farmers, often accounting for more than half their income. These income sources contribute to higher consumption, income stabilization (by offsetting agricultural losses in bad years), and are used for financing on-farm investments (Walker and Ryan 1990; Reardon et al., 1998). Low human, financial, and physical assets confine many small farm households to low-productivity, low-growth nonfarm activities from which there are few pathways out of poverty. In this environment, the policy challenge becomes one of equipping poor households with the means to move from these "refuge" nonfarm jobs to more remunerative ones. To do so, they require a variety of private assets such as education and start-up funds, and public assets such as roads and electricity and information about how to access dynamic market segments. Gender, caste, and social status can restrict access by the poor to the most lucrative nonfarm activities in some settings. Many of these investments in human capital are also helpful for small farmers seeking exit strategies out of agriculture.

While investments in human capital formation can assist small farm households diversify into nonfarm activities both locally and through migration, opportunities are also conditioned by the rate of growth of the nonagricultural sector. In stagnant countries with low per capita incomes, productive nonfarm opportunities are limited, and government needs to be careful not to encourage too rapid a rate of migration to urban areas. Appropriate macroeconomic policies and public investments are also needed to stimulate economic growth.

5.7. Targeting the vulnerable

Agricultural growth centered on small farms can make deep inroads into poverty and hunger in many poor countries. But this would not be enough to eliminate poverty and vulnerability to production and market shocks. There is also need for effective safety net programs in times of crisis. There have been real advances in recent years in targeting and delivering assistance more effectively, often by involving local communities in the design and implementation of targeted programs, which leads to programs that are primarily demand-driven and hence reflect local needs and constraints.

6. Conclusions

Small farms are seriously challenged today in ways that make their future precarious. Marketing chains are changing and are becoming more integrated and more demanding of quality and food safety. This is creating new opportunities for higher-value production for farmers who can compete and link to these markets. The danger for many small farms is that they are not yet positioned to compete and access these markets and many will simply be left behind. In developing countries, small farmers also face unfair competition from farmers in richer countries in many of their export and domestic markets. The viability of many is further undermined by the continuing shrinkage of their average size. And the spread of HIV/AIDS is further eroding the number of productive farm family workers, and leaving many children as orphans with limited knowledge about how to farm.

If most small farmers are to have a viable future, then there is need for a concerted effort by governments, nongovernmental organizations (NGOs), and the private sector to create a more enabling economic environment for their development. This must include assistance in forming effective marketing organizations,

targeted agricultural research and extension, revamping financial systems to meet small farm credit needs, improved risk management policies, tenure security and efficient land markets, and when all else fails, targeted safety net programs. In addition, the public sector needs to invest in the provision of basic infrastructure, health, education, and other human capital to improve market access and to increase the range of nonfarm opportunities available to small farm households, including permanent migration to urban areas. These interventions are possible and could unleash significant benefits in the form of pro-poor agricultural growth. But they do not seem very likely at the moment and current trends are moving in the opposite direction. For example, research and extension for small farms is declining, credit for small farms has virtually disappeared, and donor and government investment in crucial rural infrastructure is stagnant at best. The question remains: Is there a future for small farms?

References

Arndt, C., P. Hazell, and S. Robinson, "Economic Value of Climate Forecasts for Agricultural Systems in Africa," in M. V. K. Sivakumar, ed., *Climate Prediction and Agriculture* (World Meteorological Organization, International START Secretariat: Washington, DC, 2000).

Carter, M. R., and D. Wiebe Keith, "Access and Capital as Impact on Agrarian Structure and Productivity in Kenya," *American Journal of Agricultural Economics* 72, no. 5 (1990), 1146–1150.

Deininger, K., and L. Squire, "New Ways of Looking at Old Issues: Inequality and Growth," *Journal of Development Economics* 57 (1998), 59–287.

Eastwood, R., J. Kirsten, and M. Lipton, "Premature Deagriculturalization?" in *Land Inequality and Rural Dependency in Limpopo Province, South Africa*, Draft paper (2003).

Gabre-Madhin, E. Z., *Market Institutions, Transaction Costs, and Social Capital in the Ethiopian Grain Market*, Research Report No. 124 (International Food Policy Research Institute (IFPRI): Washington, DC, 2001).

Hazell, P. B. R., "The Appropriate Role of Agricultural Insurance in Developing Countries," *Journal of International Development* 4, no. 6 (1992), 567–581.

Hazell, P., C. Pomareda, and A. Valdés, eds., *Crop Insurance for Agricultural Development Issues and Experience* (Johns Hopkins University Press: Baltimore, 1986).

Hazell, P., and A. Roell, *Rural Growth Linkages: Household Expenditure Patterns in Malaysia and Nigeria*, Research Report No. 41 (International Food Policy Research Institute: Washington, DC, 1983).

Heltberg, R., "Rural Market Imperfections and the Farm Size-Productivity Relationship: Evidence from Pakistan," *World Development* 26, no. 10 (1998), 1807–1826.

Jayne, T. S., T. Yamano, M. T. Weber, D. Tschirley, R. Benfica, A. Chapoto, and B. Zulu, "Smallholder Income and Land Distribution in Africa: Implications for Poverty Reduction Strategies," *Food Policy* 28 (2003), 253–275.

Kherallah, M., C. Delgado, E. Gabre-Madhin, N. Minot, and M. Johnson, *Reforming Agricultural Markets in Africa* (Johns Hopkins University Press: Baltimore, 2002).

Kindness, H., and A. Gordon, *Agricultural Marketing in Developing Countries: The Role of NGOs and CBOs*, Policy Series No. 13 (Social and Economic Development Department, Natural Resources Institute. University of Greenwich: London, UK, 2002).

Knox, A., R. Meinzen-Dick, and P. Hazell, "Property Rights, Collective Action and Technologies for Natural Resource Management: A Conceptual Framework," in *Innovation in Natural Resource Management; The Role of Property Rights and Collective Action in Developing Countries*, R. Meinzen-Dick, A. Knox, F. Place, and B. Swallow (Johns Hopkins University Press: Baltimore, 2002).

Mearns, R., *Access to Land in Rural India*, Policy Research Working Paper No. 2123 (The World Bank, South Asia Region, Rural Development Sector Unit: Washington DC, 1999).

Mellor, J. W., *The New Economics of Growth: A Strategy for India and the Developing World* (Cornell University Press: Ithaca, NY, 1976).

Narayanan, S., and A. Gulati, *Globalization and the Smallholders: A Review of Issues, Approaches and Implications*, Discussion Paper No. 50 (Markets and Structural Studies Division, International Food Policy Research Institute: Washington, DC, 2003).

Otsuka, K., and F. Place, eds., *Land Tenure and Natural Resource Management: A Comparative Study of Agrarian Communities in Asia and Africa* (Johns Hopkins University Press: Baltimore and London, 2002).

Ravallion, M., and G. Datt, "Why Has Economic Growth Been More Pro-Poor in Some States of India Than Others?" *Journal of Development Economics* 68 (2002), 381–400.

Reardon, T., K. Stamoulis, A. Balisacan, M. E. Cruz, J. Berdegue, and B. Banks, "Rural Nonfarm Income in Developing Countries," Special Chapter in *The State of Food and Agriculture, 1998*. Rome, Food and Agricultural Organization of the United Nations (1998), pp. 283–356.

Reardon, T., C. P. Timmer, C. Barrett, and J. Berdegue, "The Rise of Supermarkets in Africa, Asia, and Latin America," *American Journal of Agricultural Economics* 85, no. 5 (2003), 1140–1146.

Rosegrant, M., and P. Hazell, *Transforming the Rural Asian Economy: The Unfinished Revolution* (Oxford University Press: Hong Kong, 2000).

Skees, J., P. Hazell, and M. Miranda, *New Approaches to Crop Yield Insurance in Developing Countries*, EPTD Discussion Paper No. 55. (Environment and Production Technology Division, International Food Policy Research Institute (IFPRI): Washington, DC, 1999).

Van Zyl, J., J. Kirsten, and H. P. Binswanger, *Agricultural Land Reform in South Africa: Policies, Markets, and Mechanisms* (Oxford University Press: Cape Town, 1996).

Walker, T. S., and J. G. Ryan, *Village and Household Economies in India's Semi-Arid Tropics* (Johns Hopkins University Press: Baltimore, 1990).

World Bank, *Global Economic Prospects: Realizing the Development Promise of the Doha Agenda in 2004* (World Bank: Washington, DC, 2003).

A revival of large farms in Eastern Europe—how important are institutions?

Ulrich Koester*

Abstract

Contrary to expectations, large farms are still the dominant form of farm organization in most countries of the Commonwealth of Independent States and are important in some Central and Eastern countries. This paper analyzes the reasons for these failed expectations, focusing on the experience of Russia. The framework of institutional economics is applied to explain the reasons why the conventional approach, based on economies of scale and on-farm transaction costs, does not apply to the evolution of farm size in the east (and even the west). Specific institutions, which are partly embedded and informal, and legal, but loosely enforced, suppress the birth of family farms in Russia and stimulate the growth of large farms, leading to holdings that are not observed in developed market economies. Hence, survival and growth of large farms are not necessarily based on comparative advantage of these farm organizations in a market economy environment.

JEL classification: Q12, Q15, Q18

Keywords: transformation; Russia; embedded institutions; large farms

1. Introduction

The discussion of farm sizes is an old theme in agricultural economics. Can anything new be added to the debate? Hopefully, yes. The ongoing debate seems to have focused on four issues.

First, the measurement of farm sizes is still an unsolved problem (Lund, 1983; Lund and Price, 1998). The amount of arable or agricultural land per farm is certainly an inadequate indicator of farm size: the value of sales or value added per farm is better, albeit not ideal. Yet, this indicator has not been used in empirical work, due to lack of data.

Second, opinions about the optimal farm size differ because this optimum is difficult to define because opinions about the objective function of farmers may differ, and because the same determinants can affect farm size in different ways across different farms. Further, while on-farm transaction costs are generally considered to be among the most important determinants of farm size, their importance depends largely on the farm operator's ability to monitor and to enforce labor contracts. Another important determinant, economies of size, depends in turn on the pattern of production.

Third, in spite of the difficulties mentioned above, according to Western economic wisdom and experience, farm sizes in the transitional economies of the Central and Eastern Europe Countries (CEECs) and the Commonwealth of Independent States (CIS) in 1990 were considered to be too large.

Finally, family farms, which are the dominant mode of production in most market economies, were expected to emerge rapidly in these transitional countries.

However, contrary to expectations, large farms are still the dominant form of farm organization in most countries of the CIS (Lerman et al., 2002) and are important in some CEECs. Large holdings have emerged in Russia, while in East Germany, a more stable economy, large farms have survived, and the number of large private farms has even increased in recent years.

This paper analyses the reasons for these failed expectations, focusing on the experience of Russia. The

* University of Kiel, Germany.

Table 1

Land use shares (in percent) by type of farms in Russia, 1990–2001

	Total farm land			Arable land		
	Farm enterprises	Individual farms	Household plots	Farm enterprises	Individual farms	Household plots
1990	98.1	0.0	1.8	97.9	0.0	2.0
1991	96.8	0.6	2.6	96.5	0.6	2.4
1992	92.3	3.3	4.4	93.3	3.7	3.1
1993	90.4	5.0	4.5	91.4	5.5	3.1
1994	89.9	5.2	4.8	91.0	5.8	3.2
1995	89.4	5.4	5.2	90.4	6.0	3.6
1996	89.1	5.7	5.2	90.0	6.4	3.6
1997	88.3	6.2	5.4	89.0	7.2	3.8
1998	87.4	7.0	5.6	88.2	8.2	3.7
1999	86.4	7.2	6.4	87.2	8.7	4.1
2000	86.1	7.9	6.0	85.7	9.4	4.9
2001	n. a.	n. a.	n. a.	85.1	9.1	5.8

n. a. = not available.

Source: Calculated from Russian Cadastre Service data. Table quoted from Serova, 2003.

following section provides a brief description of the evolution of farm size and structure in Russia. In this section, the reasons why the conventional approach, based on economies of scale and on-farm transaction costs, does not apply to the evolution of farm size in the east (and even the west) are also examined. In the main section of the paper the importance of institutions[1] for the determination of farm sizes and their development is investigated.

The term "institution" is broad, and this paper follows Williamson's (2000) classification, which distinguishes four levels of institutions. First-level institutions are embedded and are shaped by informal rules, customs, cultural beliefs, norms, traditions, and religion. Second-level institutions include the institutional environment, such as laws and property rights. These are the formal rules of the game. Third-level institutions concern the way the game is played, aligning governance structure with transactions. Whereas second-level institutions are crucial for *ex ante* decisions, third-level institutions are concerned with assessing and sometimes modifying *ex post* decisions.

Finally, fourth-level institutions concern the rules for resource allocation and employment.

2. The evolution of farm structure and farm size in Russia

Unfortunately, information on the agrarian structure in transitional economies is not easily available. The information presented here is mainly based on Lerman (2003) and Lerman et al. (2002).

The use of land in Russia has not changed much over the past decade. Individual or family farms cultivated less than 8% of total farm land and less than 10% of arable land in 2001 (see Table 1). While the number of these private farms increased in the first years of transition (Table 2 and Figure 1), it has stagnated and even declined since 1996. The share of farm enterprises in use as arable land has declined from 97.9% in 1990 to 85.1% in 2001, or about 13 percentage points. The share of household plots in use as arable land has increased significantly, from 2.0% in 1990 to 5.8% in 2001, and the number of household plots has slightly increased.

The average size of private farms in Russia is smaller than the average size of scale-efficient family farms in the United States and the EU, while the average size of agricultural enterprises is much larger than the average size of efficient Western family farms. Surprisingly,

---

[1] The terms institutions and organizations are used as suggested by North (1990). Institutions are rules that may be set officially or may have evolved unofficially; they make human behavior predictable. Organizations are groups of individuals bound by common objectives, and are comparable to the players in a game.

Table 2
Relative magnitude of farm-enterprise sector and individual sector in Russian agriculture (percent)

	1995			2000		
	Farm enterprises	Private farms	Household plots	Farm enterprises	Private farms	Household plots
Number of units	26,900	280,100	16.3 mill.	27,600	261,700	16.0 mill.
Average size, ha	5,700	43	0.36	5,400	58	0.38
Agricultural land[a]	81.7	5.0	4.7	80.0	7.4	5.6
Agricultural production	50.2	1.9	47.9	43.4	3.0	53.6
Agricultural labor[b]	60	40		49	10	41

[a] "Other users" complete the sum to 100%.

[b] Very rough estimates based on data from two sources: total number of employed in agriculture from *Rossiiskii statisticheskii ezhegodnik 2002* and number of agricultural workers in farm enterprises from *Sel'skoe khozyaistvo v Rossii 2002*; employment in the individual sector estimated by difference. For 2000, employment in private farms obtained from *Agricultural Activity of Private Farms in Russia 2000*, Goskomstat, Moscow (2001); household plot employment estimated by difference.

Source: Table provided by Zvi Lerman, The Hebrew University, Rehovot, Israel and the World Bank, Washington, DC.

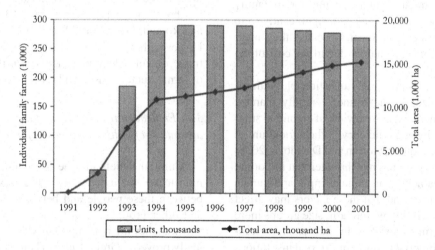

Figure 1. Number of individual farms in Russia and land area used by them. *Source:* Csaki et al. (2002, p. 57).

the number of these large enterprises increased between 1995 and 2000, while the average size declined by only 5 percentage points. Hence, the share in arable land held by enterprises declined only slightly (by 2.8 percentage points) over this period.

The organization of the agricultural sector has changed significantly in Russia over the past 5 to 6 years and, starting in 1997, and now includes new companies called agro-holdings. These are collectives of several juridical entities where one is the mother enterprise and the others have to accept the mother's decisions.[2] The mother enterprise prepares the consolidated or common financial statements of the holdings, coordinates the flows of financial resources and commodities, and may have the right to hire and fire managers and specialists in the subsidiary companies. In 2003, 13 of these holdings encompassed more than 100,000 ha each, with one reaching 500,000 ha. Some of them hold a high share on regional markets (up to 50%), and even on national markets (up to 12%). According to estimates, in 2002, 30 to 40 holdings in Russia included about 6% of all agricultural farms and contributed between 10% and 20% of total agricultural production.

In a random sample it was found that the 16 interviewed companies had 36,000 ha of arable land, meadows, and pasture on average (Rylko, 2001). The

---

[2] Information provided by Täuber, German Embassy, Moscow, 2003.

agricultural land plots of the interviewed companies ranged from 2,000 ha to one company that owned 19 collective farms in two regions totaling about 150,000 ha. These companies are, therefore, much larger than the typical former collective farm size of 4,000 to 8,000 ha (Rylko, 2001). It is estimated that these large holdings control 5–6% of the agricultural land, but the share is much higher in those oblasts (regions) with fertile land. To the author's knowledge there is no comparable development in any market economy.

2.1. The traditional approach to farm size

Large farms are thought to be inferior to family farms, first because they do not gain significantly from economies of scale, and, second because they incur higher transaction costs mainly due to the cost of supervising hired labor.

These hypotheses are highly questionable with respect to farm sizes in the former centrally planned economies. In their extensive review of empirical studies on productivity and efficiency in the agriculture of transformation countries, Gorton and Davidova (2003) found that all these studies may have left out important determinants, namely, management input and human and other resources, and that these factors may affect variation in productivity within farm size groups more than between farm size groups. In addition, most studies neglect the important impact of external institutions on the efficiency of farm size and on the evolution of the farm size structure.

The transition from plan to market requires considerable adjustment to a drastically changed economic environment. Adjustment pressures in the agricultural sector in most of these countries were delayed for a short while so that the need for structural change seemed apparent at least to many agricultural researchers. To assess the effects of external institutions on the evolution of farm structure and farm sizes in selected transitional countries, it is necessary to investigate how family farms or single-owner farms emerge and grow, as well as how large-scale farms survive. What institutions affect the birth of family farms? Why do family farms survive as they are, and why do they change so slowly? What institutions contribute to the inefficiency of large-scale farms? Why do new large farms get established, and why do they survive if they are inefficient?

This paper examines the hypothesis that institutions, which are generally country-specific, determine birth, survival, and the death of specific farm sizes and farm organizations.

3. Organizations and institutions on the farm level

3.1. Embedded institutions

Embedded institutions (first-level institutions according to Williamson, 2000) include informal rules, customs, traditions, norms, and religion. The importance of these institutions in the transformation process has been widely accepted (Greif, 1994; Dewatripont and Roland, 1996; World Bank, 1997; Bardhan, 2001). However, the importance of institutions for agricultural transformation, and in particular, for the development of farm structure has been less elaborated.

3.2. Embedded institutions and the comparative advantage of individual types of farms

"Cultural beliefs are the ideas and thoughts common to several individuals that govern interaction—between these people and between them, their gods, and other groups and differ from knowledge in that way that they are not empirically discovered or analytically proved" (Greif, 1994). There is ample evidence that embedded institutions differ among societies and that they have significant implications for economic performance. Embedded institutions have an influence on the way people behave and how they interact with each other, how they collect and deal with information, and how willing they are to change. Moreover, embedded institutions may have a strong impact on policy makers and influence how they prefer to design policies. Such cultural beliefs as part of embeddedness are important for "would-be" or "could-be" private farmers, for their survival, for the survival of large farms, and for the birth of new large holdings.

People's behavior depends not only on economic incentives, as generally assumed in neoclassical economics, but also on embedded institutions. To quote Alan Greenspan, Chairman of the Federal Reserve Board of the United States, capitalism is not human nature, but, as the Russian disaster indicated, "not nature at all, but culture" (quoted in Pfaff, 1999). Hence, the reaction to changes in the economic environment depends on the given culture in a society. Mental models are the micro-level basis of culture because they describe the underlying beliefs that influence the way people behave (Lindsay, 2000), and how they think the world works. Mental models are crucial for understanding the willingness of people and a society to change.

However, mental models cannot measure human behavior, and therefore they can only suggest the effects of embedded institutions on the structure of farms.

Table 3 provides an overview of some specific first-level institutions that might be of importance on the farm and policy level. Openness of a society is a main determinant of prosperity, as openness determines willingness to change. Rural societies in Russia were largely isolated from urban regions and even more so from other countries. Hence, they were not well prepared to react to changes in the economic environment. This situation was aggravated by the structure of human capital in rural regions, where the average level of education was generally less than in urban areas and out-migration after the onset of transition eroded the stock of human capital even further. Societies, which

are less open, tend to be more risk averse than others. This point is of special relevance in transition countries, as the environment was quite uncertain and there were no developed insurance markets. Thus, it is not surprising that most first-generation private farmers were not former employees. A survey in Russia revealed that 75% of early private farmers in Russia were ex-urbanites, and only 5–7% were former members of state and collective farms. "Romantics of the rural way of life" and demobilized military personnel accounted for 20% of private farmers (Wegren and Durgin, 1997). Thus, outsiders were the first generation of new farmers in Russia. It is difficult to assess which had the most impact on this outcome: embedded institutions or rational economic behavior, especially when the less risky alternative of working on the household plots under the umbrella of the large farm cannot be singled out exactly.

Some societies are more risk averse than others. However, the nature of agriculture, with its exposure to the elements and the long delay between investment and returns, means that farmers always have to bear risk. It is reasonable to assume that the willingness to bear risk is also dependent on education and personal experience during childhood and work. In planned economies, workers were not educated or

Table 3
Embedded institutions and the comparative advantage of individual types of farms

Farm level	
Against private farms	In favor of large farms
Not open to change[a]	Not open to change
Lack of trust	Strong belief in comparative advantage of large farms
Preference against being self-employed	Social responsibility
Preference for leisure	Belief in specific role of the state
Attitude with respect to risk	Production oriented and less profit oriented
Insufficient understanding of formal rules	Corruption and nepotism
Preference for collective action as compared to self-reliance	
Attitude against to land ownership and land sales	
Corruption and nepotism	
Policy level	
Private farms	Large farms
Mistrust in individualism	Mistrust in functioning of food markets
	Political influence
	Belief in comparative advantage of large-scale agriculture

Notes: [a] Human capital in rural areas had already eroded before the transition started. The rural population was less educated than the average person in the country and less prepared to change.

trained to be entrepreneurs nor undertake risky activities. Hence, the number of potential entrepreneur-farmers in a transitional country is likely to be limited, at least in the short term. For example, some of these societies, being risk averse, seem to be unwilling to take credit. In addition, some societies may be culturally averse to incurring debt or reluctant to do so. It is considered as "something which one should not do" because it indicates living beyond one's means. This cultural belief may explain why so many new farmers in these countries had to give up farming after a few years. Even in eastern Germany, where the potential of a new farming generation seemed to be high, 42% of the new single-owner farms had negative net investments, and 34% had to accept a negative change in equity (Koester, 2000). These individuals were either unable to become good farmers, because they lacked the necessary skills and became indebted, or they tended to live beyond their means. Therefore, changing from a planned to a market economy may likely trigger an exodus of farmers from the business and expansion of those farms that remain. The willingness to incur debt is one prerequisite for any successful restructuring.

World Bank interviews in Russia 1994 (Lerman et al., 2002) found that 42% of respondents were not willing to become a private farmer because they did not wish to change their lifestyle, while 56% were afraid of risk. The attitude toward risk is also related to culture, education, experience in dealing with risky situations, and the availability of risk-reducing institutions.

Prospective farmers might also shy away from setting up a private farm due to corruption and nepotism, which are at least partially embedded institutions. Corruption creates uncertainty and enhances the risk of starting a new activity, in particular if long-term engagement is needed. Unfortunately, most former socialist countries were and still are prone to corruption. Pervasive corruption reduces trust, but trust is a necessary ingredient of a market economy.[3,4] Even if there

is a stable and reliable legal framework in place, trust is needed in order to exploit the potential of productive interactions. Hence, lack of trust reduces the division of labor in an economy. This aspect could have been quite important for the creation of new private farms, more so, as markets were not functioning well (see below) and new farmers had to rely on discretionary decisions by bureaucrats, policymakers, and managers of large farms. Lack of trust and badly functioning markets help to explain why small-scale enterprises were the backbone of the recovery of the economy, but not so in agriculture. New farms in agriculture need long-term investment, which is often specific. Hence, some of the new assets have to be produced to order, implying a high risk for the producer of the asset concerning the willingness of the purchaser to pay, and also a high risk for the purchaser, as he does not know the exact quality of the asset. In contrast, it is easier for small-scale enterprises to expand in sectors that do not rely on asset specificity as much as agriculture and that do not need as much long-term investment.

Farmers in the Western market economies own at least some share of the land they cultivate. Indeed, the survival of many farms in these countries is only possible because farmers own land, and are, therefore, able to survive for as long as a generation even if they lose equity year after year. Potential successors of small farms seem to be inclined to run the farm because they consider the activity on the family farm as somewhat special, as the farm may have been in the hands of the family for generations and as farmers, they think in generations and stick to the soil. There are indications that embedded institutions with respect to landownership point in the opposite direction in Russia. This has important implications for the initial mode of privatization, which would not matter much if land were highly mobile. However, experience has shown that many new owners are not willing to sell or lease out their land. Land seems to have a specific value in addition to being an asset. Even if the owners wanted to, they may not be able to sell or lease out their land because they do not possess a title, (which proves ownership) or because buyers are not creditworthy, or because land is highly

---

[3] "Corruption is a contractual relationship between economic agents for the abuse of position for private gains" (Reja and Talvitie, 2000).

[4] There is much empirical evidence in the institutional literature in both economics and sociology that sustainable rural communities in a contemporary global economy need to develop both horizontal (within the community) and vertical (outside of the community) bridging ties (Woolcock, 1998; Flora and Flora, 1993). However,

these ties require a level of trust that was certainly not fostered during the 70 years of Soviet authoritarian rule. Given a lack of trust, households, as rational economic actors, will devote most of energy on the development and maintenance of highly dense networks of trusted family and friends (O'Brien, 2002, pp. 169–173).

fragmented and users of individual plots can hardly cultivate them.

The negative attitude toward private ownership in Russia is clearly expressed in interviews. About 90% of respondents in a survey conducted in Russia (Serova, 2000) disagreed with the concept of land reform and seemed to be against private land ownership. Interviews in Novosibirsk and Shitomir revealed that only 33% of the farmers were willing to mortgage their land (Schulze et al., 1999). Owners seem to be afraid of losing their land because land may be considered an important asset in risk hedging. Given the constraints on the land market due to the mental models of landowners and the rural population, it is difficult for the sector to adjust to the rapidly changing environment during the transition period. If, in addition, the initial land allocation is inefficient, this situation can be exacerbated.

Some rural people have a specific attitude with respect to land ownership. A survey conducted in Novosibirsk province revealed that 78.6% of respondents working in agriculture disapproved of selling and buying farmland (Schulze et al., 1999). This may partly explain why land may remain idle in these countries, in spite of rural unemployment. All the above-mentioned constraints to changes in land ownership accentuate the importance of the original farm structure. Taking into account embedded institutions, which are relevant for managing a farm, does not mean that family farms are necessarily inferior to large-scale farms. It only suggests that specific policy actions might be needed to overcome these embedded institutions. Williamson (2000) estimates, somewhat pessimistically, that it may take up to 100 to 1,000 years to change these institutions. Economists could play a role in accelerating this process, as education and dissemination of information will likely shorten the needed time horizon.

3.3. Embedded institutions and the survival of large farms

Embedded institutions in the form of juridical entities also play a significant role in the management of large-scale farms. Some societies strongly emphasize kinship. People in charge of hiring, monitoring, granting licenses, etc., favor their relatives. This fact has implications for managing a farm that relies on many wage earners. In a functioning market economy, managers are expected to monitor and enforce labor contracts, and to assess the performance of employees. If the manager is not the owner, he is the agent of the owner and may have a juridical role. At the same time the manager is the principal of the worker. This dual role may lead to corruption and may hinder the efficiency of this farm type.

Corruption and nepotism play a larger role if large farms are connected with household farms as in Russia. Farm workers can shift their effort from the mother farm to the household farm and can even expropriate the large farm. Given their small share in land use, the only way household farms can produce a large share in total agricultural production is through their easy access to farm inputs from the large farm. The prevailing law even allows large farms to sell inputs at below market price to the household farms, and thus encourages cross-subsidization. There is ample evidence that household farms even receive inputs and services at zero prices. The relationship between large farms and household farms negatively affects the incentive to become a private farmer. Those individuals who opt for this alternative forego a fairly secure and predictable environment in exchange for one with a high degree of economic and political uncertainty.

Managers of large farms believe in their comparative advantage, and hence hinder those who want to leave the farm. Moreover, managers were used to feeling socially responsible for the employees on the farm. Their objective was, and still is, not just to maximize profit. Instead, they are still often production-oriented, and believe in a specific role of the state, namely to accept social responsibility for the survival of the large farms.

Mental models of policymakers are also often very important in explaining the survival of large farms. First, policymakers are widely convinced that they have to intervene in markets for food security reasons. Moreover, they believe in the comparative advantage of large farms and the need to set production targets. Hence, they are inclined to bail out large firms if they face financial problems, allowing most large farms to survive even if they are not profitable. Consequently, the comparative advantage of private farms is negatively affected in two ways: first, when large farms do not go out of business they reduce the supply of land that would be available otherwise, and second, access to soft budgets affects the behavior of employees who operate a household farm, encouraging them to exploit the mother farm (Koester, 1999).

3.4. The institutional environment and the comparative advantage of individual types of farms

According to Williamson, second-level institutions are formal rules (constitutions, laws, property rights), which make individual behavior predictable. Markets are the most important second-level institutions for the comparative advantage of individual types of farms. When these markets, in particular, those for land and rural credit, do not function, structural change in agriculture is impeded. Private farms, which may start with a minimum land endowment, cannot lease or buy additional land because of restrictive land legislation and poorly functioning markets, in particular with regard to the problems of incomplete information. The potential buyer or long-term lessee has insufficient information when estimating future returns of an asset and the supplier of these factors has insufficient information when assessing creditworthiness. Good legislation would help, but is not sufficient. In general, inadequate second-level institutions increase transaction costs and thus reduce the division of labor in the economy. Private farms have to accept lower farm gate prices for their produce and have to pay higher prices for their inputs. As transaction costs are partly related to the volume of exchange, large farms incur lower transaction costs per unit of purchase or sale than private farms.

Large farms are also favored by the Russian tax system (second-level institution), which is nontransparent and inconsistent. It is almost impossible for anyone to comply fully with the law. Larger farms, which are backed by regional or central administration, can take advantage of tax authorities' discretion in prosecuting tax violations.

3.5. Governance and the comparative advantage of individual types of farms

Third-level institutions are about governance, which affects the enforcement of formal rules set in second-level institutions (the legal framework). It is important for decision makers to know what happens if the terms of a contract are violated or affected by changes in the economic or political environment. Third-level institutions play a major role in determining the comparative advantage of individual types of farm in transition countries. First, market activities may become less profitable than subsistence production, not only because of adverse macroeconomic conditions (high inflation, an inefficient tax system, etc) but also because a weak legal framework increases uncertainty for farmers. Second, nontransparent markets increase information costs in the markets for agricultural produce. As these costs per unit of output sold or per unit of input bought are smaller for larger than smaller volumes, this uncertainty increases the comparative advantage of large farms over small family farms. Uncertainty due to weak enforcement of contracts affects the setting up of new family farms even more. Access to credit is generally crucial for prospective farmers, whereas existing farms may survive without any access to credit. Third, poor governance favored existing large collective farms or their successors because the government has often intervened strongly and inconsistently in the markets. Large farms also have a better network of contacts and are, therefore, better informed than the small family farms. Finally, large farms often received favorable treatment in the form of allocation of fuel and other inputs, or credit by bureaucrats who consider these farms important for local food security and the rural social infrastructure.

3.6. Institutions concerning resource allocation and employment, and the comparative advantage of individual types of farms

Neoclassical theory assumes that decision makers at the farm level assess alternatives with respect to maximizing an objective function. It is unlikely that farm managers in transition countries use such a procedure. First, they may face incomplete information to a much more severe extent than their colleagues in market economies do. Second, they suffer from certain deficits in education, e.g., they may not be familiar with basic concepts such as opportunity costs and marginal analysis. Finally, they may take into consideration nonmonetary factors, e.g., the preference for subsistence production, taking care of those who are in social need, etc. Hence, the outcome of the decision-making process with the same given constraints may be quite different in market economies than in transition countries.

It can be assumed that managers of family farms lack, even more than managers of large-scale collective farms, these necessary elements of rational decision making. It is possible that, under *working* market conditions, family farms might be superior to large-scale farms, but that the environment existing during the transition period has given a comparative advantage to large farms.

In conclusion, the comparative advantage of individual types of farms depends on various institutions. Privatization is but one of them and may not be the determining advantage in the performance of the farm. Therefore, it is not surprising that an investigation of the impact of land privatization on sector performance does not lead to conclusive results (Lerman, 1998).

3.7. Institutions and the recent changes in the farm structure

The Russian farm sector has undergone significant restructuring over the last two years (Rylko, 2001), and a new type of farm has emerged with new outside operators. Unfortunately, there are no official records of land transactions (Serova, 2002a), with the exception of sale and rent of state-owned land included in official statistics; yet, the bulk of transactions is conducted between private agents (Serova, 2002b). Serova reveals that in a small study of three Russian regions up to one third of all farms increased their area planted by three- to seven-fold, and that external operators (processors, traders, oil companies, etc.) have become more active in the land market.

Interviewers of 16 such enterprises received the following answers to their inquiry on farmers' motives for expansion: "we got tired of non-payback by farms and decided to control the whole production chain," "we wanted to receive the necessary quantity of inexpensive quality raw material on a timely basis," "we thought that agriculture was a good place to put money in." One operator expressed what was on many others' minds: "We don't see any reason why agriculture in Russia cannot be a highly profitable business. You only need new assets, new technology, new management, and new people" (Rylko, 2001). Fortunately, Rylko and others also report on the development of some of these enterprises. Their history reveals quite clearly that the recent and ongoing development cannot be explained

with the help of neoclassical theory alone, but needs to be supplemented with the framework of institutional economics.

3.8. Embedded institutions and the birth of agricultural holdings

Embedded institutions seem to have played a major role in this context. As already discussed above, many of the large farms have not been able to adjust to the changed economic environment but have benefited from soft budget constraints and other support from policymakers. In spite of this support many large farms became gradually insolvent, productivity went down, workers did not receive wages, and shareholders did not get lease payments. Such enterprises were weak, and an easy prey for taking over by outside operators, especially considering the extent of their political support. The support took the form of subsidies and decisions to accept the creation of new enterprises and leaving behind the highly indebted old ones. This procedure was at least tolerated or even promoted by the officials. The most outstanding model was applied in Belgorod oblast (Rylko, 2001), where a special decree by the Governor allowed the transfer of all bad farm debts of insolvent collective farms (about one third of all farms) to the oblast budget. At the same time, the insolvent farms were assigned to strong nonagricultural and agricultural enterprises and to private farmers. It would have been possible to partition the collective farms and to create small family farms, but existing first-, second-, and third-level institutions prevented such an alternative.

Private ownership of land has been possible in Russia since 1991, and according to federal legislation all transactions of land have been permitted since 1993 (Serova, 2002). However, there persisted high uncertainty concerning the stability of existing land regulations. The law on mortgage in 1998 forbade the use of land as security, and thus constrained landowner rights. It was only in 2002 that the State Duma accepted a land law and ended the long-lasting uncertainty on the land market. Up to this point agricultural land had been mainly transferred in the form of lease contracts. However, the procedure for leasing out is extremely burdensome, implying several administrative steps. It is well known that paying bribes and kickbacks can speed up the process. One of the interviewed enterprises even mentioned that it gave personal computers to the local office of the land committee to try to speed up the lengthy process of registration. It is quite clear that private farmers who might have been interested in smaller areas are less able to pay competitive bribes. Moreover, they would incur higher search costs per ha of land transacted in order to find out the magnitude of a successful bribe. Landowners seem to prefer handing over land to large operators due to lack of trust in prospective private farmers. This lack of trust is partly embedded, but also supported by weak third-level institutions. Farmers may suffer from delayed payments and even from theft. Large operators are better positioned to collect debt and to secure themselves against theft. It is reported that some large farms even use armed groups with automatic guns to protect the crops on the fields. Family farmers cannot resort to such methods in order to compensate for weak third-level institutions, i.e., weak enforcement of the law.

3.9. The effect of second- and third-level institutions

Second-level institutions may also favor the creation of large farms. First, agriculture is exempt from income or other taxes; it only incurs a land tax, which is not related to farm profits, but to cadastre values of the land. Outside operators of agricultural enterprises can, therefore, save on taxes by shifting the profits from nonagriculture activities to farm activities. Second, poorly functioning land markets allows the new operators, who often enjoy regional monopoly powers, to suppress lease prices. Although rental prices are very low compared to those in other countries, they are high compared to what the former farm operator is used to. Third, badly functioning credit markets improve the comparative advantage of the new large farms. External operators that had profits to invest, proven credit worthiness, and political influence, have a comparative advantage in the land market. Transfer of a huge number of total farms instead of smaller farm units was also in the interest of many owners. Given the state of the social security system and the social infrastructure in rural areas, weak landowners prefer to lease to those potential leasers who are able to provide some social services and are considered to be reliable. Again, the weak second- and third-level institutions favored outside operators.

4. Challenges ahead

This analysis suggests that the creation of very large enterprises in Russian agriculture and their ongoing growth was not a reflection of the comparative advantage of these farms in market economies, but a consequence of embedded institutions and an inadequate institutional framework. What might be the macroeconomic and long-term implications of this development?

Russian agricultural output has increased over the past decade, which may be due to either an increase in yields or a decline in theft, or both. Indeed, this development is not surprising, since outside operators, who expected to make profits in farming, invested heavily. As a result, the share of profit-making enterprises has increased. However, the long-term effects might be highly negative.

First, the new enterprises move to highly capital-intensive production and release workers. In the absence of alternative employment, the danger of the creation of a class of landless unemployed workers in rural areas seems to be real. Given the present level of factor endowment and productivity in Russia, the shadow price of labor is low, which indicates the profitability of less capital-intensive activities. If large-scale farms release workers it is likely due to high wages, labor market legislation, and low productivity of labor. The old collective farms could not move to capital-intensive production because they had less access to credit and felt committed to preserving the social well-being of the labor force.

Second, whether the present trend of enlarging the large farms and creating new farms is reversible in the near future is an interesting point. Even if more efficient markets could improve the competitiveness of family farms, a path-dependency in the evolution of the agrarian structure would have been created. Family farms can only develop if some of the large farms are subdivided. However, as these farms have market power and political clout, it is likely that they will continue to operate in the future.

Third, the creation of these farms affects the political markets in the regions. This point can be illustrated by the recently proposed change in the agricultural tax system. The Russian parliament intended to introduce a profit tax in agriculture, as in other sectors. Not surprisingly, the new large farm operators opposed this legislation and succeeded in keeping control over decisions on whether the old system of only a land tax or the new system should be applied in the region.

In summary, the survival of large farms and the creation of new large holdings in Russia is not a reflection of market forces, but of the specific institutional environment. It may be that the short-term overall economic effects of the present changes are positive, but there is a real danger of long-run negative economic and political effects, such as rural unemployment among a new class of landless people, and even social unrest.

Acknowledgments

The author acknowledges the very helpful comments by Ken Thomson, Aberdeen, Bernhard Bruemmer, Goettingen, and by the editors David Colman and Nick Vink.

References

Bardhan, P., "Institutions, Reforms and Agricultural Performance," in K. G. Stamoulis, ed., *Current and Emerging Issues for Economic Analysis and Policy Research* (Food and Agriculture Organization of the United Nations: Rome, 2001), pp. 137–166.

Csaki, C., J. Nash, V. Matusevich, and H. Kray, "Food and Agricultural Policy in Russia," *Progress to Date and the Road Forward*, World Bank Technical Paper No. 523, Washington, DC (2002).

Dewatripont, M., and G. Roland, "Transition as a Process of Large-Scale Institutional Change," *Economics of Transition* 4, no. 1 (1996), 1–29.

Flora, C. B., and J. L. Flora, "Entrepreneurial Social Infrastructure: A Necessary Ingredient," *The Annals of the American Academy of Political and Social Science* 529 (1993), 48–58.

Gorton, M., and S. Davidova, "Farm Productivity and Efficiency in the CEE Applicant Countries: A Synthesis of Results," *Agricultural Economics* (2003), forthcoming.

Greif, A., "Cultural Beliefs and the Organization of Society: A Historical and Theoretical Reflection on Collectivist and Individualist Societies," *Journal of Political Economy* 102, no. 5 (1994), 912–950.

Koester, U., "The Evolving Farm Structure in East Germany," in C. Csaki and Z. Lerman, eds., *Structural Change in the Farming Sectors in Central and Eastern Europe*, ECSSD Environmentally and Socially Sustainable Development, Working Paper No. 23, Washington, DC (2000), pp. 49–66.

Koester, U., "Die Bedeutung der Organisationsstruktur landwirtschaftlicher Unternehmen für die Entwicklung des ukrainischen Agrarsektors," in S. von Cramon-Taubadel and L. Striewe, eds., *Die Transformation der Landwirtschaft in der Ukraine* (Wissenschaftsverlag Vauk: Kiel, 1999), pp. 189–212.

Lerman, Z., C. Csaki, and G. Feder, "Land Policy and Changing Farm Structures in Central Eastern Europe and Former Soviet Union," *World Bank Technical Paper*, Washington, DC, 2002.

Lerman, Z., "Does Land Reform Matter? Some Experiences from the Former Soviet Union," *European Review of Agricultural Economics* 25, no. 3 (1998), 307–330.

Lerman, Z., personal communication (2003).

Lindsay, S., "Culture, Mental Models, and National Prosperity," in L. E. Harrison and S. P. Huntington, eds., *Culture Matters. How Values Shape Human Progress* (Basic Books: New York, 2000), pp. 282–295.

Lund, P., "The use of alternative measures of farm size in analysis of size and efficiency," *Journal of Agricultural Economics* 1 (1983), 100–110.

Lund, P., and R. Price, "The Measurement of the Average Farm Size," *Journal of Agricultural Economics* 1 (1998), 100–110.

North, D. C., *Institutions, Institutional Change and Economic Performance* (Cambridge University Press: Cambridge, U.K., 1990).

O'Brien, D. J., "Land Privatization in Rural Russia," *Economic Systems* 26 (2002), 169–173.

Pfaff, W., *Economics Hatch a Disaster*, Boston Globe 30 August 1999.

Reja, B., and A. Talvitie, "The Industrial Organization of Corruption: What Is the Difference between Asia and Africa," Presentation at the Annual Meeting of the International Society for New Institutional Economics, Tübingen 22–24 September 2000.

Rylko, D. N., "New Agricultural Operators, Input Markets and Vertical Sector Coordination," USAID Basis Project (2001), Unpublished paper.

Schulze, E., P. Tillack, O. Dolud, and S. Bukin, "Eigentumsverhältnisse landwirtschaftlicher Betriebe und Unternehmen in Russland und in der Ukraine," Discussion Paper No. 18 (Institute of Agricultural Development in Central and Eastern Europe, Halle, 1999).

Serova, E., "Public Opinion Concerning Russia's Agrarian Reforms," in A. L. Norsworthy, ed., *Russian Views of the Transition in the Rural Sector* (The World Bank: Washington, DC, 2000), pp. 67–86.

Serova, E., *Budget Subsidizing of Medium-Term Credits in Agriculture* (Analytical Centre Agrifood Economy: Moscow, 2002a).

Serova, E., *The New Regime of Taxing Agricultural Producers* (Analytical Centre Agrifood Economy: Moscow, 2002b).

Serova, E., *Russian Agri-Food Economy: Today and Tomorrow* (2003), forthcoming.

Täuber, A., "Arbeitsgemeinschaft Agrar, Kooperationsbeirat Ostausschuss," E-mail-Information (2003).

Wegren, S. K., and F. A. Durgin, "The Political Economy of Private Farming in Russia," *Comparative Economic Studies* XXXIX, no. 3–4 (1997), 1–24.

Williamson, O., "The New Institutional Economics: Taking Stock, Looking Ahead," *Journal of Economic Literature* XXXVIII (2000), 595–613.

Woolcock, M., "Social Capital and Economic Development: Towards a Theoretical Synthesis and Political Framework," *Theory and Society* (1998).

World Bank, "The State in a Changing World," *World Development Report 1997*, Washington (1997).

Total factor productivity growth in agriculture:
a Malmquist index analysis of 93 countries, 1980–2000

Tim J. Coelli*, D. S. Prasada Rao**

Abstract

In this paper we examine the levels and trends in agricultural output and productivity in 93 developed and developing countries that account for a major portion of the world population and agricultural output. We make use of data drawn from the Food and Agriculture Organization of the United Nations and our study covers the period 1980–2000. Due to the nonavailability of reliable input price data, the study uses data envelopment analysis (DEA) to derive Malmquist productivity indices. The study examines trends in agricultural productivity over the period. Issues of catch-up and convergence, or in some cases possible divergence, in productivity in agriculture are examined within a global framework. The paper also derives the shadow prices and value shares that are implicit in the DEA-based Malmquist productivity indices, and examines the plausibility of their levels and trends over the study period.

JEL classification: D24, O13, O47, Q10

Keywords: total factor productivity growth; Malmquist index; data envelopment analysis; agriculture; catch-up; convergence; shadow prices

1. Introduction

Productivity growth in agriculture has been the subject matter for intense research over the last five decades. Development economists and agricultural economists have examined the sources of productivity growth over time and of productivity differences among countries and regions over this period. Productivity growth in the agricultural sector is considered essential if agricultural sector output is to grow at a sufficiently rapid rate to meet the demands for food and raw materials arising out of steady population growth. During the 1970s and 1980s a number of major analyses of cross-country differences in agricultural productivity were conducted, including Hayami and Ruttan (1970, 1971), Kawagoe and Hayami (1983, 1985), Kawagoe et al. (1985), Capalbo and Antle (1988), and Lau and Yotopoulos (1989).

The majority of these studies used cross-sectional data on approximately 40 countries to estimate a Cobb–Douglas production technology using regression methods. The focus was generally on the estimation of the production elasticities and the investigation of the contributions of farm scale, education, and research in explaining cross-country labor productivity differentials.[1]

In the past decade, the number of papers investigating cross-country differences in agricultural productivity levels and growth rates has expanded significantly. This is most likely driven by three factors. First, the availability of some new panel data sets, such as that produced by the Food and Agriculture Organization of the United Nations (FAO). Second, the development of new empirical techniques to analyze this type of data, such as the data envelopment analysis (DEA) and stochastic frontier analysis (SFA) techniques, described in Coelli et al. (1998). Third, a desire to assess

---

* *Centre for Efficiency and Productivity Analysis, School of Economics, Brisbane, QLD, 4072 Australia.*

** *University of Queensland, Brisbane, QLD, 4072 Australia.*

[1] Lau and Yotopolous (1989) also estimated a translog functional form so as to illustrate the restrictions inherent in the Cobb–Douglas production technology.

Table 1
Analyses of inter-country agricultural total factor productivity (TFP) growth, 1993–2003

Paper	Method	Years	Countries
Fulginiti and Perrin (1993)	CD	1961–85	18 LDC
Bureau et al. (1995)	DEA & Fisher	1973–89	10 DC
Fulginiti and Perrin (1997)	DEA	1961–85	18 LDC
Craig et al. (1997)	CD	1961–90	98
Lusigi and Thirtle (1997)	DEA	1961–91	47 Africa
Fulginiti and Perrin (1998)	CD (VC)	1961–85	18 LDC
Rao and Coelli (1998)	DEA	1980–95	97
Arnade (1998)	DEA	1961–93	70
Fulginiti and Perrin (1999)	DEA & CD	1961–85	18 LDC
Martin and Mitra (1999)	Translog	1967–92	49
Wiebe et al. (2000)	CD	1961–97	110
Chavas (2001)	DEA	1960–94	12
Ball et al. (2001)	Fisher (EKS)	1973–93	10 DC
Suhariyanto et al. (2001)	DEA	1961–96	65 Asia/Africa
Suhariyanto and Thirtle (2001)	DEA	1965–96	18 Asia
Trueblood and Coggins (2003)	DEA	1961–91	115
Nin et al. (2003)	DEA	1961–94	20 LDC

the degree to which the Green Revolution, and other programs, have improved agricultural productivity in developing countries.

In Table 1 we list 17 studies that have been conducted in the last decade. Certain comments can be made about these papers. First, the majority of these papers use FAO panel data, spanning the 1960s through the 1980s. Of these 17 papers, 11 utilize DEA, five estimate Cobb–Douglas production functions, one estimates a translog production function, and one uses the Fisher index.[2] In terms of country coverage, five papers focus on less developed countries (LDCs), two analyze small groups of developed countries (DCs), three papers look at Asia or Africa or the two combined, while the remaining seven study a mix of countries. Four of these latter seven papers cover a large number of countries, ranging from 70 countries in Arnade (1998) to 110 countries in Wiebe et al. (2000).

One of the recurring themes in the reported results in many of these papers is that less developed countries exhibit technological regression while the developed countries show technological progress. For example, Fulginiti and Perrin (1997) studied 18 LDCs and found that 14 of these countries showed a decline in agricultural productivity over the period 1961–1985.

Such results indicate a divergence in agricultural productivity. However, these results appear to be in sharp contrast to the trends in the manufacturing sector and gross domestic product level productivity, which show signs of convergence (Barro and Sala-i-Martin, 1991; Maddison, 1995). Furthermore, they do not appear to be in accordance with crop-level evidence coming out of many developing countries in the past few decades, especially in Southeast Asia.

The principal aim of this study is to provide up-to-date information on agricultural total factor productivity (TFP) growth over the past two decades (1980–2000) for 93 of the largest agricultural producers in the world. It should be noted that the study by Wiebe et al. (2000) does analyze total factor productivity (TFP) growth for 110 nations over the 1961–1997 period; however, it does use the Cobb–Douglas production function, which introduces a number of restrictive assumptions, such as, constant production elasticities (and hence input shares) across all countries, Hicks-neutral technical change, plus the requirement that crop and livestock outputs be aggregated into a single output measure. The analysis in the present study uses the DEA technique to calculate the Malmquist TFP index numbers. This method does not make any of the above assumptions. However, it is susceptible to the effects of data noise, and can suffer from the problem of

---

[2] Two of these papers use two techniques.

"unusual" shadow prices, when degrees of freedom are limited.

This issue of shadow prices is important, and is one that is not well understood among authors who apply these Malmquist DEA methods. A major advantage cited in support of the use of DEA in measuring productivity growth, is that these methods do not require any price data. This is a distinct advantage, because in general, agricultural input price data are seldom available and such prices could be distorted due to government intervention in most developing countries. However, an important point needs to be added here. Even though the DEA-based productivity measures may not *explicitly* use *market price* information, they do *implicitly* use *shadow price* information, derived from the shape of the estimated production surface. This issue is described in some detail in Coelli and Rao (2001), who show that one can use these shadow prices to calculate *shadow shares* information, to help shed light on the factors influencing these productivity growth measures. Hence, an important contribution of this paper is to demonstrate the feasibility of explicitly identifying the implicit shadow shares and to study regional variation and trends in these shares over time.

In our view, this shadow share information can provide valuable insights into why various authors have obtained widely differing TFP growth measures for some countries, when applying these Malmquist DEA methods. This has been particularly evident when the applications have involved panel data sets containing small groups of countries, and the countries included in each data set differ from study to study.

The remainder of this paper is organized into sections. In Section 2 the DEA and Malmquist TFP index methods are described, while in Section 3 the data that are used are presented. The empirical results are presented and discussed in Section 4. Concluding comments are made in the final section.

2. Methodology

In this paper total factor productivity (TFP) is measured using the Malmquist index methods described in Färe et al. (1994) and Coelli et al. (1998, Chapter 10). This approach uses data envelopment analysis (DEA) methods to construct a piece-wise linear production frontier for each year in the sample. Hence, a brief description of DEA methods is provided prior to a description of the Malmquist TFP calculations.

2.1. Data envelopment analysis (DEA)

DEA is a linear programming methodology, which uses data on the input and output quantities of a group of countries to construct a piece-wise linear surface over the data points. This frontier surface is constructed by the solution of a sequence of linear programming problems—one for each country in the sample. The degree of technical inefficiency of each country (the distance between the observed data point and the frontier) is produced as a by-product of the frontier construction method.

DEA can be either input-orientated or output-orientated. In the input-orientated case, the DEA method defines the frontier by seeking the maximum possible proportional reduction in input usage, with output levels held constant, for each country. While, in the output-orientated case, the DEA method seeks the maximum proportional increase in output production, with input levels held fixed. The two measures provide the same technical efficiency scores when a constant returns-to-scale (CRS) technology applies, but are unequal when variable returns to scale (VRS) is assumed. In this paper a CRS technology is assumed (the reasons for this are outlined in the Malmquist discussion below). Hence the choice of orientation is not a big issue in this case. However, an output orientation has been selected because it would be fair to assume that, in agriculture, one usually attempts to maximize output from a given set of inputs, rather than the converse.[3]

Given data for N countries in a particular time period, the linear programming (LP) problem that is solved for the ith country in an output-orientated DEA model is as follows:

$$\max_{\phi,\lambda} \phi,$$

$$\text{st} \quad -\phi y_i + Y\lambda \geq 0,$$
$$x_i - X\lambda \geq 0,$$
$$\lambda \geq 0, \quad \quad \quad (1)$$

---

[3] There are some obvious exceptions to this. For example, where dairy farmers are required to fill a particular output quota, and attempt to do this with minimum inputs.

where

- y_i is a $M \times 1$ vector of output quantities for the ith country;
- x_i is a $K \times 1$ vector of input quantities for the ith country;
- Y is a $N \times M$ matrix of output quantities for all N countries;
- X is a $N \times K$ matrix of input quantities for all N countries;
- λ is a $N \times 1$ vector of weights; and
- ϕ is a scalar.

Observe that ϕ will take a value greater than or equal to 1, and that $\phi - 1$ is the proportional increase in outputs that could be achieved by the ith country, with input quantities held constant. Note also that $1/\phi$ defines a technical efficiency (TE) score that varies between 0 and 1 (and that this is the output-orientated TE score reported in our results).

The above LP is solved N times—once for each country in the sample. Each LP produces a ϕ and a λ vector. The ϕ-parameter provides information on the technical efficiency score for the ith country and the λ-vector provides information on the *peers* of the (inefficient) ith country. The peers of the ith country are those efficient countries that define the facet of the frontier against which the (inefficient) ith country is projected.

The DEA problem can be illustrated using a simple example. Consider the case where there are a group of five countries producing two outputs (e.g., wheat and beef). Assume for simplicity that each country has identical input vectors. These five countries are depicted in Figure 1. Countries A, B, and C are efficient countries because they define the frontier. Countries D and E are inefficient countries. For country D the technical efficiency score is equal to

$$TE_D = 0D/0D', \tag{2}$$

and its peers are countries A and B. In the DEA output listing this country would have a technical efficiency score of approximately 70% and would have nonzero λ-weights associated with countries A and B. For country E the technical efficiency score is equal to

$$TE_E = 0E/0E', \tag{3}$$

and its peers are countries B and C. In the DEA output listing this country would have a technical efficiency

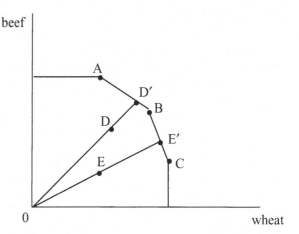

Figure 1. Output-orientated DEA.

score of approximately 50% and would have nonzero λ-weights associated with countries B and C. Note that the DEA output listing for countries A, B, and C would provide technical efficiency scores equal to one and each country would be its own peer. For further discussion of DEA methods see Coelli et al. (1998, Chapter 6).

2.2. The Malmquist TFP index

The Malmquist index is defined using distance functions. Distance functions describe a multi-input, multi-output production technology without the need to specify a behavioral objective (such as cost minimization or profit maximization). Both input distance functions and output distance functions may be defined. An input distance function characterizes the production technology by looking at a minimal proportional contraction of the input vector, given an output vector. An output distance function considers a maximal proportional expansion of the output vector, given an input vector. Only an output distance function is considered in detail in this paper. However, input distance functions can be defined and used in a similar manner.

A production technology may be defined using the output set, $P(x)$, which represents the set of all output vectors, y, which can be produced using the input vector, x. That is,

$$P(x) = \{y : x \text{ can produce } y\}. \tag{4}$$

It is assumed that the technology satisfies the axioms listed in Coelli et al. (1998, Chapter 3).

The output distance function is defined on the output set, $P(x)$, as:

$$d_o(x, y) = \min\{\delta: (y/\delta) \in P(x)\}. \quad (5)$$

The distance function, $d_o(x,y)$, will take a value that is less than or equal to 1 if the output vector, y, is an element of the feasible production set, $P(x)$. Furthermore, the distance function will take a value of unity if y is located on the outer boundary of the feasible production set, and will take a value greater than one if y is located outside the feasible production set. DEA-like methods are used to calculate the distance measures in this study. These are discussed shortly.

The Malmquist TFP index measures the TFP change between two data points (e.g., those of a particular country in two adjacent time periods) by calculating the ratio of the distances of each data point relative to a common technology. Following Färe et al. (1994), the Malmquist (output-orientated) TFP change index between period s (the base period) and period t is given by

$$m_o(y_s, x_s, y_t, x_t) = \left[\frac{d_o^s(y_t, x_t)}{d_o^s(y_s, x_s)} \times \frac{d_o^t(y_t, x_t)}{d_o^t(y_s, x_s)}\right]^{1/2}, \quad (6)$$

where the notation $d_o^s(x_t, y_t)$ represents the distance from the period t observation to the period s technology. A value of m_o greater than 1 will indicate positive TFP growth from period s to period t while a value less than one indicates a TFP decline. Note that equation (6) is, in fact, the geometric mean of two TFP indices. The first is evaluated with respect to period s technology and the second with respect to period t technology.

An equivalent way of writing this productivity index is

$$m_o(y_s, x_s, y_t, x_t)$$
$$= \frac{d_o^t(y_t, x_t)}{d_o^s(y_s, x_s)}\left[\frac{d_o^s(y_t, x_t)}{d_o^t(y_t, x_t)} \times \frac{d_o^s(y_s, x_s)}{d_o^t(y_s, x_s)}\right]^{1/2}, \quad (7)$$

where the ratio outside the square brackets measures the change in the output-oriented measure of Farrell technical efficiency between periods s and t. That is, the efficiency change is equivalent to the ratio of the technical efficiency in period t to the technical efficiency in period s. The remaining part of the index in equation (2) is a measure of technical change. It is the geometric mean of the shift in technology between the two periods, evaluated at x_t and also at x_s.

Following Färe et al. (1994), and given that suitable panel data are available, the required distance measures for the Malmquist TFP index are calculated using DEA-like linear programs. For the ith country, four distance functions are calculated in order to measure the TFP change between two periods, s and t. This requires the solving of four linear programming (LP) problems. Färe et al. (1994) assume a constant returns-to-scale (CRS) technology in their analysis. The required LPs are:

$$\left[d_o^t(y_t, x_t)\right]^{-1} = \max_{\phi, \lambda} \phi,$$
$$\text{st} \quad -\phi y_{it} + Y_t\lambda \geq 0,$$
$$x_{it} - X_t\lambda \geq 0,$$
$$\lambda \geq 0, \quad (8)$$

$$\left[d_o^s(y_s, x_s)\right]^{-1} = \max_{\phi, \lambda} \phi,$$
$$\text{st} \quad -\phi y_{is} + Y_s\lambda \geq 0,$$
$$x_{is} - X_s\lambda \geq 0,$$
$$\lambda \geq 0, \quad (9)$$

$$\left[d_o^t(y_s, x_s)\right]^{-1} = \max_{\phi, \lambda} \phi,$$
$$\text{st} \quad -\phi y_{is} + Y_t\lambda \geq 0,$$
$$x_{is} - X_t\lambda \geq 0,$$
$$\lambda \geq 0, \quad (10)$$

and

$$\left[d_o^s(y_t, x_t)\right]^{-1} = \max_{\phi, \lambda} \phi,$$
$$\text{st} \quad -\phi y_{it} + Y_s\lambda \geq 0,$$
$$x_{it} - X_s\lambda \geq 0,$$
$$\lambda \geq 0. \quad (11)$$

Note that in LPs (10) and (11), where production points are compared to technologies from different time periods, the ϕ parameter need not be greater than or equal to 1, as it must be when calculating standard output-orientated technical efficiencies. The data point could lie above the production frontier. This will most likely occur in LP (11) where a production point from period t is compared to technology in an earlier period, s. If technical progress has occurred, then a value of $\phi < 1$ is possible. Note that it could also possibly occur in LP (10) if technical regress has occurred, but this is less likely.

One issue that must be stressed is that the returns-to-scale properties of the technology are very important in TFP measurement. A CRS technology is used in this study for two reasons. First, given that the analysis involves the use of aggregate country-level data, it does not appear to be sensible to consider a VRS technology. That is, how is it possible for a *sector* to achieve scale economies? For example, the index of crop output for India and the U.S. are similar, but their average farm sizes are quite different. Hence, what can be sensibly concluded if a VRS technology is estimated and it is reported that these countries face decreasing returns to scale? The use of a VRS technology when the summary data are expressed on an "average per farm" basis may be sensible, since the scale economies of the "average farm" could be discussed, but when dealing with aggregate data (as is the case in this study) the use of a CRS technology is the only sensible option.

In addition to the above comment regarding the use of aggregate data, a second argument for the use of a CRS technology is applicable to both firm-level and aggregate data. Grifell-Tatjé and Lovell (1995) use a simple one-input, one-output example to illustrate that a Malmquist TFP index may not correctly measure TFP changes when VRS is assumed for the technology. Hence, it is important that a CRS technology be used in calculating Malquist TFP indices using DEA. Otherwise, the resulting measures may not properly reflect the TFP gains or losses resulting from scale effects.

3. Data

The present study is based on data drawn from the AGROSTAT system of the Statistics Division of the Food and Agricultural Organization in Rome. It is possible to access and download all the necessary data from the Web site of the FAO.[4] The following are some of the main features of the data series used.

3.1. Country coverage

The study includes 93 countries. These are the top 93 agricultural producers in the world, which account for roughly 97% of the world's agricultural output as well as 98% of the world's population.[5] The countries included in the study are distributed over all the regions of the world, as follows:

Africa	26 countries
North America	2 countries
South and Central America	19 countries
Asia	23 countries
Europe	20 countries
Australasia	3 countries

Data for the USSR, Czechoslovakia, and Yugoslavia in the 1990s could not be obtained due to changes in the political systems in Eastern Europe. Data for the newly formed countries for the most recent period are available but no corresponding data are available before 1990. Inclusion of USSR in the period before 1990 and replacing it with a large number of smaller countries may introduce some aggregation and scale issues. Hence these countries are omitted from the analysis.

3.2. Time period

Results are presented for the period 1980 to 2000. The initial intention was to study the 1960–2000 period; however, the analysis has been restricted to this shorter period since labor force data were not readily available for the years 1960–1979 from the FAO or the ILO sources. These years will be included in the subsequent stages of the project when appropriate labor data are obtained.[6]

---

[4] The authors are grateful to the FAO for maintaining an excellent site and for their generosity in making valuable data series available on the Internet.

[5] Ordering of the countries and estimates of country shares of agricultural output are drawn from Table 3.2 in Rao (1993). The original aim was to include 100 countries but three countries had to be dropped since we could not build the output series for those countries. An additional four countries, USSR, Czechoslovakia, Yugoslavia, and Ethiopia, are dropped due to data-related problems.

[6] The study period complements the periods covered in some of the earlier studies, which usually cover the 1960s and 1970s.

3.3. Output series

Due to the problems of degrees of freedom associated with the application of DEA methods, the present study uses two output variables, viz., crops and livestock output variables. The output series for these two variables are derived by aggregating detailed output quantity data on 185 agricultural commodities. The following steps are used in the construction of data.

For the year 1990, output aggregates are drawn from Table 5.4 in Rao (1993). These aggregates are constructed using international average prices (expressed in U.S. dollars) derived using the Geary–Khamis method (see Rao, 1993, Chapter 4 for details) for the benchmark year 1990.[7] Thus the output series for 1990 are at constant prices, expressed in a single currency unit.

The 1990 output series are then extended to cover the study period 1980–2000 using the FAO production index number series for crops and livestock separately.[8] The series that are derived using this approach are essentially equivalent to the series constructed using 1990 international average prices and the actual quantities produced in different countries in various years.

Tables of the output aggregates for the 93 countries for the years 1980 and 2000 are available from the authors on request. These tables demonstrate the differences in output mix across different countries. There are many countries that are mainly producers of crops, some countries are mainly livestock producers, while the remaining countries have a fair balance between crops and livestock.[9] A point to note here is the concept of output used in the study. Consistent with the definition of the FAO production index, the output concept used here is the output from the agriculture sector, net of quantities of various commodities used as feed and seed.[10] This is the reason for not including feed and seed in the input series.

Another point regarding the output series that is important to remember is the fact that the output series are based on 1990 international average prices. So the output series would change when the base is shifted from 1990 to another period, thus potentially influencing the final results. In this study it was decided that it is more appropriate to use 1990 prices as the basis for the study spanning 1980 to 2000 rather than using 1980 or 2000 international average prices.

3.4. Input series

Given the constraints on the number of input variables that can be used in a DEA analysis, this analysis considers only six input variables. Details of these variables are given below.

Land: This variable covers arable land, land under permanent crops as well as the area under permanent pasture. Arable land includes land under temporary crops (double-cropped areas are counted only once), temporary meadows for mowing or pasture, land under market and kitchen gardens, and land temporarily fallow (less than 5 years). Land under permanent crops is the land cultivated with crops that occupy the land for long periods and need not be replanted after each harvest. This category includes land under flowering shrubs, fruit trees, nut trees, and vines but excludes land under trees grown for wood or timber. Land under permanent pasture is the land used permanently (5 years or more) for forage crops, either cultivated or growing wild.

Tractors: This variable covers the total number of wheel and crawler tractors, but excludes garden tractors, used in agriculture. It is important to note that only the number of tractors is used as the input variable with no allowance made to the horsepower of the tractors.[11] This aspect will be examined in future work.

Labor: This variable refers to the economically active population in agriculture. This population is defined as all persons engaged or seeking employment in an economic activity, whether as employers,

---

[7] The Geary–Khamis international average prices are based on prices (in national currency units) and quantities of 185 agricultural commodities in 103 countries.

[8] See the 1997 FAO Production Yearbook for details regarding the construction of production index numbers.

[9] The DEA method employed here is specially suited to this type of situations. The method benchmarks countries against countries with similar output and input mixes.

[10] The output concept used here is consistent with the concept used in some of the earlier inter-country comparison studies (see Kawagoe and Hayami [1985] and Hayami and Ruttan [1970]).

---

[11] Assuming that farming in developing countries is on fragmented land, average horsepower of tractors in these countries could be significantly lower than those used in countries with large farms using highly mechanized farming techniques. This could understate the productivity levels and changes in developing countries.

own-account workers, salaried employees, or unpaid workers, assisting in the operation of a family farm or business. The economically active population in agriculture includes all economically active persons engaged in agriculture, forestry, hunting, or fishing. This variable obviously overstates the labor input used in agricultural production, where the extent of overstatement depends upon the level of development of the country.[12]

Fertilizer: Following other studies (Hayami and Ruttan 1970; Fulginiti and Perrin 1997) on inter-country comparison of agricultural productivity, fertilizer is measured as the sum of nitrogen, phosphate, and potasn contained in the commercial fertilizers consumed. This variable is expressed in thousands of metric tons.

Livestock: The livestock input variable used in the study is the sheep equivalent of five categories of animals used in constructing this variable. The categories considered are: buffalo, cattle, pig, sheep, and goat. Numbers of these animals are converted into sheep equivalents using conversion factors: 8.0 for buffalo and cattle; 1.00 for sheep, goat, and pig.[13] Chicken numbers are not included in the livestock figures.

Irrigation: In this study, the area under irrigation is used as a proxy for the capital infrastructure associated with the irrigation of farmlands.[14]

4. Results and discussion

The results of the DEA and TFP calculations are summarized in this section. Given that there are 21 annual observations on 93 countries, there is a lot of computer output to describe. The calculations involved the solving of $93 \times (21 \times 3 - 2) = 5{,}673$ LP problems.

Table 2
Means of technical efficiency for the continents, 1980–2000

Continent	Countries	1980	1990	2000
Africa	1–26	0.700	0.746	0.804
North America	27, 37	1.000	1.000	1.000
South America	28–36, 38–47	0.888	0.888	0.911
Asia	48–70	0.681	0.707	0.739
Europe	71–90	0.859	0.871	0.907
Australasia	91–93	1.000	1.000	1.000
Mean	1–93	0.784	0.806	0.842

There are thousands of pieces of information on the efficiency scores and peers of each country in each year. Furthermore, measures of technical efficiency change, technical change, and TFP change for each country in each pair of adjacent years have been calculated.

Hence, by necessity only a selection of the results are presented in this paper. Information on the means of the measures of technical efficiency change, technical change, and TFP change for each country (over the 21-year sample period) and the mean changes between each pair of adjacent years (over the 93 countries) are provided. Furthermore, means for certain groups of countries and plots of the TFP trends of some selected groupings of countries are presented. In addition to this, a table of peers for all countries in the first year (1980) and in the final year (2000) is provided.[15] Each of these sets of results is now discussed in turn.

Average technical efficiency scores in 1980 and 2000 are reported in Table 2 for the six regions and the full sample. Note that the average technical efficiency score of 0.784 in 1980 implies that these countries are, on average, producing 78.4% of the output that could be potentially produced using the observed input quantities.[16] It is interesting to note that those regions with the lowest mean technical efficiency scores in 1980—Asia and Africa—also achieved the largest increases in mean technical efficiency over the sample period. This provides evidence of catch-up in these countries, which was not found in many of the studies listed in Table 1. This is most likely due to the fact that the data in this study span the past two decades, while the

---

[12] There could be a significant percentage of the labor force (as defined here) in disguised unemployment.

[13] The conversion figures used in this study correspond very closely with those used in the 1970 study of Hayami and Ruttan.

[14] This irrigation variable was not included in an earlier analysis of the 1980–1995 data (see Rao and Coelli, 1998). In the present study, the DEA analysis was run with this variable included and also excluded. It was interesting to note that the (unweighted) mean TFP growth increased from 1.1% to 1.3% when this variable was excluded. This is not surprising, given that there has been significant investment in irrigation infrastructure in many countries over the past two decades, especially in Asian countries.

---

[15] These can obviously change from year to year, but it is not feasible to present this information for every year.

[16] This figure should be interpreted with care. No attempt has been made to adjust the data for differences in climate, soil quality, labor quality, etc.

Table 3
Peers from DEA, 1980 and 2000

	Country	Peers												Count*	
		1980						2000							
1	Algeria	10	62	31	39	57		8	97	40	34	93		0	0
1	Algeria	79	38	72	56	9	82	93	33	82	89	39		0	0
2	Angola	92	33	62	93	38		18	62					0	0
3	Burundi	65	18					3						0	0
4	Cameroon	4						4						4	2
5	Chad	5						5						3	1
6	Egypt	6						6						0	4
7	Ghana	93	92	33	62	24		7						0	2
8	Guinea	18	93	39	33			33	24	18	7			0	0
9	Cote d'Ivoire	9						9						17	16
10	Kenya	10						10						1	1
11	Madagascar	93	30	39	33			39	33	62	18			0	0
12	Malawi	65	93	9	18	4	33	18	65	9	93			0	0
13	Mali	10	5	4	33	32		9	62	33	17	32		0	0
14	Morocco	43	61	38	56	30	9	30	28	9	56	93		0	0
15	Mozambique	79	93	33	82			33	93	18				0	0
16	Niger	16						18	33					0	0
17	Nigeria	18	33	4	5	93		17						0	3
18	Rwanda	18						18						11	10
19	Senegal	33	4	18	65			9	33	18	4			0	0
20	South Africa	72	79	61	56	38	9	28	43	38				0	0
21	Sudan	33	39	30	93			33	39	18	62			0	0
22	Tanzania	93	62	34	33	44		39	33	24	62	18		0	0
23	Tunisia	9	56	93	38			82	38	56	9	78		0	0
24	Uganda	24						24						1	4
25	Burkina Faso	18	33	5				24	5	4	10	33		0	0
26	Zimbabwe	9	92	72	61	79	38	9	92	72	93	17	82	0	0
27	Canada	27						27						0	1
28	Costa Rica	38	92	79	61	9	30	28						0	6
29	Cuba	30	92	82	79	61		17	39	89	33	82	46	0	0
30	Dominican Republic	30						30						17	7
31	El Salvador	31						31						3	0
32	Guatemala	32						32						1	1
33	Haiti	33						33						21	16
34	Honduras	34						34						1	0
35	Mexico	30	56	61	79	38		28	43	56	30	9		0	0
36	Nicaragua	44	30	79	9	61		62	39	33	46	92	30	0	0
37	United States	37						37						1	1
38	Argentina	38						38						18	11
39	Bolivia	39						39						3	6
40	Brazil	44	79	61	9	72		38	9	72	61	92		0	0
41	Chile	56	9	61	30	38		38	28	56	43	73	61	0	0
42	Colombia	9	44	92	30	79		42						0	0
43	Ecuador	43						43						3	5
44	Paraguay	44						44						6	0
45	Peru	9	38	44	30	43		38	9	62	30	43		0	0
46	Uruguay	46						46						0	3
47	Venezuela	44	92	72	9	79	38	46	38	30	82	92		0	0
48	Bangladesh	93	59	18				48						0	1
49	Myanmar	93	18	33	65			65	48	93	33			0	0
50	Sri Lanka	61	93	56				93	65	6				0	0

(Continued)

Table 3
(Continued)

	Country	Peers 1980						Peers 2000						Count*	
51	China	59	79	93	33			28	93	30				0	0
52	India	65	82	31	93	30		56	6	82	65	93		0	0
53	Indonesia	53						61	9	93	65			0	0
54	Iran	56	9	61	30	38		43	38	61	9	56		0	0
55	Iraq	9	56	93	43	38		78	61	38	9			0	0
56	Israel	56						56						17	12
57	Japan	57						59	82	56				0	0
58	Cambodia	93	33	18	65			24	93	7	33	18		0	0
59	Korea Rep	59						59						4	2
60	Laos	93	65	33	18			60						0	0
61	Malaysia	61						61						17	11
62	Mongolia	62						62						3	7
63	Nepal	65	93	18	59	33		93	33	6	65			0	0
64	Pakistan	31	30	65	93	82		65	82	33	30			0	0
65	Philippines	65						65						10	7
66	Saudi Arabia	61	30	31	33	93		28	93	61				0	0
67	Syria	67						78	56	9	61	38		0	0
68	Thailand	93	65	30	33	9		56	61	93				0	0
69	Turkey	9	61	81	56	93		9	61	72	78	81	93	0	0
70	Vietnam	82	59	93	33			93	59	6				0	0
71	Austria	71						71						0	2
72	Bel-Lux	72						72						10	7
73	Bulgaria	61	56	92	38	79		73						0	1
74	Denmark	56	37	82	72			74						0	0
75	Finland	82	89	79				72	89	82	79			0	0
76	France	76						76						1	2
77	Germany	82	89	79	72			61	79	76	72	27		0	0
78	Greece	38	79	61	81	56		78						0	5
79	Hungary	79						79						22	3
80	Ireland	80						80						0	0
81	Italy	81						81						4	2
82	Netherlands	82						82						12	13
83	Norway	89	79	82				89	82					0	0
84	Poland	79	93	89	72			61	89	93	71			0	0
85	Portugal	89	56	33	30	93	38	78	82	93	56	9		0	0
86	Romania	56	61	82	30	92	79	38	9	82	56			0	0
87	Spain	61	38	81	79	56		38	81	37	56	76		0	0
88	Sweden	79	89	72	82			79	82	89	72			0	0
89	Switzerland	89						89						6	6
90	United Kingdom	79	56	81	72	76	38	71	72	93				0	0
91	Australia	91						91						0	0
92	New Zealand	92						92						9	4
93	Papua N. Guin.	93						93						26	19

* The *count* is the peer count. That is, the number of times that firm acts as a peer for another firm.

majority of these other studies consider the 1960–1985 period.

This information on changes in average technical efficiency only tells the "catch-up" part of the productivity story. TFP change can also appear in the form of technical change (or frontier shift). The means of the measures of technical efficiency change, technical change and TFP change for each country (over the

Table 4

Mean technical efficiency change, technical change, and TFP change, 1980–2000

	Country	Efficiency Change	Technical Change	TFP Change
51	China	1.044	1.015	1.060
58	Cambodia	1.024	1.033	1.057
1	Algeria	1.033	1.013	1.046
3	Burundi	1.015	1.030	1.046
66	Saudi Arabia	1.031	1.010	1.042
2	Angola	1.061	0.978	1.037
17	Nigeria	1.016	1.020	1.037
20	South Africa	1.014	1.023	1.037
60	Laos	1.022	1.011	1.034
27	Canada	1.000	1.033	1.033
74	Denmark	1.009	1.022	1.032
28	Costa Rica	1.003	1.026	1.028
62	Mongolia	1.000	1.028	1.028
37	U.S.	1.000	1.026	1.026
85	Portugal	1.019	1.007	1.026
91	Australia	1.000	1.026	1.026
29	Cuba	1.005	1.020	1.025
21	Sudan	1.016	1.008	1.024
48	Bangladesh	1.007	1.017	1.024
70	Vietnam	1.027	0.997	1.024
64	Pakistan	1.012	1.011	1.023
86	Romania	1.008	1.015	1.023
7	Ghana	1.010	1.012	1.022
12	Malawi	1.013	1.009	1.022
82	Netherlands	1.000	1.022	1.022
19	Senegal	1.008	1.013	1.021
84	Poland	1.015	1.007	1.021
89	Switzerland	1.000	1.021	1.021
40	Brazil	1.001	1.019	1.020
54	Iran	1.013	1.008	1.020
73	Bulgaria	1.014	1.006	1.020
76	France	1.000	1.020	1.020
15	Mozambique	1.031	0.988	1.019
23	Tunisia	1.011	1.008	1.018
36	Nicaragua	1.014	1.004	1.018
49	Myanmar	1.008	1.011	1.018
78	Greece	1.007	1.010	1.017
14	Morocco	1.004	1.012	1.016
35	Mexico	1.000	1.015	1.015
45	Peru	1.011	1.004	1.015
9	Cote d'Ivoire	1.000	1.014	1.014
42	Colombia	1.001	1.013	1.014
52	India	1.008	1.006	1.014
71	Austria	1.000	1.014	1.014
90	U.K.	1.001	1.013	1.014
77	Germany	1.003	1.011	1.013
6	Egypt	1.000	1.012	1.012
39	Bolivia	1.000	1.011	1.011
41	Chile	0.998	1.013	1.011
75	Finland	1.002	1.009	1.011

(Continued)

Table 4

(Continued)

	Country	Efficiency Change	Technical Change	TFP Change
80	Ireland	1.000	1.011	1.011
30	Dominican Republic	1.000	1.010	1.010
63	Nepal	1.010	1.000	1.010
87	Spain	1.009	1.001	1.010
4	Cameroon	1.000	1.009	1.009
69	Turkey	1.005	1.004	1.009
81	Italy	1.000	1.009	1.009
26	Zimbabwe	0.997	1.011	1.008
31	El Salvador	1.000	1.008	1.008
65	Philippines	1.000	1.008	1.008
47	Venezuela	0.997	1.009	1.006
10	Kenya	1.000	1.005	1.005
32	Guatemala	1.000	1.005	1.005
56	Israel	1.000	1.004	1.004
61	Malaysia	1.000	1.004	1.004
92	New Zealand	1.000	1.004	1.004
22	Tanzania	1.013	0.990	1.003
34	Honduras	1.000	1.003	1.003
43	Ecuador	1.000	1.003	1.003
79	Hungary	1.000	1.003	1.003
88	Sweden	0.992	1.012	1.003
50	Sri Lanka	1.004	0.998	1.002
57	Japan	0.993	1.009	1.002
46	Uruguay	1.000	1.000	1.000
11	Madagascar	1.008	0.990	0.998
16	Niger	0.995	1.004	0.998
25	Burkina Faso	0.990	1.007	0.997
72	Bel-Lux	1.000	0.996	0.996
59	Korea Republic	1.000	0.995	0.995
68	Thailand	0.994	1.000	0.995
83	Norway	0.986	1.010	0.995
93	Papua N. Guin.	1.000	0.992	0.992
67	Syria	0.982	1.007	0.989
44	Paraguay	1.000	0.984	0.984
13	Mali	0.982	1.001	0.983
53	Indonesia	0.978	1.003	0.981
24	Uganda	1.000	0.977	0.977
55	Iraq	0.968	1.008	0.976
38	Argentina	1.000	0.973	0.973
18	Rwanda	1.000	0.967	0.967
8	Guinea	1.006	0.958	0.964
33	Haiti	1.000	0.957	0.957
5	Chad	1.000	0.947	0.947
	Mean	1.005	1.006	1.011

21-year sample period) are presented in Table 4. Tables 5 and 6, respectively, show the unweighted and weighted annual averages (averaged over the 93 countries) of efficiency change, technical change, and TFP change. Table 7 shows the regional averages of changes

Table 5

Annual mean technical efficiency change, technical change, and TFP change, 1980–2000

Year*	Efficiency Change	Technical Change	TFP Change
1981	1.021	0.966	0.987
1982	0.993	1.027	1.020
1983	0.999	0.997	0.996
1984	1.023	0.990	1.012
1985	0.993	1.023	1.016
1986	1.011	0.988	0.999
1987	0.991	0.985	0.976
1988	1.012	1.048	1.060
1989	1.007	0.987	0.993
1990	0.995	1.025	1.020
1991	0.996	1.018	1.014
1992	1.009	0.979	0.987
1993	1.023	0.979	1.001
1994	1.010	0.986	0.995
1995	0.994	1.030	1.023
1996	1.020	1.039	1.059
1997	1.009	0.980	0.989
1998	0.997	1.033	1.030
1999	0.989	1.044	1.033
2000	1.006	1.003	1.009
Mean	1.005	1.006	1.011

* Note that 1981 refers to the change between 1980 and 1981, etc.

Table 6

Weighted annual mean technical efficiency change, technical change, and TFP change, 1980–2000

Year*	Efficiency Change	Technical Change	TFP Change
1981	1.017	1.011	1.028
1982	0.990	1.028	1.018
1983	1.012	0.985	0.996
1984	1.022	1.022	1.044
1985	1.008	1.021	1.030
1986	1.003	0.996	1.000
1987	0.991	1.008	0.999
1988	1.018	1.007	1.025
1989	1.008	1.005	1.013
1990	0.990	1.029	1.019
1991	1.008	1.011	1.019
1992	1.035	0.995	1.030
1993	1.030	0.981	1.010
1994	1.029	1.011	1.041
1995	0.984	1.048	1.031
1996	1.028	1.010	1.038
1997	1.030	1.011	1.041
1998	1.002	1.011	1.012
1999	0.983	1.039	1.022
2000	0.996	1.019	1.015
Mean	1.009	1.012	1.021

* Note that 1981 refers to the change between 1980 and 1981, etc.

Table 7

Weighted means of annual technical efficiency change, technical change, and TFP change for the continents, 1980–2000

Continent	Countries	Efficiency Change	Technical Change	TFP Change
Africa	1–26	1.006	1.007	1.013
North America	27, 37	1.000	1.027	1.027
South America	28–36, 38–47	1.000	1.006	1.006
Asia	48–70	1.019	1.010	1.029
Europe	71–90	1.002	1.011	1.014
Australasia	91–93	1.000	1.018	1.018
Mean	1–93	1.009	1.012	1.021

in efficiency and TFP. Table 8 shows the changes in TFP for groups of countries classified by their technical efficiency score in the initial period 1980.

In Table 3 we can identify all those countries that define the frontier technology for the years 1980 and 2000 (in the vicinity of their observed output and input mixes). The table shows that there are 39 and 45 countries that are on the frontier in 1980 and 2000, respectively. Only four countries, Niger, Indonesia, Japan, and Syria, which were on the frontier in 1980, were no longer in the frontier in 2000. Table 3 also provides a list of countries that define the best practice (peers) for each of the countries that are not on the frontier. It is interesting to observe the changes in the sets of peer countries over the two periods. For example, in 1980 Cuba had the Dominican Republic, the Netherlands, Malaysia, New Zealand, and Hungary as its peers. However, in 2000 only the Netherlands remained in the peer country set, the other countries in the new set being Nigeria, Bolivia, Switzerland, Haiti, and Uruguay. Sets of peer countries defining best practice for countries in Asia seem to be relatively stable over the study period.

Table 8

Weighted means of annual technical efficiency change, technical change, and TFP change for efficient and inefficient countries, 1980–2000

Efficiency Level in 1980	Efficiency Change	Technical Change	TFP Change
$TE = 1$	0.998	1.013	1.012
$0.6 < TE < 1$	1.003	1.012	1.015
$TE < 0.6$	1.025	1.011	1.036
Mean	1.009	1.012	1.021

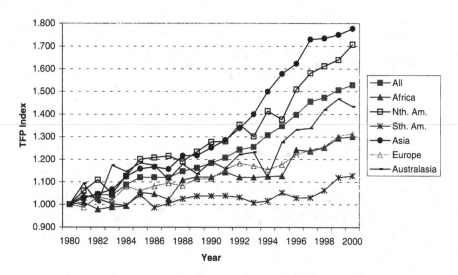

Figure 2. Cumulative TFP indices.

The last two columns of Table 3 show the number of times each of the efficient countries on the frontier appear as a peer for the technically inefficient countries. Countries that do not appear as a peer for any other country may be considered to be on the frontier due to the unique nature of their input and output mixes. For example, Australia does not appear as a peer for any country in 1980. In contrast, Papua New Guinea appears as a peer for 26 countries in 1980.

Table 4 shows the mean technical efficiency change, technical change, and TFP change for the 93 countries over the period 1980 to 2000. Countries in the table are presented in descending order of the magnitude of the TFP changes. The table shows China and Cambodia as the two countries with maximum TFP growth. China shows a 6.0% average growth in TFP, which is due to 4.4% percent growth in technical efficiency, and 1.5% growth in technical change.[17] Australia, United States, and India, respectively, exhibit TFP growth rates of 2.6, 2.6, and 1.4%. The unweighted average (across all countries) growth in TFP is 1.1%.

Tables 5 and 6 show the annual average technical efficiency change, technical change, and TFP change using, respectively, unweighted (where each country has the same weight) and weighted (where each country change is weighted by the country's share in total

agricultural output). These tables show the effect of using weights on the annual averages derived. Unweighted averages show only 1.1% growth in TFP whereas the weighted TFP growth over the period is 2.1%. The results show that the use of unweighted averages understates the changes in TFP and in its components. Another implication of this difference is that TFP growth has been higher in countries with a higher share of global agricultural output. It seems reasonable to argue that for purposes of assessing regional and global performance a weighted average (across countries) of annual growth rates is more appropriate.

Tables 5 and 6 show that over the whole period there has been no technological regression though for some individual years there has been some evidence of technological regression. The extent of technological regression seems to be less serious when weighted average changes are considered.

Table 7 provides measures of annual changes in technical efficiency, technical change, and TFP change by different regions. Asia posted the highest TFP growth of 2.9% (mainly due to efficiency change growth of 1.9%) followed by North America (consisting of the United States and Canada), Australasia, Europe, Africa, and South America. South America has posted the lowest growth rate of 0.6%, followed by Africa with 1.3% growth in TFP. A surprising result is that over the period 1980–2000, these results show no evidence of global or regional technological regression. This is in

---

[17] This result appears to be consistent with some of the recent studies on Chinese economic growth (Maddison, 1997).

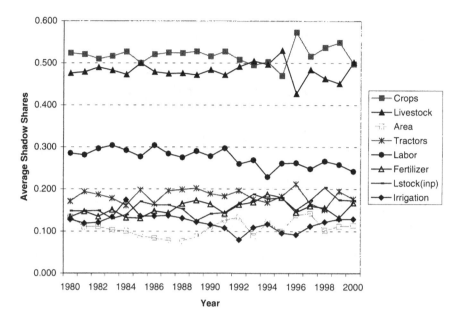

Figure 3. Mean shadow shares.

Table 9
Annual mean shadow shares, 1980–2000

Year	Outputs		Inputs					
	Crops	Livestock	Area	Tractors	Labor	Fertilizer	Livestock	Irrigation
1980	0.524	0.476	0.134	0.171	0.285	0.134	0.148	0.128
1981	0.521	0.479	0.111	0.194	0.281	0.147	0.148	0.119
1982	0.510	0.490	0.111	0.187	0.296	0.135	0.149	0.121
1983	0.517	0.483	0.103	0.178	0.304	0.151	0.130	0.134
1984	0.527	0.473	0.101	0.161	0.292	0.132	0.139	0.174
1985	0.500	0.500	0.088	0.198	0.277	0.131	0.170	0.136
1986	0.521	0.479	0.084	0.165	0.304	0.148	0.162	0.136
1987	0.525	0.475	0.078	0.196	0.284	0.143	0.162	0.137
1988	0.524	0.476	0.077	0.199	0.275	0.165	0.153	0.131
1989	0.528	0.472	0.087	0.202	0.290	0.173	0.125	0.122
1990	0.516	0.484	0.110	0.189	0.278	0.164	0.141	0.116
1991	0.528	0.472	0.126	0.183	0.297	0.142	0.144	0.108
1992	0.508	0.492	0.133	0.197	0.260	0.163	0.166	0.080
1993	0.495	0.505	0.089	0.176	0.269	0.169	0.188	0.109
1994	0.503	0.497	0.121	0.168	0.229	0.188	0.177	0.117
1995	0.470	0.530	0.100	0.183	0.261	0.180	0.179	0.096
1996	0.573	0.427	0.138	0.212	0.262	0.145	0.150	0.092
1997	0.516	0.484	0.142	0.166	0.248	0.160	0.173	0.112
1998	0.537	0.463	0.102	0.151	0.266	0.155	0.204	0.122
1999	0.549	0.451	0.112	0.195	0.258	0.132	0.174	0.129
2000	0.498	0.502	0.114	0.176	0.242	0.167	0.173	0.129
Mean	0.519	0.481	0.108	0.183	0.274	0.154	0.160	0.121

Table 10
Mean shadow shares, 1980–2000

Country	Outputs		Inputs					
	Crops	Livestock	Area	Tractors	Labor	Fertilizer	Livestock	Irrigation
Algeria	0.393	0.607	0.000	0.025	0.222	0.449	0.244	0.060
Angola	0.227	0.773	0.000	0.003	0.276	0.222	0.250	0.249
Burundi	1.000	0.000	0.000	0.159	0.044	0.065	0.731	0.000
Cameroon	0.308	0.692	0.030	0.254	0.349	0.015	0.000	0.352
Chad	0.149	0.851	0.000	0.483	0.160	0.092	0.001	0.263
Egypt	0.785	0.215	0.657	0.081	0.120	0.130	0.012	0.000
Ghana	0.455	0.545	0.000	0.091	0.191	0.246	0.281	0.192
Guinea	0.553	0.447	0.001	0.188	0.359	0.452	0.000	0.000
Cote d'Ivoire	0.992	0.008	0.000	0.264	0.407	0.084	0.109	0.136
Kenya	0.002	0.998	0.045	0.287	0.166	0.121	0.014	0.368
Madagascar	0.531	0.469	0.007	0.180	0.608	0.205	0.000	0.000
Malawi	0.817	0.183	0.006	0.438	0.381	0.000	0.158	0.016
Mali	0.112	0.888	0.049	0.092	0.342	0.074	0.006	0.438
Morocco	0.620	0.380	0.012	0.198	0.408	0.186	0.197	0.000
Mozambique	0.536	0.464	0.000	0.011	0.000	0.256	0.731	0.002
Niger	0.295	0.705	0.001	0.600	0.150	0.125	0.023	0.101
Nigeria	0.568	0.432	0.019	0.241	0.473	0.032	0.019	0.215
Rwanda	0.721	0.279	0.160	0.216	0.118	0.258	0.058	0.190
Senegal	0.656	0.344	0.000	0.235	0.621	0.069	0.050	0.026
South Africa	0.566	0.434	0.000	0.401	0.214	0.129	0.179	0.077
Sudan	0.191	0.809	0.005	0.267	0.536	0.191	0.000	0.000
Tanzania	0.479	0.521	0.005	0.121	0.518	0.180	0.005	0.172
Tunisia	0.814	0.186	0.037	0.080	0.361	0.370	0.153	0.000
Uganda	0.550	0.450	0.165	0.043	0.057	0.498	0.017	0.219
Burkina Faso	0.000	1.000	0.066	0.254	0.111	0.050	0.060	0.459
Zimbabwe	0.704	0.296	0.098	0.185	0.455	0.071	0.017	0.175
Canada	0.751	0.249	0.001	0.000	0.539	0.112	0.214	0.134
Costa Rica	0.260	0.740	0.047	0.413	0.293	0.025	0.158	0.064
Cuba	0.142	0.858	0.102	0.325	0.087	0.382	0.104	0.000
Dominican Republic	0.114	0.886	0.136	0.413	0.186	0.043	0.139	0.083
El Salvador	0.056	0.944	0.440	0.213	0.118	0.021	0.136	0.071
Guatemala	0.015	0.985	0.039	0.219	0.141	0.007	0.263	0.331
Haiti	0.023	0.977	0.118	0.343	0.045	0.359	0.004	0.131
Honduras	0.078	0.922	0.034	0.281	0.249	0.199	0.017	0.221
Mexico	0.471	0.529	0.000	0.272	0.320	0.169	0.237	0.002
Nicaragua	0.136	0.864	0.009	0.252	0.221	0.178	0.025	0.315
United States	0.844	0.156	0.000	0.105	0.641	0.043	0.069	0.141
Argentina	0.630	0.370	0.021	0.113	0.469	0.297	0.089	0.010
Bolivia	0.286	0.714	0.021	0.140	0.350	0.339	0.029	0.121
Brazil	0.917	0.083	0.143	0.126	0.331	0.138	0.000	0.262
Chile	0.559	0.441	0.013	0.397	0.261	0.155	0.174	0.000
Colombia	0.441	0.559	0.000	0.356	0.353	0.016	0.000	0.275
Ecuador	0.688	0.312	0.160	0.210	0.423	0.154	0.034	0.018
Paraguay	0.821	0.179	0.035	0.141	0.346	0.321	0.000	0.157
Peru	0.447	0.553	0.000	0.197	0.369	0.281	0.153	0.000
Uruguay	0.030	0.970	0.036	0.039	0.172	0.523	0.000	0.231
Venezuela	0.381	0.619	0.107	0.314	0.145	0.200	0.003	0.231
Bangladesh	0.874	0.126	0.550	0.426	0.021	0.003	0.000	0.000
Myanmar	0.867	0.133	0.137	0.311	0.358	0.194	0.000	0.000
Sri Lanka	1.000	0.000	0.257	0.127	0.549	0.067	0.000	0.000

(Continued)

Table 10
(Continued)

Country	Outputs		Inputs					
	Crops	Livestock	Area	Tractors	Labor	Fertilizer	Livestock	Irrigation
China	0.133	0.867	0.022	0.222	0.000	0.017	0.740	0.000
India	0.710	0.290	0.328	0.156	0.445	0.070	0.000	0.001
Indonesia	1.000	0.000	0.021	0.293	0.422	0.006	0.258	0.000
Iran	0.855	0.145	0.084	0.237	0.360	0.221	0.098	0.000
Iraq	0.949	0.051	0.073	0.265	0.347	0.230	0.085	0.000
Israel	0.658	0.342	0.472	0.099	0.173	0.032	0.223	0.000
Japan	0.298	0.702	0.564	0.000	0.019	0.004	0.389	0.023
Cambodia	0.771	0.229	0.036	0.246	0.537	0.181	0.000	0.000
Korea Republic	0.710	0.290	0.629	0.089	0.140	0.023	0.086	0.034
Laos	0.492	0.508	0.102	0.053	0.640	0.204	0.000	0.000
Malaysia	0.818	0.182	0.253	0.118	0.218	0.066	0.302	0.043
Mongolia	0.003	0.997	0.000	0.248	0.127	0.354	0.040	0.231
Nepal	0.580	0.420	0.594	0.221	0.142	0.044	0.000	0.000
Pakistan	0.320	0.680	0.325	0.401	0.227	0.048	0.000	0.000
Philippines	0.767	0.233	0.237	0.328	0.280	0.030	0.124	0.000
Saudi Arabia	0.137	0.863	0.000	0.260	0.092	0.006	0.643	0.000
Syria	0.956	0.044	0.003	0.235	0.351	0.248	0.162	0.001
Thailand	0.940	0.060	0.121	0.073	0.608	0.156	0.041	0.000
Turkey	1.000	0.000	0.104	0.021	0.377	0.306	0.030	0.161
Vietnam	0.543	0.457	0.718	0.182	0.000	0.026	0.002	0.073
Austria	0.866	0.134	0.051	0.036	0.159	0.181	0.414	0.158
Bel-Lux	0.452	0.548	0.078	0.031	0.261	0.030	0.110	0.489
Bulgaria	0.770	0.230	0.036	0.402	0.331	0.166	0.064	0.000
Denmark	0.234	0.766	0.025	0.000	0.508	0.000	0.460	0.008
Finland	0.034	0.966	0.000	0.000	0.017	0.000	0.865	0.118
France	0.928	0.072	0.104	0.030	0.529	0.041	0.173	0.123
Germany	0.204	0.796	0.016	0.000	0.070	0.000	0.832	0.082
Greece	1.000	0.000	0.033	0.025	0.238	0.279	0.237	0.187
Hungary	0.633	0.367	0.181	0.105	0.174	0.104	0.250	0.186
Ireland	0.080	0.920	0.000	0.215	0.057	0.000	0.000	0.728
Italy	0.975	0.025	0.146	0.001	0.209	0.285	0.073	0.285
Netherlands	0.029	0.971	0.277	0.052	0.438	0.103	0.093	0.036
Norway	0.006	0.994	0.005	0.000	0.002	0.000	0.808	0.185
Poland	0.836	0.164	0.028	0.036	0.352	0.008	0.468	0.109
Portugal	0.531	0.469	0.085	0.005	0.128	0.690	0.074	0.018
Romania	0.597	0.403	0.155	0.246	0.195	0.353	0.051	0.000
Spain	0.966	0.034	0.000	0.040	0.261	0.291	0.107	0.301
Sweden	0.162	0.838	0.000	0.000	0.075	0.035	0.750	0.140
Switzerland	0.042	0.958	0.021	0.002	0.120	0.507	0.228	0.123
United Kingdom	0.951	0.049	0.000	0.415	0.258	0.018	0.110	0.200
Australia	0.510	0.490	0.000	0.240	0.586	0.013	0.110	0.051
New Zealand	0.015	0.985	0.013	0.169	0.382	0.097	0.078	0.260
Papua N. Guin.	0.906	0.094	0.297	0.116	0.022	0.015	0.122	0.429
Mean	0.519	0.481	0.108	0.183	0.274	0.154	0.160	0.121

contrast to the work of Fulginiti and Perrin (1997) who report technical regression in a group of 18 developing countries over the period 1961–1985. Another interesting feature is the predominance of efficiency change (or "catch-up") as a source of TFP growth. Both in Asia

and Africa efficiency change is the principal source of TFP growth.

Figure 2 shows cumulative TFP indices from 1980 to 2000 for the different regions. From the figure it is evident that Asia has the highest cumulative growth by

2000, followed by North America and Europe. Asia has a higher cumulative growth than the global growth in TFP. Africa and South America remain as the bottom groups.

Table 8 shows the average annual changes for groups of countries classified by their technical efficiency scores in 1980. The first group, consisting of 39 countries on the frontier in 1980, posted only 1.2% growth in TFP driven by a 1.3% growth in technical change. In contrast, those countries that had an efficiency score between 0.6 and 1, posted a 1.5% growth in TFP mainly driven by 0.3% growth in technical change and 1.2% growth in technical efficiency. However, the bottom group of countries, with a technical efficiency score of less than 0.6, posted an impressive 3.6% growth in TFP mainly driven by 2.5% growth in technical efficiency and 1.5% growth in technical change. These results indicate a degree of catch-up due to improved technical efficiency along with growth in technical change.

While the results in Tables 7 and 8 are very encouraging in terms of the catch-up and convergence shown by many countries, a feature of concern is the low TFP growth experienced by a number of countries in Africa and South America. These are the two continents with the highest population growth during 1980–2000, which suggests that food security will remain an issue on these continents for some time yet.

Figure 3 summarizes our estimated shadow shares obtained from the DEA frontiers used in computing the Malmquist TFP indices. Summary information on these shares is also given in Tables 9 and 10. The top two series in Figure 3 represent the value shares for crops and livestock (both sum to unity) over the study period. These shares appear to be fairly steady over the period, with crops accounting for more than 50% of the total output in most years.

The six series graphed at the bottom of Figure 3 represent the shadow input shares resulting from the application of the DEA methodology. The figure serves to demonstrate the plausibility of the input shares derived here. The average labor share shows a steady decline from 28.5% in 1980 to 24.2% in 2000. The share of land, aggregated over all the countries, seems to be quite stable at around 11%. While the share of tractors remained essentially the same, the shares of fertilizer and livestock have shown small increases.

Table 10 shows the country-specific output and input shares underlying the TFP indices reported here. These shares are averaged over the study period from 1980 to 2000. These shadow shares seem to be quite meaningful. For example, India shows 71% share for crops and 29% for livestock confirming the importance of crops in India. Similarly, in the Netherlands the share of livestock is shown to be 97.1%. Similar livestock shares are shown for Norway (99.4%), Switzerland (95.1%), and Finland (96.6%).

The last six columns of Table 10 show the shares of the six inputs. These shares also appear to be meaningful and consistent with the general factor endowments enjoyed by these countries. For example, the shadow shares of labor are quite high in countries like the United States (64.1%), Canada (53.9%), and Australia (58.6%). Labor shares are also quite high in those countries where labor is abundant and agriculture is very labor intensive. India and Indonesia, respectively, have shadow labor shares of 44.5 and 42.2%, respectively. In countries where land is a limiting factor its shadow share is quite high. For example, in the

Table 11
Mean shadow shares for the continents, 1980–2000

Continent	Outputs		Inputs					
	Crops	Livestock	Area	Tractors	Labor	Fertilizer	Livestock	Irrigation
Africa	0.501	0.499	0.052	0.208	0.294	0.176	0.128	0.143
North America	0.798	0.203	0.001	0.053	0.590	0.078	0.142	0.138
South America	0.342	0.658	0.077	0.251	0.257	0.200	0.082	0.133
Asia	0.669	0.331	0.245	0.200	0.280	0.110	0.140	0.025
Europe	0.515	0.485	0.062	0.082	0.219	0.155	0.308	0.174
Australasia	0.477	0.523	0.103	0.175	0.330	0.042	0.103	0.247
Mean	0.519	0.481	0.108	0.183	0.274	0.154	0.160	0.121

Table 12

Comparison of mean TFP change when average DEA shadow prices used as shares in a Tornqvist index, 1980–2000

	Country	Malmquist	Tornqvist	Difference
5	Chad	0.947	0.984	−0.037
38	Argentina	0.973	1.004	−0.031
18	Rwanda	0.967	0.995	−0.028
53	Indonesia	0.981	1.005	−0.024
24	Uganda	0.977	0.997	−0.020
8	Guinea	0.964	0.983	−0.019
33	Haiti	0.957	0.973	−0.016
93	Papua N. Guin.	0.992	1.007	−0.015
77	Germany	1.013	1.028	−0.015
61	Malaysia	1.004	1.019	−0.015
50	Sri Lanka	1.002	1.017	−0.015
92	New Zealand	1.004	1.019	−0.015
46	Uruguay	1.000	1.015	−0.015
72	Bel-Lux	0.996	1.010	−0.014
44	Paraguay	0.984	0.998	−0.014
13	Mali	0.983	0.997	−0.014
22	Tanzania	1.003	1.017	−0.014
43	Ecuador	1.003	1.016	−0.013
59	Korea Republic	0.995	1.007	−0.012
45	Peru	1.015	1.027	−0.012
23	Tunisia	1.018	1.028	−0.010
6	Egypt	1.012	1.022	−0.010
39	Bolivia	1.011	1.020	−0.009
67	Syria	0.989	0.997	−0.008
56	Israel	1.004	1.011	−0.007
87	Spain	1.010	1.016	−0.006
4	Cameroon	1.009	1.015	−0.006
75	Finland	1.011	1.016	−0.005
65	Philippines	1.008	1.013	−0.005
57	Japan	1.002	1.007	−0.005
40	Brazil	1.020	1.025	−0.005
54	Iran	1.020	1.025	−0.005
83	Norway	0.995	0.999	−0.004
80	Ireland	1.011	1.015	−0.004
41	Chile	1.011	1.014	−0.003
88	Sweden	1.003	1.006	−0.003
26	Zimbabwe	1.008	1.011	−0.003
19	Senegal	1.021	1.024	−0.003
14	Morocco	1.016	1.019	−0.003
81	Italy	1.009	1.011	−0.002
71	Austria	1.014	1.016	−0.002
47	Venezuela	1.006	1.007	−0.001
11	Madagascar	0.998	0.999	−0.001
74	Denmark	1.032	1.033	−0.001
73	Bulgaria	1.020	1.020	0.000
9	Cote d'Ivoire	1.014	1.014	0.000
82	Netherlands	1.022	1.022	0.000
16	Niger	0.998	0.998	0.000
7	Ghana	1.022	1.021	0.001
2	Angola	1.037	1.036	0.001

(Continued)

Table 12

(Continued)

	Country	Malmquist	Tornqvist	Difference
69	Turkey	1.009	1.008	0.001
30	Dominican Republic	1.010	1.009	0.001
90	U.K.	1.014	1.012	0.002
42	Colombia	1.014	1.012	0.002
27	Canada	1.033	1.031	0.002
76	France	1.020	1.018	0.002
28	Costa Rica	1.028	1.025	0.003
35	Mexico	1.015	1.012	0.003
34	Honduras	1.003	1.000	0.003
32	Guatemala	1.005	1.002	0.003
91	Australia	1.026	1.023	0.003
31	El Salvador	1.008	1.004	0.004
79	Hungary	1.003	0.999	0.004
49	Myanmar	1.018	1.013	0.005
63	Nepal	1.010	1.005	0.005
37	U.S.	1.026	1.021	0.005
64	Pakistan	1.023	1.018	0.005
10	Kenya	1.005	1.000	0.005
12	Malawi	1.022	1.017	0.005
68	Thailand	0.995	0.990	0.005
52	India	1.014	1.009	0.005
85	Portugal	1.026	1.021	0.005
36	Nicaragua	1.018	1.012	0.006
25	Burkina Faso	0.997	0.990	0.007
21	Sudan	1.024	1.016	0.008
66	Saudi Arabia	1.042	1.032	0.010
17	Nigeria	1.037	1.027	0.010
78	Greece	1.017	1.007	0.010
15	Mozambique	1.019	1.009	0.010
55	Iraq	0.976	0.965	0.011
89	Switzerland	1.021	1.009	0.012
60	Laos	1.034	1.021	0.013
86	Romania	1.023	1.010	0.013
51	China	1.060	1.047	0.013
84	Poland	1.021	1.007	0.014
48	Bangladesh	1.024	1.009	0.015
20	South Africa	1.037	1.019	0.018
70	Vietnam	1.024	1.003	0.021
1	Algeria	1.046	1.025	0.021
29	Cuba	1.025	1.000	0.025
58	Cambodia	1.057	1.031	0.026
62	Mongolia	1.028	0.997	0.031
3	Burundi	1.046	0.972	0.074
	Mean	1.011	1.011	

Netherlands the land share is 27.7%. In Japan and Israel the land shares are, respectively, 56.4% and 47.2%. These large shares for land also reflect the scarcity of land resulting from increasing urbanization of agricultural land.

Table 13
Comparison of weighted mean TFP change when average DEA shadow prices used as shares in a Tornqvist index for the continents, 1980–2000

Continent*	Countries	Malmquist	Tornqvist	Difference
Africa	1–26	1.013	1.016	0.003
North America	27, 37	1.027	1.022	−0.005
South America	28–36, 38–47	1.006	1.015	0.009
Asia	48–70	1.029	1.025	−0.004
Europe	71–90	1.014	1.015	0.001
Australasia	91–93	1.018	1.021	0.003
Mean	1–93	1.021	1.021	0.000

Shares of other factors, including fertilizers, tractors, livestock, and irrigation are also plausible and appear to support the general scarcities of these resources in different countries. We find that the general trends in these shares over time and differences across countries appear to support the discussion in Ruttan (2002) where various constraints to productivity growth in world agriculture are identified. Table 11 summarizes the shadow share information by continents. The Asian continent has the highest input share associated with land whereas North America and Europe have large shares for labor, livestock, and irrigation inputs.

As one final exercise, we have taken the average shadow share estimates from the bottom of Table 11 and used them as fixed shares in the calculation of the Tornqvist TFP index numbers for each country.[18] These Tornqvist TFP indices are reported in Table 12, along with the original Malmquist TFP indices from the final column of Table 4. The differences between these two columns of indices are reported in the final column of Table 12. This table has been sorted by the size of this difference. The reported differences are quite large in some cases, with 40 countries reporting differences of 1% per annum or more. These differences may be rationalized in two ways. Either the shadow shares for some countries are not well estimated (due to the dimensionality problem in DEA) or the shadow shares are well estimated, but they differ significantly from the sample average, because of country-specific factors, such as land scarcity, labor abundance, etc. For many countries, the observed difference may well be a combination of these two factors, to varying degrees.

Finally, the country-level information in Table 12 is summarized in Table 13 for our six regions.

---

[18] The Tornqvist index is described in chapter 4 in Coelli et al. (1998).

The largest difference occurs for South and Central America, where the average TFP growth measure increases from 0.6% to 1.5% per annum. This is not a minor difference, and emphasizes the key point that TFP indices depend crucially upon the prices that are used—be they market prices or shadow prices.

5. Conclusions

This paper presents some important findings on levels and trends in global agricultural productivity over the past two decades. The results presented here examine the growth in agricultural productivity in 93 countries over the period 1980 to 2000. The results show an annual growth in TFP of 2.1%, with efficiency change (or catch-up) contributing 0.9% per year and technical change (or frontier shift) providing the other 1.2%. There is little evidence of the technological regression discussed in a number of the papers listed in Table 1. This is most likely a consequence of the use of a different sample period and an expanded group of countries. In terms of individual country performance, the most spectacular performance is posted by China with an average annual growth of 6.0% in TFP over the study period. Other countries with strong performance are, among others, Cambodia, Nigeria, and Algeria. The United States has a TFP growth rate of 2.6% whereas India has posted a TFP growth rate of only 1.4%.

Turning to the performance of various regions, Asia is the major performer with an annual TFP growth of 2.9%. Africa seems to be the weakest performer with only 0.6% growth in TFP. Examining the question of catch-up and convergence, we find that those countries that were well below the frontier in 1980 (with technical efficiency coefficients of 0.6 or below) have a TFP growth rate of 3.6%. This is in contrast to a low 1.2% growth for those countries that were at the frontier in 1980. These results indicate a degree of catch-up in productivity levels between high-performing and low-performing countries. These results are of interest since they indicate an encouraging reversal (during 1980–2000 period) in the phenomenon of negative productivity trends and technological regression reported in some of the earlier studies for the period 1961–1985.

Though the results are quite plausible and meaningful, the authors are quite conscious of the data limitations and the need for further work in this area. Future work could include: (i) an examination of the robustness of the results to shifts in the base period for the computation of output aggregates; (ii) the inclusion of pesticides, herbicides, and purchased feed and seed in the input set; (iii) an investigation of the effects of land quality, irrigation, and rainfall; and (iv) utilization of parametric distance functions to study the robustness of the findings to the choice of methodology.

References

Arnade, C., "Using a Programming Approach to Measure International Agricultural Efficiency and Productivity," *Journal of Agricultural Economics* 49 (1998), 67–84.

Ball, V. E., J. C. Bureau, J. P. Butault, and R. Nehring, "Levels of Farm Sector Productivity: An International Comparison," *Journal of Productivity Analysis* 15 (2001), 5–29.

Barro, R., and X. Sala-i-Martin, "Convergence across States and Regions," *Brookings Paper on Economic Activity* (1991), p. 107.

Bureau, C., R. Färe, and S. Grosskopf, "A Comparison of Three Nonparametric Measures of Productivity Growth in European and United States Agriculture," *Journal of Agricultural Economics* 46 (1995), 309–326.

Capalbo, S. M., and J. M. Antle, eds., *Agricultural Productivity: Measurement and Explanation* (Resources for the Future: Washington, DC: 1988).

Chavas, J. P., "An International Analysis of Agricultural Productivity," in L. Zepeda, ed., *Agricultural Investment and Productivity in Developing Countries* (FAO: Rome, 2001).

Coelli, T. J., and D. S. P. Rao, "Implicit Value Shares in Malmquist TFP Index Numbers," *CEPA Working Papers No. 4/2001* (School of Economics, University of New England: Armidale, 2001), pp. 27.

Coelli, T. J., D. S. P. Rao, and G. E. Battese, *An Introduction to Efficiency and Productivity Analysis* (Kluwer Academic Publishers: Boston, 1998).

Craig, B. J., P. G. Pardey, and J. Roseboom, "International Productivity Patterns: Accounting for Input Quality, Infrastructure, and Research," *American Journal of Agricultural Economics* 79 (1997), 1064–1077.

Färe, R., S. Grosskopf, M. Norris, and Z. Zhang, "Productivity Growth, Technical Progress and Efficiency Changes in Industrialised Countries," *American Economic Review* 84 (1994), 66–83.

Fulginiti, L., and R. Perrin, "Prices and Productivity in Agriculture," *Review of Economics and Statistics* 75 (1993), 471–482.

Fulginiti, L. E., and R. K. Perrin, "LDC Agriculture: Nonparametric Malmquist Productivity Indexes," *Journal of Development Economics* 53 (1997), 373–390.

Fulginiti, L. E., and R. K. Perrin, "Agricultural Productivity in Developing Countries," *Journal of Agricultural Economics* 19 (1998), 45–51.

Fulginiti, L. E., and R. K. Perrin, "Have Price Policies Damaged LDC Agricultural Productivity?" *Contemporary Economic Policy* 17 (1999), 469–475.

Grifell-Tatjé, E., and C. A. K. Lovell, "A Note on the Malmquist Productivity Index," *Economics Letters* 47 1995), 169–175.

Hayami, Y., and V. Ruttan, "Agricultural Productivity Differences among Countries," *American Economic Review* 40 (1970), 895–911.

Hayami, Y., and V. Ruttan, *Agricultural Development: An International Perspective* (Johns Hopkins University Press: Baltimore, 1971).

Kawagoe, T., and Y. Hayami, "The Production Structure of World Agriculture: An Intercountry Cross-Section Analysis," *Developing Economies* 21 (1983), 189–206.

Kawagoe, T., and Y. Hayami, "An Intercountry Comparison of Agricultural Production Efficiency," *American Journal of Agricultural Economics* 67 (1985), 87–92.

Kawagoe, T., Y. Hayami, and V. Ruttan, "The Intercountry Agricultural Production Function and Productivity Differences among Countries," *Journal of Development Economics* 19 (1985), 113–132.

Lau, L., and P. Yotopoulos, "The Meta-Production Function Approach to Technological Change in World Agriculture," *Journal of Development Economics* 31 (1989), 241–269.

A., Lusigi, and C. Thirtle, "Total Factor Productivity and the Effects of R&D in African Agriculture," *Journal of International Development* 9 (1997), 529–538.

Maddison, A., *Monitoring the World Economy: 1820–1992* (OECD: Paris, 1995).

Maddison, A., *Chinese Economic Performance in the Long Run* (OECD: Paris, 1997).

Martin, W., and Mitra, D., "Productivity Growth and Convergence in Agriculture and Manufacturing," *Agriculture Policy Research Working Papers No. 2171* (World Bank: Washington, DC, 1999).

Nin, A., C. Arndt, and P. V. Preckel, "Is Agricultural Productivity in Developing Countries Really Shrinking? New Evidence Using a Modified Nonparametric Approach," *Journal of Development Economics* 71 (2003), 395–415.

Rao, D. S. P., *Intercountry Comparisons of Agricultural Output and Productivity* (FAO: Rome, 1993).

Rao, D. S. P., and T. J. Coelli, "Catch-up and Convergence in Global Agricultural Productivity, 1980–1995," *CEPA Working Papers No. 4/98* (Department of Econometrics, University of New England: Armidale, 1998), pp. 25.

Ruttan, V. W., "Productivity Growth in World Agriculture: Sources and Constraints," *Journal of Economic Perspectives* 16 (2002), 161–184.

Suhariyanto, K., and C. Thirtle, "Asian Agricultural Productivity and Convergence," *Journal of Agricultural Economics* 52 (2001), 96–110.

Suhariyanto, K., A. Lusigi, and C. Thirtle, "Productivity Growth and Convergence in Asian and African Agriculture," in P. Lawrence and C. Thirtle, eds., *Asia and Africa in Comparative Economic Perspective* (Palgrave: London, 2001), pp. 258–274.

Trueblood, M. A., and J. Coggins, *Intercountry Agricultural Efficiency and Productivity: A Malmquist Index Approach* (World Bank: Washington, DC, 2003) mimeo.

Wiebe, K., M. Soule, C. Narrod, and V. Breneman, *Resource Quality and Agricultural Productivity: A Multi-Country Comparison* (USDA: Washington, DC, 2000) mimeo.

Is small beautiful? Farm size, productivity, and poverty in Asian agriculture

Shenggen Fan*, Connie Chan-Kang

Abstract

Small farms characterize agriculture in Asia. With the fragmentation of land holdings, the average size of farms fell in the region, while the number of small-size holdings increased significantly. These small-scale farmers play an important role for food security and poverty alleviation. However, whether and how these small farms can survive under globalization is a hotly debated topic. In particular, the traditional claim that "small is beautiful," which is based on empirical observation that small farms present higher land productivity than large farms, is being challenged. It has been shown that a positive relationship also exists between farm size and labor productivity (and therefore income). To help these small farms prosper under increasing globalization, the governments have to change the "business as usual" attitude. Innovative land reform, for example, is crucial to secure property rights to farmers and to increase farm size. Equally important is the reform of public institutions in order to help small farmers to have access to credit, marketing, and technology. Moreover, promoting diversification in the production of high-value commodities can play an important role in raising the small-holders' income. Finally, policies that facilitate urban–rural migration and promote the development of the rural nonfarm sector are essential to help alleviate poverty among small-farm households and among the rural poor in general.

JEL classification: J22, N65, O13, Q15

Keywords: farm size; productivity; poverty; Asia; smallholders

1. Introduction

The debate on the relationship between farm size and productivity in Asia has gone through a complete circle. In the 1960s small farms were regarded as being efficient because they could fully use their resources, particularly family labor, and they could monitor their production activities more closely. In the 1970s and 1980s, however, as many Asian countries moved rapidly toward industrialization and urbanization, small farms were regarded as a major obstacle in this process. On the one hand, industrialization leads to greater demand for labor from rural areas, which is in conflict with labor-intensive small farm practices. On the other hand, by providing cheaper modern inputs such as machinery, industrialization

made an increase in farm size possible by relaxing labor constraints during the peak season. Therefore, there was a call for larger farms in the 1970s and 1980s. In the 1990s, however, "the small is beautiful" view was once again revived. In the past decade, agricultural production has become more diversified into high-value commodities, for example, from grains to cash crops, and from crops to livestock and horticultural products, in which small farms may have comparative advantages. Moreover, large farms and input-intensive practices (fertilizer, pesticides, machinery) have led to the degradation of natural resources and the environment. When these externalities are considered, large farms may no longer be viewed as efficient.

About 55% of the world's population lives in Asia, 58% of which depends on agriculture for a livelihood. However, the Asian region holds only 20% of the world's agricultural land. Moreover, the average size of holdings continues to fall in several countries of the

* International Food Policy Research Institute, Washington, DC, USA.

region with the fragmentation of land holdings. Today, in most countries of Asia, the average land holding ranges from only 1 to 2 hectares, well below the world average of 3.7 hectares per person. At the same time, the number of small-size holdings has increased significantly (Pookpakdi, 1992). As the world has become increasingly globalized, whether these small farms can survive is a hotly debated topic. The objectives of this paper are to review the evidence between farm size and productivity, and between farm size and poverty, and to synthesize whether and how small farms can prosper under increasing trade liberalization. This article focuses on five Asian countries, namely China, India, Thailand, Japan, and South Korea.

The paper is organized in five parts. Section 2 describes the changes in farm size over the past several decades and explores the reasons behind these changes. Section 3 reviews the level and the rate of change in production and productivity. The links between production growth, productivity, and farm size are explored in Section 4, while Section 5 assesses how small farmers can prosper under globalization, and offers policy options on how governments can help small farms avoid the adverse effects of trade liberalization and globalization.

2. Changes in farm size

Farm size varies substantially among countries in Asia, and has changed dramatically over the last several decades. Table 1 presents the average farm size and its change over time in China, India, Thailand, Japan, and South Korea.[1] Small farms characterize the agricultural sector in the five Asian countries. In the early 1990s, the average farm size in Thailand was the largest, at 3.36 hectares, compared with 1.55 hectares in India, and 1.37 and 1.23 hectares in Japan and South Korea, respectively. The average Chinese farm was significantly smaller, averaging 0.43 hectares.

The determination of farm size and its change over time is complex. This includes factors such as history, institutions, economic development, the development of the nonfarm sector (both in rural and urban areas),

---

[1] The size and number of agricultural holdings are typically surveyed in agricultural census. To our knowledge, Thailand's and India's latest agricultural census were conducted in 1993 and 1990/91 respectively.

Table 1
Average farm size

	China	India	Thailand	Japan	Korea
	(Hectares per Farm)				
1950		2.20			
1960		2.70	3.47	1.00	2.06
1970		2.30		1.01	0.88
1977		2.00			
1980	0.56	1.84	3.70	1.17	1.02
1985	0.51	1.69		1.23	1.11
1990	0.43			1.37	1.19
1991		1.55		1.37	1.23
1992				1.38	1.26
1993			3.36	1.39	1.29
1994				1.40	1.30
1995	0.41			1.47	1.32
1996				1.47	1.32
1997				1.48	1.34
1998				1.49	1.35
1999	0.40			1.50	1.37
2000				1.55	1.37
2001				1.56	1.39
2002				1.57	1.46

Sources: China: Statistical Yearbook of China (SSB), various issues. India: 1950, 1960, and 1970 from FAO's supplement to the World Census of Agriculture; 1977 from FAO's World Census of Agriculture; 1980, 1985, and 1991: data downloaded from IndiaStat.
Thailand: All data from FAO's supplement to the World Census of Agriculture.
Japan and Korea: 1960 and 1970 from FAO's supplement to the World Census of Agriculture; 1980–2002 from the Korean Ministry of Agriculture and Forestry.

land and labor markets, and policies related to land tenure and property rights. Among these factors, land policy, institutions, and legislation have been the most influential.

2.1. Japan

Immediately after World War II, drastic agricultural land reform was implemented in Japan. Land reform in Japan demolished a class structure based on landholding. Landlords were no longer supreme and rural society was restructured, so the rural population became supportive of the ruling conservative party. But land reform had little effect on agricultural production. Land ownership was transferred from landlords to tillers of the soil, and small tenant farmers became small owner-cultivators, with no apparent change in farm size. The traditional agricultural production structure from pre-war Japan remained (Kawagoe, 1999).

The year 1961 marked a turning point in Japan's agricultural sector with the passing of the Basic Law of Agriculture, which supported an expansion in the size of farms to balance farmers' incomes with those of city workers (Ukawa, 1995). Since then, several amendments to this law as well as various policies aimed at increasing the scale of farms have been passed, which promoted an increase in the size of farms (Kajii et al., 1988). The major reason for the slow increase in Japanese farm size, however, is the heavy subsidy on agricultural production, which artificially raises farmers' incomes; the high proportion of farmer's incomes, which comes from the nonfarm sector (82% in 2001, MAFF); and the extremely high land prices.

2.2. South Korea

In contrast to Japan, the average size of a farm in South Korea declined drastically after World War II, from 2.06 hectares in 1950 to 0.88 hectares in 1960. This significant drop was a result of the Land Reform Program, which marked the beginning of the modern agricultural system in South Korea. Prior to 1950, land was highly concentrated: a small number of landlords owned most South Korean farmland. Most agricultural workers were tenants and paid rent to the landlords, sometimes in excess of 50% of gross revenue. With the Land Reform Program, the government established a maximum farm size of 3 hectares and bought any farmland in excess of that limit. In addition, the government procured farmland owned by nonfarmers. The purchased land was redistributed to small or landless farmers (Kim, 1992). Since the 1970s, the total area of farmland has been declining in South Korea as a result of the increased demand for land by the non-agricultural sector. However, the decline in the number of farm households over the same period was greater (Kim, 1992). As a result, the average area per farm household increased marginally, from 0.93 hectare in 1970 to 1.46 hectares in 2002.

2.3. Thailand

The average farm size in Thailand has remained constant—at around 3.4 hectares—for the past several decades.[2] Historically, the King owned all the land in Thailand. The concept of individual land ownership was introduced by the King Chulalongkorn in 1874, and by 1901 formal land title could be obtained. In 1975, the Parliament passed the Agricultural Land Reform Act, which aimed at distributing land owned by private owners to landless and tenant farmers. However, the implementation of the reform was constrained by the military coups in 1976 and 1977. Although over 70 areas of the country were elected as Land Reform Areas by 1979, the Agricultural Land Reform Office was unsuccessful in acquiring land for redistribution as large landholders, wealthy aristocrats, businessmen, and senior military officers opposed the designation of the Land Reform Areas. Today, many landowners still hold legal title to their land, and many farmers still lack legal title of ownership (Cabrera 2002).

2.4. India

In contrast to the trends in Japan and South Korea, the average farm size decreased in India and China. Table 1 shows that India's average farm size declined from 2.20 hectares in 1950 to 1.55 hectares in 1991. But this average masks a large variation among different subgroups and a dramatic change in farm size structure over time (Table 2). The total number of rural households almost doubled in India from 63.5 million in 1953–1954 to 116.4 million in 1991–1992. Despite this huge increase, the number of landless households remained constant at about 14 million between 1953–1954 and 1991–1992. However, the share of rural households that was landless declined significantly over time, from 23% in the mid 1950s to 11% in the early 1990s (Table 2). On the other hand, the number of landed households (owning greater than 0.01 acres) increased dramatically (from 48.8 million in 1953–1954 to 103.3 million in 1991–1992), showing that despite rapid population growth, most rural households have been able to acquire at least a small amount of land. Of the landed households, the vast majority owns 5 acres or less, and most of the increase in farm numbers has occurred among the submarginal

---

[2] This review of Thailand's land policy and institutions draws heavily on information obtained from http://www.1upinfo.com/country-guide-study/thailand/thailand94.html.

Table 2
Size distribution of ownership holdings, India

Size Category of Holding (acres)	Number and Percentage[a] of Total Ownership Holdings (millions)				
	Year				
	1953/54	1961/62	1971/72	1981/82	1991/92
Landless (<0.01)	14.67	8.47	7.56	10.64	13.09
	(23.1)	(11.7)	(9.6)	(11.3)	(11.3)
Submarginal (0.01–0.99)	15.36	23.58	27.61	34.61	46.69
	(24.2)	(32.6)	(35.2)	(36.9)	(40.1)
Marginal (1.0–2.49)	8.88	11.48	13.91	17.30	23.88
	(14.0)	(15.9)	(17.7)	(18.4)	(20.5)
Small (2.5–4.99)	8.57	10.92	12.14	13.96	15.62
	(13.5)	(15.1)	(15.5)	(14.7)	(13.4)
Medium (5.0–14.99)	11.15	13.00	12.98	13.78	14.09
	(17.5)	(18.0)	(16.6)	(14.7)	(12.1)
Large (>15.0)	4.91	4.95	4.18	3.73	3.04
	(7.7)	(6.8)	(5.3)	(4.0)	(2.6)
Total	63.53	72.47	78.37	93.86	116.41
	(100)	(100)	(100)	(100)	(100)

Note: [a]Percentages are in parentheses.
Source: Thorat et al. (2003).

and marginal holdings. Taken together, these three categories accounted for 74% of all landed holdings in 1991–1992, up from 52% in 1953–1954. At the same time the number of large farms (15 acres and above) has fallen, from 4.9 million in 1953–1954 to 3 million in 1991–1992. Many of these changes appeared to have occurred most rapidly after 1981–1982 (Table 2).

The decline in the number of large farms suggests that much of the growth in the number of smaller farms may have resulted from the subdivision of large farms. Thorat et al. (2003) found that the share of the total land area owned by large farms declined from 52.5% in 1953–1954 to 26.7% in 1991–1992, while the average size of large farms dropped from 32.7 acres to 25.5 acres over that same period. Despite the reallocation of land from large to smaller farms, the distribution of owned land has barely improved when measured by the Gini coefficient, which fell from 0.75 in 1953 to 0.71 in 1991, with all the change occurring between 1953 and 1960 (Thorat et al., 2003).

2.5. China

China has experienced three major land policy and institutional reforms since the establishment of the People's Republic in 1949. The first land reform, which

was characterized by the confiscation of land from landlords, and redistribution to landless poor farmers, was completed in 1953. This was soon followed by a second land reform, in which government policies promoted the development of large, collective operations. Consequently by 1956 most of China's agricultural production was done on a collective basis. Under this system, land ownership was vested in a collective that usually consisted of around 200 families. Beginning in 1958, the central government promoted an even larger scale of production in agriculture. Advanced cooperatives were merged into communes. At the height of the Commune movement in 1958–1959, the *average* communal unit had grown to 5,000 households covering 10,000 acres. Virtually all production means other than agricultural labor were owned by communes. Failures in the commune system, along with a great natural disaster, which lasted for 3 years (1959–1961), led the government to implement an adjustment and consolidation policy after 1961. Production was decentralized into a smaller production unit called a production team—a subunit of the commune consisting of only 20–30 neighboring families. Under this form of organization, a farmer's income was not closely related to production effort, and virtually all input and output markets were controlled by the government. Moreover, market transactions of major agricultural products outside of the procurement system were restricted. Market exchanges of land between different production units in the collective system were also outlawed. This system characterized Chinese agriculture throughout the 1960s and most of the 1970s.

In the late 1970s, the third land reform reestablished family farming in China agricultural sector after years of collectivized agriculture (Chen et al., 1998). By 1984, more than 99% of production units had adopted the Household Production Responsibility System (HRS). Under the HRS, farmers had freedom of decision making on major production and marketing activities, but were not given ownership of the land allocated to them. Instead, they were granted user rights. In theory, the collectives or farmers in the same village jointly owned the land. Although the HRS contributed to rapid growth in agricultural production during the initial stage of reforms, land was fragmented due to equal distribution of land to households on an egalitarian basis (i.e., based on household size and demographic composition).

3. Agricultural output and productivity growth

Output and productivity growth are typically used to evaluate the performance of the agricultural sector. Table 3 shows that output has grown the fastest in China (4.57% per year), followed by South Korea (3.39% per year), Thailand (3.25% per year), and India (2.93% per year) since 1961. Japan fared the worst, with agricultural output growth of only 0.58% per year over this period. There was a marked acceleration in the growth rate of agricultural output in India in the 1970s and 1980s (the so-called Green Revolution period) relative

Table 3
Production growth in agriculture

	India (%)	Japan (%)	South Korea (%)	Thailand (%)	China (%)	World (%)
1961–1969	1.23	3.23	5.3	4.01	4.81	2.75
1970–1979	2.58	1.36	6.68	4.96	3.01	2.49
1980–1989	3.77	0.88	3.27	2.41	5.31	2.4
1990–2002	2.56	−0.88	2.47	2.04	5.43	2.23
1961–2002	2.93	0.58	3.39	3.25	4.57	2.32

Note: Growth rates are exponential growth rates.
Source: FAOSTAT (2003).

Table 4
Labor productivity

	China	India	South Korea	Japan	Thailand
	(1995 US$ per person)				
1961	106.1	257.3	2,311.1	8,068.4	377.0
1970	155.9	283.0	2,953.6	10,527.5	509.8
1980	156.9	276.5	3,270.5	17,219.8	620.7
1990	236.0	352.9	7,399.2	26,664.8	773.6
2000	333.2	394.4	13,508.8	30,038.3	934.3
Growth rates (%)					
1961–1969	5.70	0.23	3.73	4.62	3.24
1970–1979	0.17	−0.01	3.76	5.29	2.58
1980–1989	4.09	1.92	8.39	4.30	2.50
1990–2000	3.79	1.59	6.15	1.98	1.61
1961–2000	2.64	1.31	4.72	3.80	2.38

Note: Labor productivity is defined as the ratio of agricultural GDP to economically active population in agriculture.
Source: Calculated from FAOSTAT (2003) and WDI (2002).

to the 1960s, followed by a slowdown during the 1990s (the reform period). On the other hand, output growth decelerated continuously in Japan from 3.23% per year in the 1960s to 1.36% in the 1970s, 0.88% in the 1980s, and continued its slide to a negative growth of −0.88% in the 1990s. For South Korea, agricultural output grew rapidly during the 1960s (5.3% per year) and 1970s (6.68% per year) but experienced a slowdown in the 1980s (3.27% per year) and the 1990s (2.47% per year). Similar to South Korea, output growth accelerated in Thailand from the 1960s (4.01% per year) to the 1970s (4.96% per year) but slowed down in the 1980s (2.41% per year) and 1990s (2.04% per year).

In terms of labor productivity (Table 4), taken here to be the value of aggregate agricultural GDP (measured in 1995 US$) per economically active agricultural population, Japan had the highest level of labor productivity, producing more than US$30,000 (in constant 1995 US$) per person in 2000, a level that was 2.2 times higher than South Korea, 32 times higher than Thailand, 76 times higher than India, and 90 times higher than China.[3]

---

[3] Higher labour productivity in Japan can be attributed to several factors: higher mechanization, higher government financial support, and greater protection on agriculture by isolating domestic from international markets. The latter represents a welfare transfer from consumers to producers. Since agricultural GDP is measured using the domestic prices, labor productivity may be lower when international market prices are used.

There are also marked differences in the pattern of labor productivity growth among these countries. Between 1961 and 2000, South Korea and Japan experienced the fastest growth in labor productivity with 4.72% and 3.80% per year respectively; the comparable figures for China, Thailand, and India were 2.64%, 2.38, and 1.31%. Labor productivity growth deteriorated continuously in Thailand from an annual average of 3.24% in the 1960s to 2.58% in the 1970s, 2.50% in the 1980s, and 1.61% in the 1990s. After three decades of high growth, the rate of labor productivity growth dropped by half in Japan in the 1990s. In South Korea labor productivity grew at impressive rates of nearly 4% per year in the 1960s and 1970s, 8.4% in the 1980s and 6.15% in the 1990s. Finally, India and China encountered a slowdown in labor productivity growth during the 1970s and the 1990s.

The growth performance in land productivity provides a different picture (Table 5). Over the whole period of study (1961–2000) land productivity grew the fastest in China, at 2.89% per year, followed closely by South Korea, India, and Thailand, while Japan experienced a dismal rate of growth of 0.33% per year. China, India, and Thailand shared a similar growth pattern: land productivity growth slowed down in the 1970s but accelerated in the 1980s and 1990s. After an increase in land productivity growth in the 1970s, there was a decline in the rate of growth in South Korea in the 1980s and 1990s. In contrast, Japan showed an erratic growth pattern: land productivity growth declined in the 1970s, increased in the 1980s, and dropped significantly in the 1990s.

Growth in total factor productivity is a ratio of total output growth to total input growth. It is a better measure of efficiency improvements, because growth in total output and partial factor productivities can simply be achieved by using more inputs. Total factor productivity (TFP) for India grew at an average annual rate of 1.75% between 1970 and 1995 (Fan et al., 1999). In the 1970s, total factor productivity did not improve, but it grew rapidly in the 1980s, at 2.52% per annum. Since 1990, TFP growth in Indian agriculture has continued to grow, but at a slower rate of 2.29% per annum. Using district level data from India for 1970–1994, Fan and Hazell (2000) compared TFP growth between irrigated and high and low potential rainfed areas. They found that TFP grew fastest in high-potential rainfed areas during 1970–94 (3.1% per year), followed by irrigated areas (2.21% per year) and the low-potential rainfed areas (1.58% per year). TFP growth has slowed in irrigated areas since 1990, remained unchanged at nearly 4% per year in high-potential rainfed areas, and accelerated to 3.06% per year in low-potential rainfed areas.

In Thailand TFP grew at an average rate of 1.27% per year during 1971–1981, but dropped by nearly 50% in the subsequent period covering 1981 to 1995 (Mundlak et al., 2002). For the whole period of study (1971–1995), TFP grew at an average rate of 1.08% per year. The major sources of TFP growth has been improved varieties of crops and changes in output composition.

TFP grew at an average rate of 1.09% in South Korea over the period 1918–2000. The pattern of TFP growth varies markedly over time. From 1920 to 1960, TFP grew at a dismal rate of 0.09% per year (Park, 2003). Various factors contributed to this low productivity performance, including the colonization of South Korea by Japan (1918–1938), the Second World War, and the Korean War (Sharma, 1991). From 1960 to 1998, South Korean TFP grew rapidly, at an average annual rate of 2.12% per year. Park estimated that productivity growth contributed 52% of South Korean agricultural output growth over the 1918–2000 period.

Productivity analyses in the Chinese agricultural sector typically distinguishes between the pre- and the post-reform period. Beginning with the First Five Year Plan Period (1953–1957), during which large-scale

Table 5
Land productivity

	China	India	South Korea	Japan	Thailand
	(1995 US$ per hectare)				
1961	89.9	226.3	5,695.2	16,367.1	377.0
1970	139.3	280.1	7,123.5	16,446.8	512.2
1980	147.2	320.0	8,393.9	17,810.8	565.7
1990	219.0	447.6	12,071.6	21,868.6	727.3
2000	317.8	575.8	16,336.5	16,651.9	1,048.7
Growth rates (%)					
1961–1969	6.36	1.54	2.84	1.92	3.40
1970–1979	0.75	1.49	4.39	1.20	1.27
1980–1989	3.64	2.84	3.93	2.03	2.68
1990–2000	4.05	3.04	3.23	− 2.33	3.75
1961–2000	2.89	2.63	2.81	0.33	2.46

Notes: Land productivity is defined as the ratio of agricultural GDP to agricultural land.
Source: Calculated from FAOSTAT (2003) and WDI (2002).

land reform became one of the priorities of the Communist government, TFP in Chinese agriculture increased steadily, as a result of institutional and technological changes (Fan and Zhang, 2002). TFP declined by 13% per year during the Great Leap Forward (1958–1960), following the establishment of the commune system. From 1961 to 1965 (Adjustment Period), Chinese agriculture recovered through a series of adjustments made by the government, and TFP grew by an annual rate of 4.7%. However, with the Cultural Revolution (1966–1976), production was centrally controlled by the government and executed by production teams. As a result of low incentives, agricultural production was inefficient and there was almost no gain in TFP. With the First Phase of Reform (1979–1984), which was characterized by the decentralization of the agricultural production system, TFP grew by 6.2% per year. Since 1984, agricultural prices and marketing systems have been reformed, and consequently TFP has continued to increase, at an annual growth rate of 2.2%.

Over the last several decades there has been a declining trend in the growth rate of TFP in Japanese agriculture. The average annual growth rate was 2.82% from 1960 to 1968 but dropped to 1.11% in the 1969–1990 period. Kuroda (1997) attributed the sluggish growth rate in TFP after the late 1960s to a slowdown in technological progress. As the average size of farms increased in Japan and South Korea, both land and labor productivities have increased, with a much faster growth in labor than land productivity. On the other hand, in China, India, and Thailand, as farm sizes declined due to the increased rural population, land productivity has increased much faster than labor productivity. In the case of China and India, TFP continued to increase. This all suggests that there might be an inverse relationship between farm size and land productivity and TFP, but a positive relationship between farm size and labor productivity.

4. Farm size and productivity: a literature review

A popular stylized fact in development economics is that there is a strong inverse relationship between farm size and land productivity. Sen, in a seminal paper published in 1962, observed that small farmers were more productive per unit of land than large farmers. The inverse relationship is typically explained by the

difference in factor endowments between small and large farms: by using family labor small farms face lower labor transaction costs than larger farms (Raghbendra et al., 2000, Berry and Cline, 1979, Bhalla, 1979). As a result, smaller farms have higher labor/land ratios and can achieve higher yield per hectare (Feder, 1985). The inverse relationship has important implications for land policy, as it entails that any type of land reform that reduces the inequality in landholdings will have a positive effect on productivity (Lipton, 1993, Singh et al., 2002).

A significant volume of literature has been produced on the inverse relationship since Sen's paper, although no consensus on the inverse relationship has been reached. On the one hand, a body of literature supports the hypothesis that small farms produce more per unit of land than large farms (Heltberg, 1996, 1998). With the advent of the Green Revolution, however, research has also shown that the relationship diminishes or is even reversed as agriculture becomes more capital intensive. As a large proportion of the literature has focused on the relationship in India, the empirical evidence from that country is reviewed below, and factors that may explain the lack of consensus on this debate are discussed.

The literature that appears to clearly validate Sen's findings includes Mazumdar (1965), Bharadwaj (1974), Chaddha (1978), Ghose (1979), Bhalla (1979), and Carter (1984). Several explanatory factors on the inverse relationship have been advanced. Some stress that differences in the intensity of land use across farms of different sizes influence land productivity. A typical example is the study by Cornia (1985). Cornia analyzed the relationship between factor inputs, yields, and labor productivity for farms of different sizes in 15 developing countries. In all but three countries (Peru, Bangladesh, and Thailand), a negative relationship was established between farm size and land productivity. Cornia attributed the higher yields observed on small farms to greater application of inputs and to a more intensive use of land. Similarly, Banerjee (1985) observed that smaller farms in the district of Nadia in West Bengal use their land and fertilizer inputs more intensely than the larger farms. Banerjee took the analysis a step further and showed that the cost per unit of output is directly related with the size of holdings, but inversely related with the value of output. This finding implies that small farms are using their variable resources more efficiently than the bigger farms, yielding higher output per hectare.

Environmental factors also appear to affect the farm size–productivity relationship. Tadesse and Krishnamoorthy (1997) examined the level of technical efficiency across agro-ecological zones and farm size in Tamil Nadu in 1992–1993. The authors found significant differences in the level of technical efficiency among paddy farms across agro-ecological zones as well as size groups. Small and medium-sized farms achieved a higher level of technical efficiency than large holdings. Moreover, the analysis revealed that small and medium-sized paddy farms located, respectively, in the agro-climatic zones of the southern and northeastern part of the state were operating at a higher level of technical efficiency than all other farms.

With the transformation of agriculture toward a science-based approach, family labor becomes less important in shaping land productivity, while other inputs such as fertilizer play a greater role. Following the Green Revolution, farms relied increasingly on purchased inputs and capital. As these inputs require cash and credit flow, they are clearly more accessible to farmers with large landholdings. Opponents of the inverse relationship hypothesis argue that the earlier adoption of new technology by large farmers has reduced or even reversed the yield advantage of small farmers. Deolalikar (1981), for example, found that the inverse size–productivity relationship cannot be rejected at low levels of agricultural technology in India, but can be rejected at higher levels. This finding suggests that although the inverse relationship remains valid for traditional agriculture, it cannot be assumed to exist in an agriculture experiencing technical change. Hanumantha (1975) and Subbarao (1982) also found a positive relationship between farm size and productivity, and attributed this positive association to higher application of fertilizer and other cash-intensive inputs on large farms.

Bhalla and Roy (1988) suggested that the inverse relationship might be a result of differing land fertility between small and large farms. Using a comprehensive dataset with observations including 21,500 farm households from different parts of India, Bhalla and Roy confirmed the inverse relationship at the state level. However, when incorporating soil quality variables and running the regressions on more disaggregated geographical units, the authors observed that the inverse relationship weakened, and in many cases disappeared. The authors concluded that the stylized fact of a negative relationship between farm size and farm productivity might in large part be due to the omission of soil quality variables from the estimated equations. Similarly, Carter (1984) found that intra-village soil quality differences and other farm assets explain part of the size–productivity relationship in Haryana. On the other hand, Newell et al. (1997) argued that in Gujarat the inverse relationship is an interregional phenomenon: farms tend to be smaller in fertile regions and larger in less fertile regions. The authors also observed that labor per hectare is higher on small farms. Thus the inverse relationship between output per hectare and farm size is explained by regional variations in fertility and labor supply.

Managerial factors also appear to have an influence on the inverse relationship. The econometric results of Rao and Chotigeat (1981) showed that when hired labor is employed in preference to family labor and nontraditional capital is used as opposed to traditional capital, large-sized holdings are positively related with higher productivity.

Farm size is typically defined in terms of the physical size of the operational holdings. According to Sampath (1992), this conventional way of defining farm size leads to biased estimates and misleading inferences as it fails to discriminate between irrigation and nonirrigated areas. By adding the two types of land to define total area, one implicitly assumes that one unit of irrigated land has the same cropping intensity potential as one unit of nonirrigated land. Sampath demonstrated that using the conventional definition of farm size in the regression equation leads to biased estimates, resulting in the misleading inference that there are diseconomies of scale in the use of land. In contrast, using the same dataset, the author found that there are no diseconomies of scale in land use when the difference between irrigated and nonirrigated land is recognized in the econometric specification.

5. How small farms can prosper under trade liberalization

As demonstrated above, the relationship between farm size and labor productivity is not clearly established. As shown both empirically and theoretically, increased labor productivity is essential to raise a farmer's income.

Labor productivity in agriculture can be decomposed into two components, i.e., land productivity and the land-to-labor ratio:

$$Y/L = Y/A \times A/L,$$

where Y is output, L is the number of agricultural workers, and A is the total land area available for agricultural production. To increase labor productivity, and therefore farmers' income, either land productivity has to increase or the land-to-labor ratio has to improve.

This is a daunting task, as the number of small farms is still large and continues to increase over time. For example, the proportion of farms below 1 hectare rose from 45% in India in 1971–1972 to 51.4% in 1991–1992, while the share of the larger farms decreased from 5.3% to 2.6% over the same period (Table 2). Farms of less than 2 hectares constituted 78% of the total number of farms in India but contributed 41% of the national grain production. A large percentage of these small farmers are poor (Fan et al., 1999), and they constitute more than half of the nation's poor in India (Singh et al., 2002). The Chinese agricultural sector is also still dominated by very small farms with less than 1 hectare of land: 83% of Chinese farms were less than 0.6 hectares in 1997, while only 0.24% of farms were bigger than 6.6 hectares. To get the small farmers out of the "poor but efficient" trap, appropriate polices should be carefully designed.

5.1. Reforming land polices and institutions

The success of land reforms in Asia has been mixed. India's land reforms, implemented after the country's independence, consisted of introducing new regulations that place a ceiling on agricultural land, restrict leasing, regulate rents, and provide security of tenure. However, implementation has varied by region. At present, more than 10% of agricultural workers still do not have land, and for landed farmers, a large proportion have extremely small farms, not sufficient to support the farmer's family. The radical approach to land reform used by China and South Korea in the 1950s is no longer politically feasible in India. It is imperative for the government to design and implement an innovative approach to promote more efficient and equitable land policies and institutions.

On the other hand, the Chinese land tenure system provided equal access to land, which has prevented an increase in the number of rural landless poor, and also increased the efficiency of production during the initial stage of the reform in the 1980s. Unfortunately, the current system does not provide permanent ownership for farmers. The lack of property rights has hindered further development of agriculture and rural areas. For example, farmers cannot use land as collateral for accessing credit. It has also constrained an increase in farm size because farmers only have the right to use—but not to own—their land. The newly proposed Land Lease Law will allow farmers to sell and buy land use rights at market prices without interference from local villages and governments, but the legal framework to implement this policy has yet to be established. On the other hand, this new law may also increase the concentration of land ownership, and lead to large numbers of landless farmers or migrants in the urban centers. Without a sound social safety net, these landless people will fall into the poverty trap.

5.2. Reforming public institutions to serve small farms

In many Asian countries, small farms are still ignored by the government. For example, due to their limited land, it is difficult for small farms to access credit, marketing, and technology services. Given that investment in agricultural research has in the past also been biased against small farms, it is not surprising that larger farms adopted green revolution technology first. Small farms are often losers in the initial adoption stage of a new technology, since prices of agricultural products are pushed down as a result of the greater supply of products from large farms. To avoid losses for small farms in this initial stage, government may have to consider using limited subsidies on credit, inputs and new technologies, targeted to small farms. Once the new technology has been adopted, the subsidies should be removed gradually to avoid future efficiency losses.

5.3. Development of high-value commodities

Given the small size of farm holdings and limited labor movement out of the agricultural sector, land productivity must increase to increase labor productivity and farmers' income. However, due to natural constraints, the potential to increase yields in traditional crops such as rice, wheat, maize, cotton, and rapeseeds is limited. In addition, international prices of these products are low due to oversupply and heavy protection in the developed countries. A possible solution is for farmers to diversify their farming activities and to engage in the production of high-value commodities. The government has much to do to facilitate this process.

First, the government must change its funding priorities in agricultural research from traditional crops such as rice and wheat to cash crops, livestock, and post-harvest technologies. It is still the case that many developing countries spend more than half of their agricultural research expenditures on staple crops (Fan et al., 2004).

Second, the government should gear up its public investments, or should design public policy to attract private investment in transportation, retail chain stores, processing, and storage. At present, governments spend a great proportion of public investments in traditional activities such as irrigation and large crop-extension programs.

5.4. Migration

Lack of economic and employment opportunities leads to migration, either to urban areas or rural areas in other parts of the country. During period of economic booms in Japan, South Korea, and lately China and Thailand, rural–urban migration not only improved the well-being of the migrants, but also improved the land-to-labor ratio in the agricultural sector, enabling nonmigrants to raise their labor productivity and income. For example, the increase in land-to-labor ratios in Japan and South Korea, and more recently in China, was the result of the net flow of rural labor to the urban and rural nonfarm sectors. However, today there are still formal and informal institutions and policy barriers to restrict these movements. Lack of education and access to information and infrastructure is the most critical constraint. In China, many jobs in the urban areas still require urban residence and farmers are not eligible for these jobs. Even if farmers are employed, their rights are usually not protected. In addition, social services such as health care, education

of children, retirement, and unemployment benefits, to which the urban residents are entitled, are often not available to migrant farmers. All these restrictions and barriers should be removed to make large-scale migration possible.

5.5. Development of the rural nonfarm sector

Expanding off-farm employment is important for poverty alleviation. Hazell and Haggblade (1993) showed that the share of household nonagricultural income is inversely related to farm size, with landless and near-landless workers deriving between a third and two thirds of their income from off-farm sources. In India, Dev (1986) indicated that the bulk of the poor are landless or live on small farms with inadequate land to meet their own food needs. Consequently, they depend heavily on earnings from supplying unskilled wage labor to other farms or to nonfarm enterprises. Moreover, public investment in physical infrastructure (road, transportation, communication) as well as in education and health is crucial for the small farms to establish their own businesses and to access nonfarm jobs in the rural nonfarm sector.

6. Conclusion

Poverty remains essentially a rural phenomenon in Asia and most of the rural poor depend on farming for their livelihood. Agricultural production typically takes place on small holdings in the region. Moreover, the number of small farms has been increasing over time due to land fragmentation. Therefore, small-scale agriculture plays an important role for food security and poverty alleviation.

It has been argued that an inverse relationship exists between farm size and productivity. The validity of this claim and the factors causing it have been thoroughly researched. However, the empirical literature has failed to reach a consensus. The relationship between farm size and productivity appears to depend on a number factors including the difference in the intensity of land use, land fertility, and managerial factors. The viability of small-farm production is now being questioned with the ongoing process of trade liberalization, which places small farms in a disadvantaged position.

A number of policy options have been proposed to help small-scale farmers who face increasing globalization. Reforming land policies, for example, is crucial to secure property rights to farmers and to increase farm size. Equally important is the reform of public institutions in order to help small farmers have access to credit, marketing, and technology. Moreover, promoting diversification toward the production of high-value commodities can play an important role in raising smallholders' income. Finally, policies that facilitate urban–rural migration and that promote the development of the rural nonfarm sector are essential to help alleviating poverty among small-farm households and among the rural poor in general.

References

Banerjee, B. N., "Concepts of Farm Size, Resource-Use and Productivity in Agriculture—A Case Study," *Economic Affairs* 30, no. 1 (1985), 17–22.

Berry, R. A., and W. R. Cline, *Agrarian Structure and Productivity in Developing Countries* (Johns Hopkins University Press: Baltimore, 1979).

Bhalla, S., "Farm Size and Productivity and Technical Change in Indian Agriculture," in R. A. Berry, and W. R. Cline, eds., *Agrarian Structure and Productivity in Developing Countries* (John Hopkins University Press: Baltimore, 1979), pp. 141–193.

Bhalla, S. S., and P. Roy, "Mis-specification in Farm Productivity Analysis: The Role of Land Quality," *Oxford Economic Papers* 40, no. 1 (1988), 55–73.

Bharadwaj, K., "Notes on Farm Size and Productivity," *Economic and Political Weekly* 9, no. 13 (1974), A11–A24.

Cabrera, J., *Bumpy Ride for Land Reform*, Economic News and Article 64 (Chulalongkorn University, Faculty of Economics, 2002). http://www.geocities.com/econ_10330/articles.html

Carter, M. R., "Identification of the Inverse Relationship between Farm Size and Productivity: An Empirical Analysis of Peasant Agricultural Production," *Oxford Economic Papers* 36 (1984), 131–145.

Chaddha, A. N., "Farm Size and Productivity Revisited: Some Notes from Recent Experience of Punjab," *Economic and Political Weekly* 13, no. 39 (1978), A82–A96.

Chattopadhyay, M., and A. Sengupta, "Farm Size and Productivity: A New Look at the Old Debate," *Economic and Political Weekly* 32, no. 52 (1997), A172–A175.

Chen, F., and J. Davis, "Land Reform in Rural China since the Mid-1980s," *Land Reform* 2 (1998), 122–137.

Cornia, G. A., "Farm Size, Land Yields and the Agricultural Production Function: An Analysis for Fifteen Developing Countries," *World Development* 13, no. 4 (1985), 513–534.

Deolalikar, A. B., "The Inverse Relationship between Productivity and Farm Size: A Test Using Regional Data from India," *American Journal of Agricultural Economics* 63 (1981), 275–279.

Dev, S. M., "Growth in Labour Productivity in Indian Agriculture: Regional Dimensions," *Economic and Political Weekly* 21, no. 25–26 (1986), A65–A74.

Fan, S., and P. Hazell, "Should Developing Countries Invest More in Less-Favoured Areas? An Empirical Analysis of Rural India," *Economic and Political Weekly* 35, no. 17 (2000), 1455–1464.

Fan, S., P. Hazell, and S. Thorat, *Government Spending, Agricultural Growth and Poverty: An Analysis of Interlinkages in Rural India*, IFPRI Research Report 110 (International Food Policy Research Institute, 1999).

Fan, S., K. Qian, and X. Zhang, "Agricultural R&D Policy in China: An Unfinished Reform Agenda," in P. G. Pardey, J. M. Alston, and R. R. Piggott, eds., *Agricultural R&D Policy in the Developing World* (forthcoming).

Fan, S., and X. Zhang, "Production and Productivity Growth in Chinese Agriculture: New National and Regional Measures," *Economic Development and Cultural Change* 50, no. 4 (July 2002), 819–838.

Fan, S., L. Zhang, and X. Zhang, *Growth and Poverty in Rural China: The Role of Public Investments*, Research Report 125 (International Food Policy Research Institute: Washington, DC, 2002).

FAO, *Report on the 1990 World Census of Agriculture*, FAO Statistical Development Series 9 (FAO: Rome, 1997).

FAO, *Supplement to the Report on the 1990 World Census of Agriculture*, FAO Statistical Development Series 9A (FAO: Rome, 2001).

FAO, "FAOSTAT. Food and Agricultural Organization of the United Nations," Available at http://faostat.fao.org/default.htm. Accessed 10 July 2003.

Feder, G., "The Relation between Farm Size and Farm Productivity: The Role of Family Labour, Supervision and Credit Constraints," *Journal of Development Economics* 18, no. 2–3 (1985), 297–313.

Ghose, A. K., "Farm Size and Land Productivity in Indian Agriculture: A Reappraisal," *The Journal of Development Studies* 16 (1979), 27–49.

Hanumantha, R. C. H., *Technological Change and Distribution of Gains in Indian Agriculture* (MacMillan: New Delhi, 1975).

Hazell, P., and S. Haggblade, "Farm–Non Farm Growth Linkages and the Welfare of the Poor," in M. Lipton and J. van der Gaag, eds., *Including the Poor* (The World Bank: Washington, DC, 1993).

Heltberg, R., *How Rural Market Imperfections Shape the Relation between Farm Size and Productivity—A General Framework and an Application to Pakistani Data* (1996).

Heltberg, R., "Rural Market Imperfections and the Farm Size–Productivity Relationship: Evidence from Pakistan," *World Development* 26, no. 10 (1998), 1807–1826.

Indiastat, "Datanet India Private Limited," http://www.indiastat.com (accessed 2002) (2002).

Kajii, I., S. Usami, and S. Nakayasu, *Changes in Japan's Agrarian Structure* (Food and Agriculture Policy Research Center: Tokyo, 1998).

Kawagoe, T., *Agricultural Land Reform in Postwar Japan: Experiences and Issues*, Policy Working Paper 2111 (World Bank: Washington, DC, 1999).

Kim, S. H., *Farm Size and Structural Reform of Agriculture: Korea*, Extension Bulletin April 1992 (Food and Fertilizer Technology Center: Taipei, 1992).

Korea National Statistical Office, *Statistical Handbook of Korea 2002* (Republic of Korea: Korea National Statistical Office: Daejeon, 2003).

Kuroda, Y., "Research and Extension Expenditures and Productivity in Japanese Agriculture, 1960–1990," *Agricultural Economics* 16, no. 2 (1997), 111–124.

Lipton, M., "Land Reform as Commenced Business: The Evidence against Stopping," *World Development* 21, no. 4 (1993), 641–657.

MAF, *Major Statistics Related to Agricultural Industry* (Korea Ministry of Agriculture and Forestry: Gyeonggi, 2003). http://www.maf.go.kr/maf_eng/data/data1_2_05.htm

MAFF, *Abstract of Statistics on Agriculture Forestry and Fisheries in Japan* (Ministry of Agriculture, Forestry, and Fisheries of Japan: Tokyo, 2003). http://www.maff.go.jp/toukei/abstract/index.htm

Mazumdar, D., "Size and Farm Productivity Revisited: A Problem of Indian Peasant Agriculture," *Economica* 32 (1965), 161–173.

Mundlak, Y., D. F. Larson, and R. Butzer, *Determinants of Agricultural Growth in Indonesia, the Philippines, and Thailand*, Policy Working Paper 2803 (World Bank: Washington, DC, 2002).

Newell, B., K. Pandya, and J. Symons, "Farm Size and the Intensity of Land Use in Gujarat," *Oxford Economic Papers* 49 (1997), 307–315.

Park, J. K. *Source of Korea Agricultural Growth, 1918–2000*, Mimeo (2003).

Pookpakdi, A., *Sustainable Agriculture for Small-Scale Farmers: A Farming Systems Perspective*, Extension Bulletin (Food and Fertilizer Technology Center: Taipei, 1992).

Raghbendra, J., P. Chitkara, and S. Gupta, "Productivity, Technical and Allocative Efficiency and Farm Size in Wheat Farming in India: A DEA Approach," *Applied Economics Letters* 7 (2000), 1–5.

Rao, V., and T. Chotigeat, "The Inverse Relationship between Size of Land Holdings and Agricultural Productivity," *American Journal of Agricultural Economics* 63 (1981), 571–574.

Sampath, R. K., "Farm Size and Land Use Intensity in Indian Agriculture," *Oxford Economic Papers* 44, no. 3 (1992), 494–501.

Sen, A. K., "An Aspect of Indian Agriculture," *Economic Weekly* 14, no. 4–6 (1962), 243–246.

Sharma, S. C., "Technological Change and Elasticities of Substitution in Korean Agriculture," *Journal of Development Economics* 35 (1991), 147–172.

Singh, R. B., P. Kumar, and T. Woodhead, *Smallholder Farmers in India: Food Security and Agricultural Policy*, RAP publication 2002/03 (FAO Regional Office for Asia and the Pacific: Bangkok, 2002).

Subbarao, K., *Technology Gap and the Emerging Size–Productivity Relationships Following the Adoption of New Technology: An Analysis of Evidence from Northwest and Eastern India*, Unpublished paper (University of California at Berkeley, Department of Agriculture and Resource Economics, 1982).

Tadesse, B., and S. Krishnamoorthy, "Technical Efficiency on Paddy Farms of Tamil Nadu: An Analysis Based on Farm Size and Ecological Zone," *Agricultural Economics* 16 (1997), 185–192.

Thorat, S., P. Hazell, and S. Fan, "Population Growth, Land Distribution and Rural Poverty in India," Mimeo (2003).

Ukawa, H., *Crop-Livestock Integration in Hokkaido Japan, Based on Ammonia Treated Straw as Livestock Feed*, Extension Bulletin (Food and Fertilizer Technology Center: Taipei, 1995).

World Bank, *World Development Indicators 2002* (World Bank: Washington, DC, 2002) CD-ROM.

Plenary 3

Will food safety jeopardize food security?

Jean Kinsey*

Abstract

By a new definition proposed here, food safety is not just about safe food but the safe consumption of food. This draws attention to the issue of overeating as well as undernourishment. With the transition of diets and a rapid rise in obesity, worldwide, chronic illnesses join acute illness (from microbiological contamination) as health issues related to food consumption. The economics of obesity overwhelms the costs of microbiological contamination, with large impacts on private and social health care costs. In food-insecure areas, producing and consuming unsafe food does not build a healthy population nor does it allow export of food to other nations. Safety and security must work hand in hand to enhance human health.

JEL classification: I12

Keywords: food safety; food security; obesity; hunger; health costs

1. The safe consumption of food

This paper will conclude by saying no, food safety and food security reinforce each other. But first, some new definitions are in order. Food is an edible substance that will nourish the human body, provide it with energy for normal activities, and maintain or enhance its healthy state. Safe food consumption makes a person feel good in the short and the long run. It does not make you ill! This defines food by its fundamental purpose and how well it performs, rather than its physical characteristics like grams of nutrients, production technology like organically grown, or implied freedom from pathogens due to being "triple washed" or irradiated. Safe food consumption focuses on a simple but comprehensive performance standard. That is, eating (or drinking) food facilitates the health and growth of the human body.

When consuming food does not achieve this end it cannot be defined as safe food consumption. Unsafe food consumption constitutes ingesting a fast-acting poison or a set of substances that lead to debilitating diseases over a long period of time. Some

* *Applied Economics Department and The Food Industry Center, University of Minnesota, St. Paul, MN, USA.*

dangerous food substances are well known (e.g., wild mushrooms); some are unknowable until long after the damage is done (e.g., pesticides). At the University of Minnesota there was a course in food safety titled "The Dose Makes the Poison." This is an incredibly insightful title. There are many substances in this world that are harmful to human health; some of them are in the foods we eat every day. In minute quantities they are not harmful. However, there is some quantity of exposure that tips the scales; at some dose they become dangerous. That dose differs by the size, genetics, and immunities of individuals, but at some critical level food that carries potentially harmful substances interferes with nourishment and, therefore, diminishes health, normal cell growth, and bodily functions. This is consistent with the educational slogans of many nutritionists who say, "There are no bad foods, just bad diets." Rather than focus on substances in food that make it unsafe, this new definition focuses on *the safe consumption of food*. It requires the acceptance of some risk, an acknowledgement that quantity matters, and responsibility for the dietary context and needs of people in various situations and cultures.

In this paper the boundaries of the safe food consumption and public policies that might alleviate unsafe consumption will be explored. The implications for food security, which also now has two distinctly different definitions, will be considered as a function of how well food and diets enhance the health and well-being of people. Unsafe consumption of food and subsequent health issues include:

1. Foods that contain microbes in sufficient quantities to lead to short-term illness or death such as botulism or *E. coli* O157 H7.
2. Foods that contain substances that are believed to pose potential long-term health problems such as pesticide residues or bovine spongiform encephalopathy.
3. Foods that have unknown, but suspected, health consequences such as foods that have been genetically modified or irradiated.
4. Foods that contain nutrients or ingredients such as trans-fats or simple sugars that, when consumed in excess quantities, lead to chronic diseases such as diabetes, cancer, and cardiovascular heart disease.

2. Microorganisms and the safe consumption of food

Traditionally, those who study and regulate food safety concentrate their research and policy analysis on microbial contamination. Table 1 lists the ten most

Table 1
Reported foodborne illnesses from bacteria, viruses, or parasites in the United States

	Cases/Year (millions)	Hospitalization	Deaths
Norwalk-like virus	9.20	20,000	124
Campylobacter (1/1,000 cases lead to Guillain-Barre syndrome)	2.00	10,500	1,000
Salmonella	1.413*	15,600	550
Clostridium perfringens	0.25	50	10
Giardia lamblia	.200	500	1
Escherichia coli	.173	2,800	80
Listeria monocytogenes	.003*	2,500	500
Taxoplasma gondii	.113	22,600	375
Shigella	.09	1,250	14
Total reported	13.44	75,896	2,654
CDC estimated total incidents	76.00	325,000	5,000

Source: Ropeik and Gray (2002).
Note: *Adjusted from data on http://www.ers.usda.gov/data/foodborneillness/.

known and tracked pathogens leading to foodborne illnesses in the United States. The Centers for Disease Control (CDC) estimates that these pathogens represent only about 20% of the cases and hospitalizations, and less than half of the deaths actually caused by foodborne pathogens. Norwalk-like viruses generate the largest number of reported cases of foodborne illnesses per year, Taxoplasma gondii (a parasite) generates the largest number of hospitalizations, and campylobacter causes the largest number of deaths (Ropeik and Gray, 2002). Identifying a hazardous organism and the probability of it causing a foodborne illness and then deciding on an acceptable level of risk in order to set food safety standards involves long and arduous study and debate. The tasks involve science, politics, culture, and international consensus. It is one thing to say that regulations should be based on science. It is quite another to agree on the scientific evidence and how to apply it. Two organizations intimately involved in setting standards for food safety as it relates to microbial contamination are the International Commission on Microbiological Specifications for Foods (ICMSF, 2002) and Codex Alimentarius. The former is a group of scientists that assess risk and establish protocol for setting food safety objectives and standards;[1] the latter is the consensus-building arm of the United Nations that identifies international standards for food safety. Figure 1 depicts the complex decision process proposed by ICMSF to manage food safety and prevent as many foodborne illness as possible (ICMSF, 2002, p. 5).

Hazard is the measurable probability that contamination exists in an amount sufficient to cause illness. This can generally be determined by laboratory tests. Risk refers to the hazard plus the consequences that consumers will suffer when they are subjected to the hazard and become ill. To assess the risk, one needs to know the probability that people of various ages and lifestyles will become ill when exposed to various hazards in their food. This involves epidemiological data that track outbreaks of foodborne illnesses and identify their cause. Tracking such data requires some nationally agreed upon reporting system by doctors and hospitals. A national food safety management

---

[1] A Commission of the International Union of Microbiological Societies.

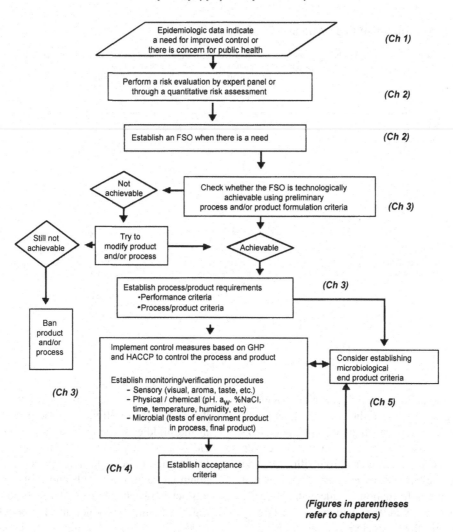

Figure 1. Food safety management scheme. *Source:* ICSMF, 2002, p. 5.

system involves identifying the hazards, the risks, and the magnitude of the problem for consumers' health (van Schothorst, 1998). It also involves identifying the magnitude of the problem for a nation's economy.

In assessing the magnitude of the problem one looks at the pain and suffering caused to individuals, the costs of health care, and the value of lost productivity due to illness and death. The value of these losses is a measure of the benefits that could be derived from eliminating, or drastically reducing, the foodborne illness. Finding policies and practices that can be used to ensure the safe consumption of food involves assessing the public and private consequences of foodborne illnesses. Yes, there is a "public good" aspect to the safe consumption

of food. The benefits are nonrival and nonexclusive. That is, the safety, health, trust, and security that one person enjoys from safe food consumption are not diminished by a neighbor's enjoyment of the same. In fact, the more healthy people there are in a society, the better off everyone is. There is a positive (negative) spillover effect from having a community of healthy (unhealthy) people. The role of governments and food industry executives is to discover the right combination of policies and practices that will work in a given economy and culture in order to deliver the optimum level of safe food consumption.

Figure 2 presents a continuum of types of policies that may be implemented to match various levels of

Continuum of Food Attributes

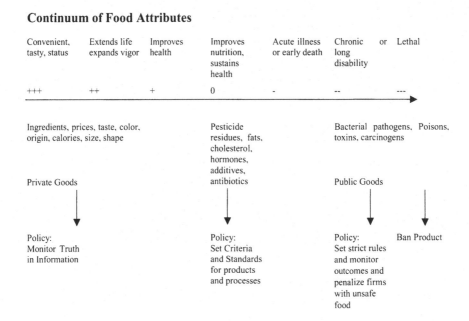

Figure 2. Food safety policy depends on nature of the risk.

hazard and risk to appropriate public policies (Kinsey, 1993). Moving from left to right along the continuum the characteristics of food and their accompanying hazards and risks increase from disappointment in the taste or convenience to acute and then chronic illnesses and even death. At the left end, policies and practices that focus on consumer information are adequate. Here the characteristics of food under consideration are largely private goods and the characteristics are transparent or can be readily made so with labeling and education. Moving to the right, food characteristics become increasingly nontransparent and the consequences of consumption more uncertain. Pesticide residues are believed to make people ill, even cause cancer, but there is little proof. Food additives, antibiotics, fats, and hormones may lead to illness in some people at some dose, but the amounts are uncertain. As you move to the far right, there are some contaminants that cause violent illnesses or immediate death (e.g. *E. coli* 0157H7 or botulism). They are not transparent and the only way to effectively deal with them is to eliminate them from the food supply. In the center of the continuum appropriate practices and policies are less clear. Here is where scientists and policymakers alike debate the standards for products and processes in order to protect people

from harm and provide some regulatory clout to monitor and penalize offending food providers. Examples in the middle of the continuum might include pesticides, trans-fatty acids, and antibiotics in animal feed. Designing public and private policies to maximize the potential for good health and longevity is a balancing act between the cost of regulation and/or building safety into a product and the benefits of healthy, productive people who might be incapacitated from a foodborne illness.

Which food safety problems are consumers most concerned about? In surveys taken by the Food Marketing Institute consumers have ranked microbes and the chance of being exposed to microbes higher than pesticides since 1995 (Figure 3, FMI, 1993–2000). The concern about microbes leads to concern about food-handling practices and spoiled food. With consumers relying on many strangers' hands to prepare their food, its freshness and how it is handled is a proxy for fear of microbes and the illnesses they can cause. Between 1995 and 1997 when FMI asked consumers to identify the most serious health hazards related to food they consistently ranked microbes first, pesticides second, antibiotics third, and biotechnology last FMI, 1993–2000.

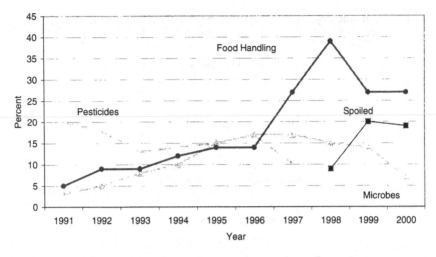

Figure 3. U.S. consumer perceptions of food safety problems. *Source:* Consumer Trends, Food Marketing Institute, Washington D.C. 1993–2000.

3. Acute illnesses and costs

The cost of foodborne illnesses caused by microbes is estimated at $6.9 to $33 billion per year (USDA). This includes direct medical costs as well as lost wages, productivity, and the estimated value of life lost to premature death. Figure 4 illustrates the types of costs and considerations included in the calculations. It makes one realize the vast number and types of costs involved in foodborne illnesses and the great loss to individuals and society. The numbers quoted herein do not even begin to include the costs to the food industry or to the public health sector.

Dollar estimates have a wide range, partly because foodborne illnesses are vastly underreported both by consumers and doctors. Most consumers who become ill think they have the flu and, though violently ill in many cases, they recover in a few days and go on about their lives. A few, less fortunate, die. There are an estimated 2,654 to 5,000 deaths per year in the United States attributable to foodborne illness.

These acute and temporary illnesses are largely preventable by good manufacturing practices (GMP) and good handling practices in the supply chain starting at the farm and ending with the consumer. Many studies have been conducted to estimate the cost of diminishing these illnesses. They typically find that safe handling and storage practices cost less than the resulting benefits. For example, Ollinger and Mueller (2003) found that a Pathogen Reduction/Hazard

Analysis and Critical Control Point program in meat and poultry plants would cost plants about 1.1 % of their total costs, adding about 1.2 cents to a pound of beef, 0.7 cents to a pound of pork, and 0.4 cents to a pound of poultry. The benefits were estimated to range from $1.9 to $171.8 billion annually. This translates into a benefit value (in terms of health cost savings) that is at least two times the cost to the industry. Lakhani (2000) estimated that the benefit–cost ratio from reducing salmonella enteritidis in shell eggs by refrigeration to be 0.65, 3.56, 2.56, and 8.87 depending on the method used to calculate the benefits. Since three of four estimates are greater than one, measures to reduce salmonella-caused foodborne illness was deemed to be worthwhile. A third example comes from an analysis of adopting HACCP programs in meat and poultry slaughterhouses in the United States using a Social Accounting Matrix (SAM) method (Golan et al., 2000). This provides a comprehensive picture of how well an entire economy fares as a result of investments in food safety. Their model showed that for every dollar saved by preventing a premature death from a foodborne illness, there is an economy-wide gain of $1.92. They also found that for every dollar of household income saved due to lower medical expenses, the whole economy lost $0.27. On the cost side, they found that for every dollar spent implementing a HACCP program, the economy gained $0.66, leading to a net increase in production output of $10.63 billion, an increase in factor payments of $6.08 billion, and an increase in household

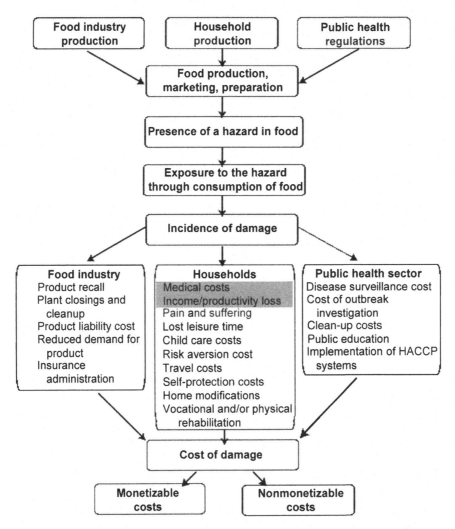

Figure 4. Foodborne disease, exposure, and types of costs. *Source:* http://www.ers.usda.gov/Briefing/FoodborneDisease/gallery/Foodborne Diseas.gif.

income of $9.38 billion (in 1993 dollars) not considering the benefits from reduced work-loss days.

Antle (2001) points out that studies indicate that consumers are willing to pay more for safer food than the losses that might incur due to illness, using the cost-of-illness approach to measure the benefits of safer food. In the real world consumers demonstrate their willingness to pay at the supermarket when they buy organic food to avoid pesticides. They pay for safer food at tax time when they support government agencies such as the Food and Drug Administration, Department of Agriculture, and state health departments who test,

inspect, and regulate the processes by which food is produced, labeled, and sold. In most developed countries consumers have come to expect their government to ensure safe food and they are generally willing to pay for it.

4. Chronic and long-term illnesses

Foodborne illnesses due to microbes are well known; cause and effect are relatively well established. The relationship between food, diet and chronic diseases,

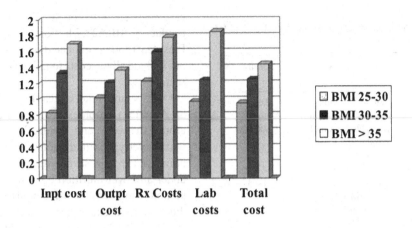

Figure 5. Health care costs index by body mass index. *Source:* Wolf (1998).

and delayed illnesses is less well established. For example, there is virtually no known link between pesticide residue in food and cancer even though most believe it to exist. Many believe that there is a link between antibiotic resistance in humans and eating meat from animals that have been routinely fed antibiotics to keep them healthy and fast growing. Some fear that feeding growth hormones to cattle or genetically modifying plants and animals will lead to human illness, though there is little to no scientific evidence to support these suspicions. The link between spongiform encephalopathy (mad cow disease) and variant creutzfeldt Jakob disease (vcjd) was confirmed using transgenic mice in 1999 (Acheson, 2001) but the time lag between exposure and illness is several years, making epidemiological evidence in humans hard to establish. Many chronic and long-term illnesses suspected of being linked to unsafe food consumption are difficult to trace to the source due to time lags and intervening genetic and environmental factors.

On the other hand, Type 2 diabetes (Knowler et al., 2002) and 20–40% of cancers in U.S. adults (Calle et al., 2003) are known to be linked to obesity and are rising at a near-epidemic rate. The rapid rise in obesity in the United States. and around the world, leads us to redefine the boundaries of safe food consumption. With obesity individual food characteristics are not the problem, but the quantity eaten—the total dose—is a problem. Just as it is the quantity of microbes in the food that leads to acute illness, it is the quantity of calories in the diet, relative to energy expended by the body, that leads to diabetes and other obesity-related complications.

In 1999, an estimated 61% of adults and 13% of children and adolescents in the United States were overweight. Adult obesity has doubled since 1980 to 24% of the population and overweight adolescents have tripled since 1980 to 15%. (FDA; CDC). Overweight children aged 2–5 years have increased from 7% to 10% since 1994. Eight percent of U.S. adults (Knowler et al., 2002) and about 4% of children in the United States have Type 2 diabetes. The rise in this non-inherited, Type 2 diabetes in children is of great concern, since diabetes is a chronic disease that absorbs over 10% of all health care dollars. It is growing along with obesity in children; it is a health care disaster in slow motion. Obese children with diabetes will absorb an increasing amount of our health care dollars for as long as they live. Figure 5 illustrates the higher cost of all types of health care for people with body mass indexes (BMI)[2] more than 30 or 35 (obese and morbidly obese). One study estimated that health care for overweight and obese people costs an average of 37% more than for people of normal weight, adding an average of $732 to the annual medical bills of every American (Connolly, 2003). This places the problem of obesity squarely in the realm of a public good (bad) and one that will take

---

[2] BMI is measured by dividing an individual's weight in kilograms by height in meters squared (weight in pounds by the height in inches squared × 703). BMI of 20–25 is considered healthy. BMI over 25 is considered overweight and over 30 is obese.

Table 2

Costs associated with the unsafe consumption of food leading to obesity: United States, 2000

Type of health care problem	Health care costs	Deaths
Microbial contamination	$6.9*–$37 billion (includes losses due to death)	2,654–5,000
Obesity-related diseases	$93–$230 billion (direct and indirect costs)	300,000
Diabetes (10% of all health care costs)	$132 billion** (direct and indirect costs)	
Ratio of obesity costs to microbial costs	Low: 93/6.9 = 13.5 High: 230/37 = 6.2	300/5 = 60
Ratio of diabetes costs to microbial costs	Low: 132/37 = 3.5 High: 132/6.9 = 19.1	

Notes: *Estimated cost based on four types of microbes: Campylobactor, Salmonella, E.-coli, Listeria. http://www.ers.usda.gov
**2003 update on www.cdc.gov/diabetes/pubs/estimates.htm

a concerted effort on the part of many agents in society to correct.

What does it cost for obesity-related diseases in the United States? Total and indirect costs were estimated to be $93 billion (Connolly, 2003) to $117 billion in 2000 (FDA). Some public officials are quoting $230 billion (T. Thompson, Secretary of Health and Human Services). Table 2 compares the costs of microbial-related foodborne illnesses to health care costs related to obesity. By any comparison you want to select, the cost of obesity is much larger than the costs of microbial contamination. The $117 billion for obesity health

care costs are 1.1% of the 2002 U.S. gross domestic product of $10,623.7 billion (Economic Report to the President) and, as indicated on Figure 4, these costs do not include all the costs to industry or the public health sector.

Obesity and related health problems are not just an American problem. Obesity is being documented around the world and, ironically, it exists side by side with poverty and undernourishment. Haddad (2003) points out that in seventy-eight developing countries under and over-nutrition coexist with 5% of the population being obese and 7% being underweight. Often this condition exists in the same household (Garrett and Ruel, 2001). Around the world it is estimated that 53% of children and 18% of the total population are undernourished, while in Australia 20% of children are overweight or obese, as are 17% of Malaysian boys, 8% of Malaysian girls, and 7% of urban Chinese children (IFIC, 2001). Most of the undernourished in 1998 were in India, China, and sub-Saharan Africa (FAO, 2002). Figure 6 shows the global prevalence of both underweight and obese adults in 2000 (WHO). On balance 8.2% are obese and 5.8% are underweight, only a 2.4 percentage point difference. In developing countries more than two thirds as many people are obese as are undernourished. In the poorest countries, 20% as many people are obese as are undernourished. The magnitude of these dual food and diet issues clearly poses new challenges for global food policy and food security.

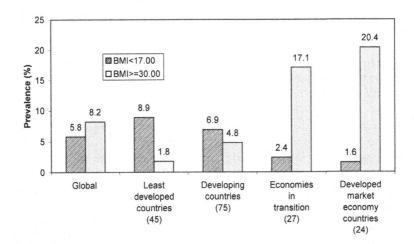

Figure 6. Global prevalence of underweight and obesity in adults for year 2000, by level of development. *Source:* Nutrition for Health and Development, A Global Agenda for Combating Malnutrition, WHO 2000. www.who.int/nut/db_bmi.htm.

5. Food security and safe food consumption

There are now two distinct definitions of food security. The traditional well-known definition refers to having enough food to maintain growth and health. The USDA defines food security as having access to enough food, at all times, for an active, healthy life (Nord, 2002). The new definition of food security refers to the production, processing, and distribution chain being secure from bioterrorists so that food cannot be *deliberately* contaminated with an agent that would make people ill, cause death, or economic chaos. Arguably, if food is produced according to good farming and manufacturing practices the chances of it being compromised by a deliberate terrorist is small but certainly not zero. Government agencies such as FDA and USDA in the United States are actively studying this new hazard, developing educational programs, and taking precautionary measures to minimize the impact of any such event. Consequently, for the purposes of this discussion I will focus on the traditional definition of food security.

The proposition here is that food security is not jeopardized by activities designed to improve the safety of food consumption. People who do not have enough to eat and are undernourished obviously benefit from more food availability, but food that makes them ill is not helpful. Making the food delivered to all people as safe and nourishing as possible should be a paramount criterion for delivering nourishment at all levels of income and caloric intake. Arguments that focus on the inability of developing countries to meet the food safety standards of countries that are potential importers and, therefore, should be allowed a lower food safety standard, end up jeopardizing the health of people everywhere, including the poor in a potentially exporting, developing nation. It is well known that hunger and poverty go together, everywhere. If resources are focused on helping poor nations to meet health and safety standards for their citizens and to be able to participate in world commerce, their incomes should rise, and food security problems can start to be alleviated.

Even in the United States, almost 11%, or 11.5 million households, were not food secure in 2001. One third of them were hungry at some time. They spent an average of 15% of their income on food per year compared to 5.5% for food-secure households (Nord,

2002). Several studies have shown that people, especially women, in these food-insecure households are also overweight (Olson, 1999; Townsend et al., 2001). Women between the ages of 19 and 55 years, who were in food-insecure households, were found to be significantly more likely to be overweight, and consumed ninety-one calories more per day that women in food-secure households (Basiotis and Lino, 2002). Based on the standard conversion of calories to body weight of 100 calories per day leading to 10 pounds of weight gained (or lost) per year, it is easy to see how those who are food insecure are more likely to be overweight. It begs the question of whether cheaper food has more calories and fat than more expensive foods, but it is a common observation that inexpensive and fast food is often higher in calories than higher priced food or food that is prepared slowly from scratch. The point is that poverty, hunger, and being overweight exist simultaneously, and that being overweight jeopardizes health, which jeopardizes the ability to work and be productive, which in turn jeopardizes the ability to earn income to buy healthy food. Therefore, safe consumption of food is compatible and consistent with food security in all parts of the world. The goal of food consumption is to nourish the body and improve health over a lifetime. If the food available is not safe or its consumption does not improve health, it does not contribute to food security.

In conclusion I quote Lawrence Haddad, "The diet transition in the developing world seems to be accelerating. It seems to be a transition towards an increased burden of chronic disease. It is increasing human costs in terms of mortality and the disease burdens. It is increasing the economic costs in terms of lower productivity. It is driven by changing preferences fuelled by growing incomes, changing relative prices, urbanization; by changing options fuelled by changes in food technology and changes in the food distribution systems; and by a legacy of low birth weights from the previous generation." He posits that there is a good case for public investment in efforts to influence the diet transition toward increasingly healthy outcomes. To do so will require us to address the dual issues of over and under consumption of food.

Food safety does not jeopardize food security; both act together to enhance human health. New definitions of both food safety and food security broaden the scope of concern and provide applied economists and policy makers with new challenges in analysis and public policy.

References

Acheson, D., "You Are What I Eat," *Food Quality* (July/August, 2001), pp. 22–33.

Antle, J., "Economic Analysis of Food Safety," in B. Gardner and G. Rausser, eds., *Handbook of Agricultural Economics*, Vol. 1B (Elsevier: New York, 2001).

Basiotis, P. P., and M. Lino, "Food Insufficiency and Prevalence of Overweight among Adult Women," *Nutrition Insights* (USDA, Center for Nutrition Policy and Promotion: Washington, DC, 2002).

Calle, E. E., C. Rodriguez, K. Walker-Thurmond, and M. J. Thun, "Overweight, Obesity, and Mortality from Cancer in a Prospectively Studied Cohort of U.S. Adults," *The New England Journal of Medicine* 348, no. 17 (2003), 1625–1638.

Centers for Disease Control, http://www.cdc.gov/nccdphp/dnpa/obesity/Trend/index.html, accessed 28 October 2004.

Connolly, C., "Health Costs of Obesity Near Those of Smoking," *Washington Post* (May 14, 2003), p. A9.

FAO, "State of Food Insecurity in the World," http://www.fao.org/DOCCREP/x8200E/x8200e03.htm#P0_0 *FDA Consumer* (March–April, 2002), p. 8.

Food Marketing Institute (FMI), *Consumer Trends*. Relevant years (1993–2000).

Garrett, J., and J. C. Ruel, "Stunted Child–Overweight Mother Pairs: An Emerging Policy Concern?" *17th International Congress of Nutrition: Annals of Nutrition and Metabolism* 45, Suppl 1 (2001), 404.

Golan, E. H., K. Ralston, P. Frenzen, and S. Vogel, "The Costs, Benefits and Distributional Consequences of Improvements in Food Safety: The Case of HACCP," in L. J. Unnevehr, ed., *Economics of HACCP: Costs and Benefits* (Eagan Press: St. Paul, MN, 2000).

Haddad, L., "Redirecting the Diet Transition: What Can Food Policy Do?" Presented at OEDC Conference, Amsterdam, The Netherlands (January 2003). (l.haddad@cigar.org)

International Food Information Council Foundation (IFIC), "Childhood 'Globesity'," *Food Insight* (January/February, 2001).

(http://ific.policy.net/proactive/newsroom/release.vtml?id=19401&PROACTIVE_ID=cecfc)

International Commission on Microbiological Specifications for Foods (ICMSF), *Micro-organisms in Foods 7: Microbiological Testing in a System for Managing Food Safety* (Kluwer Academic/Plenum: New York, 2002).

Kinsey, J., "GATT and the Economics of Food Safety," *Food Policy* (April, 1993), pp. 163–175.

Knowler, C. W., E. Barret-Connor, S. E. Fowler, R. Hamman, J. M. Lachin, E. A. Walker, and D. M. Nathan, "Reduction in the Incidence of Type 2 Diabetes with Lifestyle Intervention or Metformin," *The New England Journal of Medicine* 346, no. 6 (2002), 393–403.

Lakhani, H., "Benefit–Cost Analysis of Reducing Salmonella Enteritidis: Regulating Shell Egg Refrigeration," in L. J. Unnevehr, ed., *Economics of HACCP: Costs and Benefits* (Eagan Press: St. Paul, MN, 2000).

Nord, M., "Household Food Security in the United States," *ERS Information* (November, 2002), p. 1,3. (http://ers.usda.gov/publications/fanrr29/)

Ollinger, M., and V. Mueller, "Managing for Safer Food: The Economics of Sanitation and Process Controls in Meat and Poultry Plants," Washington DC: USDA, ERS, Agricultural Economics Report No. 817 (April 2003).

Olson, C. M. "Nutrition and Health Outcomes Associated with Food Insecurity and Hunger," *Journal of Nutrition* 131 (1999), 521S–524S.

Ropeik, D., and G. Gray, *Risk* (Houghton Mifflin, Co.: Boston, 2002), pp. 98–100.

Townsend, M. S., J. Peerson, B. Love, C. Achterberg, and S. P. Murphy, "Food Insecurity Is Positively Related to Overweight Women," *Journal of Nutrition* 131 (2001), 2880–2884.

USDA, USDA, ERS, at http://www.ers.usda.gov/data/foodborneillness/, accessed 28 October 2004.

U.S. Government Printing Office, *Economic Report to the President, 2004* H. Doc. 108–145, p. 284, available at http://www.gpoaccess.gov/eop/, accessed 26 October 2004.

Van Schothorst, M., "Principles for the Establishment of Microbiological Food Safety Objectives and Related Control Measures," *Food Control* 9, no. 6 (1998), 379–384.

WHO. http://www.who.int/nit/db_bmi.htm, accessed June 2003.

Wolf, A., "Impact of Obesity on Healthcare Delivery Costs," *American Journal of Managed Healthcare* 4, no. 3, supplement (1998), S141–S145.

Poverty amidst plenty: food insecurity in the United States

Michael LeBlanc*, Betsey Kuhn*, James Blaylock*

Abstract

The United States faces domestic food security issues that differ from those encountered by many countries. Yet, in 2001, 10.7% of U.S. households were estimated to be food insecure at some point during the year. Food security, poverty, and food insecurity are strongly linked by economic conditions. Job transitions, layoffs, and family disruptions result in periods of low income and vulnerability to food insecurity. Economic and food assistance programs have helped protect many U.S. households when the market economy has failed to do so. These programs have reduced vulnerability to falling income and food insecurity during economic downturns in the business cycle. However effective food assistance programs have been for reducing short-term vulnerability, they do not enhance a household's ability to achieve sustainable food security. Prospects for improving long-term food security are tied to the same economic forces shaping a household's income and budget, particularly those related to labor productivity and wages.

JEL classification: I-Health, Education, and Welfare

Keywords: poverty; food security; household income; labor market

The fault, Dear Brutus, is not in our stars, but in ourselves.

—William Shakespeare (Julius Caesar)

1. Introduction

The latest estimates by the Food and Agriculture Organization (FAO) indicate some 840 million people were undernourished in 1998–2000—11 million in the industrialized countries, 30 million in countries in transition, and 799 million in the developing world (FAO, 2002a). Undernourishment occurs when food intake falls below a minimum calorie (energy) requirement or when people exhibit physical symptoms caused by energy and nutrient deficiencies resulting from an inadequate or unbalanced diet.[1] A variety of factors, alone or in combination, exists when people are undernourished. Food can be physically unavailable, people can lack social or economic access to adequate food, and food utilization by the body may be inadequate.

Food security—access by all people at all times to enough food for an active healthy life—is an important objective of every nation, formalized in the "Rome Declaration." The United States faces domestic food security issues that differ from those encountered by many countries. Only a small proportion of the U.S. population is food insecure in any given year, and, in most cases, their food insecurity is occasional or episodic, not chronic. Undernourishment as a result of poverty is unusual and health effects like wasting and stunting are rare. Indeed, health problems resulting from overweight are far more widespread than health problems resulting from undernutrition.[2]

Nevertheless, not all U.S. households have achieved food security. Each year, a small proportion of the country's population is food insecure and a smaller number experience hunger at times because they cannot afford enough food. In 2001, 10.7% of U.S. households were estimated to be food insecure at some point

* *Economic Research Service, U.S. Department of Agriculture, Washington DC.*

[1] Energy and nutrient deficiencies may also result from the body's inability to use food effectively because of infection or disease.

[2] Undernutrition results from undernourishment, poor absorption, and/or poor biological use of consumed nutrients.

during the year (Nord et al., 2002). Research suggests that even the food insecurity that exists in the United States—in most cases occasional or episodic occurrences of disrupted eating patterns and reduced food intake—can have deleterious effects on nutrition, health, and children's psychosocial development and learning.

Food security and the economy are strongly linked. In countries with a high prevalence of undernourishment, a comparably high proportion of the population lives on less than US$1 per day. In higher income countries, including the United States, the relationship between relative poverty and increased risk of food insecurity and hunger is well recognized. Although poverty is undoubtedly a cause of hunger, hunger can deprive people of the strength and skill to engage in work and production. Hunger in childhood impairs mental and physical growth, limiting capacity to learn and earn. Recent estimates suggest that halving the number of undernourished by 2015 would yield over US$120 billion per year in increased income (FAO, 2002a).

Poverty and food insecurity are affected by economic conditions in the business cycle. Job transitions, layoffs, and family disruptions result in periods of low income and vulnerability to food insecurity. Government transfer programs in the United States provide an economic safety net to buffer people from the vagaries of the market, but are not typically viewed as mechanisms for permanently or sustainably lifting people out of poverty. Instead, economic growth has long been considered as the most effective instrument to reduce poverty. Over the last 20 to 40 years, however, critics have questioned the continued efficacy of growth for improving the incomes of the poor in the United States (Blank, 1997).

Since the mid-1960s considerable attention in the United States has focused on alternative approaches for alleviating poverty. The anti-poverty program of the 1960s was built on targeted measures like increased education, improvements in public and individual health, vocational training, and community development initiatives, to improve the earning capacities of individuals and communities.

Complementing the anti-poverty programs has been the growth of food-related assistance programs. Born of the Great Depression, but growing to maturity during the 1960s and 1970s, food assistance programs in the United States are meant to protect households' food security when the market economy may fail to do so. In addition to farm programs that promote crop production and lower food prices, the core food assistance programs include the Food Stamp Program, the school meals program, the Special Supplemental Nutrition Program for Women, Infants, and Children (WIC), and commodity distribution programs.

Economic and food assistance programs have helped protect many U.S. households when the market economy has failed to do so. These programs have reduced vulnerability to falling income and food insecurity during economic downturns in the business cycle. However effective food assistance programs have been for reducing short-term vulnerability, they do not enhance a household's ability to achieve sustainable food security. Prospects for improving long-term food security are tied to the same economic forces shaping a household's income and budget, particularly those related to labor productivity and wages.

2. Poverty in the United States

Each society defines poverty in its own terms. Conventional measures of poverty count the number of people below the poverty line and define the poverty rate as the proportion of the total population below the poverty line. Poverty is, therefore, a normative concept, not a statistical one and setting the poverty level requires a judgment about social norms.

In 1968, the U.S. government adopted an official definition of poverty that it uses to publish statistics on income and set eligibility standards for public programs. The official definition of poverty compares a family's cash income with an estimate of its needs. Family needs are calculated as a function of the number of family members and their ages and sex. At the heart of the original definition of poverty was the economy food plan, the least costly of four nutritionally adequate food plans designed by the Department of Agriculture. It was determined from the Department of Agriculture's 1955 Household Food Consumption Survey that families of three or more persons spent approximately one third of their after-tax money income on food. Poverty thresholds for families of three or more persons were set at three times the cost of the economy food plan.

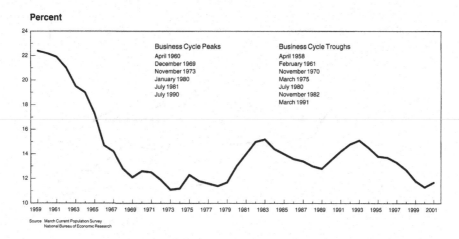

Figure 1. Poverty status of persons (1959–2001). *Source:* U.S. Bureau of the Census, Statistical Abstract of the United States.

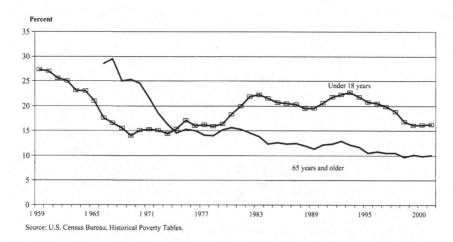

Figure 2. Poverty status by age (1959–2001). *Source:* U.S. Bureau of the Census, Poverty in the United States.

During the rapid growth of the postwar era, poverty fell dramatically and the United States enjoyed its longest period of uninterrupted growth. From 1959 through 1973 poverty declined from over 22% to 11% (Figure 1). So precipitous was the decline in poverty that it gave rise to hopes of eliminating poverty. As incomes grew rapidly many poor and near-poor families were lifted out of poverty into the middle class. Indeed, the entire income distribution was moved toward higher income levels. Since 1973, however, the U.S. poverty rate has increased and then fluctuated around a narrow range.

The poor are not homogeneous. Poverty rates differ significantly by race, sex, and household head. In general, the relative disparities among demographic subgroups have persisted over the last 30 years. The only exception is that the poverty rate for children now exceeds the poverty rate for the elderly (Figure 2).[3] Over the last two decades, the elderly have experienced rising incomes and declining poverty.

There is a dramatic difference in poverty rates for black and white populations (Figure 3). In the mid-1960s the black rate was nearly 4 times the white poverty rate. From 1966 through 1997 the average black poverty rate was about 2.5 times greater than

---

[3] Many U.S. federal income and food assistance programs (School lunch, WIC, TANF) target children or families with children.

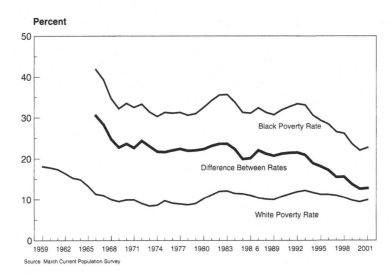

Figure 3. Poverty status by race (1959–2001). *Source:* U.S. Bureau of the Census, Poverty in the United States.

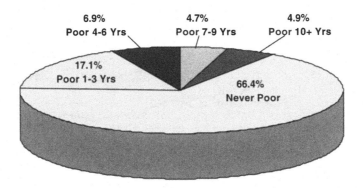

Figure 4. The extent of poverty among Americans (1979–1991). *Source:* Blank, 1997.

white poverty rate. More recently, the differences between the black and white poverty rates have been decreasing as white poverty rates have increased and the black poverty rate has fallen to about 27%. During the 1990s the mean difference between black and white poverty rates was 19%.

Popular notions often cast the poor in the United States as a persistently poor underclass where those in poverty typically remained poor from year to year. Evidence suggests poverty is a much more dynamic condition (Duncan, 1984). Individuals with persistently low incomes are not predominantly an "underclass" of young adults living in large urban areas. Rather, the designation persistently poor falls disproportionately on blacks, on the elderly, and on those living in rural areas and in the South. The persistently poor are

more sharply defined by these demographic characteristics than those found to be poor in a given year (Figures 4–7).

3. Measuring food security

The relationship between poverty and food security has been long recognized. For many years, however, constructive discussions about the level and distribution of food security and hunger were hampered by the lack of an adequate measurement and monitoring methodology. There are many methods for measuring food insecurity, each with different strengths and weaknesses.[4] Alternative approaches can generally be

_____

[4] FAO suggests using a suite of approaches. See FAO (June 2002b).

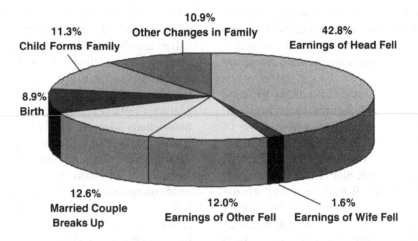

Figure 5. Reasons why poverty spells begin. *Source:* Blank, 1997.

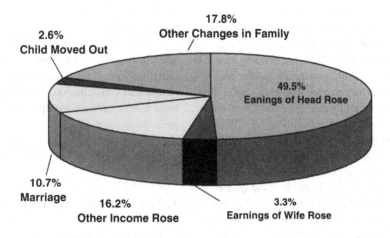

Figure 6. Reasons why poverty spells end. *Source:* Blank, 1997.

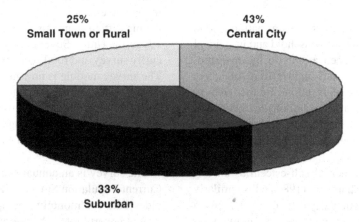

Figure 7. Composition of poor in the United States by geographic location (1993). *Source:* Blank, 1997.

characterized in three ways: those comparing estimates of dietary energy availability or intakes with energy requirements; those measuring nutritional outcomes; and those measuring perceptions of food insecurity and hunger.

There have been substantial shifts in the thinking about the definition of food security itself (Maxwell, 2001), principally away from concern with issues of global and national food supply adequacy and toward household and individual food access. The 1975 World Food Conference definition reflected policy concerns with national food self-sufficiency and proposals for world food stocks and import stabilization schemes. That era also witnessed the establishment of institutions, the World Food Council and the FAO Committee on Food Security, to promote a policy agenda aimed at augmenting food supply. It was evident, however, that widespread hunger could and did coexist with the presence of adequate food supply.

Sen (1981) is often credited with expressing ideas that helped move the issue of food access to the forefront. Sen's work gave an economic and philosophical voice to ideas that were embraced by nutrition planners and empirically supported in field studies (Berg, 1973; Joy, 1973; Levinson, 1974; Kielman et al., 1983). In Sen's view, it is more useful to define food security as being foremost a problem of food access, with food production at best a route to food entitlement. Most current definitions of food security begin with the individual entitlement, though recognizing the complex interlinkages between the individual, the household, the community, the nation, and the international community.

In addition to a paradigm shift from national food supply adequacy and toward household and individual food access, there has been a shift in the measurement of food insecurity and hunger from objective indicators to subjective perception. In the poverty literature there has been a long-standing distinction between "the conditions of deprivation" referring to objective analysis, and "feelings of deprivation," related to the subjective (Townsend, 1974). Kabeer (1998), for example, identifies lack of self-esteem as an element of poverty, and Chambers (1989) talks similarly of self-respect. It is particularly difficult to establish a metric for an experiential state of well-being like food security. The predicament is that states of mind are involved.

Most conventional approaches to food security have relied on what is viewed as objective (actually physical) measurement. These measures include target levels of consumption (Siamwalla and Valdes, 1980); consumption of less than 80% of World Health Organization average required daily calorie intake (Reardon and Matlon, 1989); or more generally, a timely, reliable and nutritionally adequate supply of food (Staaz, 1990).

Physical definitions present two major difficulties. First, the notion of nutritional adequacy is itself problematic. For any individual, nutritional adequacy depends on age, health, size, workload, environment, and behavior. Estimates of calorie requirements for average activity patterns in average years are subject to constant revision (Payne, 1990). Adding adaptation strategies complicates the calculation. Estimating precise calorie needs for different groups in the population is, therefore, difficult.

A second problem arises because qualitative aspects are omitted from the kind of quantitative measures listed earlier. The issues here include food quality (European Commission, 1988) consistency with local food habits, cultural acceptability, and human dignity, even autonomy and self-determination. The implication is that nutritional adequacy is a necessary but not sufficient condition for food security.

4. Food security in the United States

The United States measures household food security with a survey of the behaviors and experiences that are thought to characterize households in the United States having difficulty meeting their food needs. During the 1990s, the Unites States developed and tested a food security survey and food security scale for domestic use. The survey module is now in regular use in household surveys both for research and monitoring purposes. A large, nationally representative food security survey is fielded annually, and findings are published as a statistical series by the U.S. Department of Agriculture (Nord et al., 2002).

The survey is an annual supplement to the monthly Current Population Survey (CPS), the survey that provides data for monthly unemployment statistics and annual poverty rates. A nationally representative sample of about 43,000 households responds to questions about food expenditures, use of federal and community

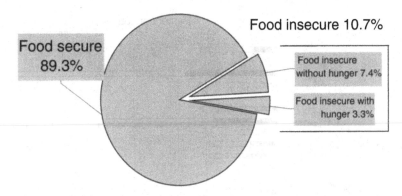

Figure 8. Prevalence of food insecurity and hunger in the United States, 2001. *Source:* Nord et al. (2002).

food programs, and whether they are consistently able to meet their food needs. Other surveys in the United States also utilize the food security survey module for monitoring and research.

The U.S. Food Security Scale is a "direct" experiential measure of the severity of household food stress or food deprivation. This approach contrasts with indirect indicators such as measures of household resources (generally income) or measures of outcomes of inadequate food access such as nutritional anthropometry. It is based on self-reported behaviors, experiences, and conditions collected by interviewing one member of each household using a standardized survey instrument, the U.S. Food Security Survey Module.

The food security status of a household is assessed by its responses to 18 questions about food-related behaviors, experiences, and conditions that are known to characterize households having difficulty meeting their food needs. The questions cover a range of food deprivation. For example, the least severe question asks whether household members worried if their food would run out before they got money to buy more; the most severe question asks whether any child in the household did not eat for a whole day because there was not enough money for food. Each question identifies a lack of money or other resources to obtain food as the reason for the condition. All questions are referenced to the previous 12 months.

Responses to the 18 questions are combined into a scale using nonlinear statistical methods based on the Rasch measurement model. The scale provides a continuous, graduated measure of the severity of food

deprivation across the range of severity encountered in U.S. households. Based on their food security scale scores, households are classified into three categories for monitoring and statistical analysis of the food security status of the population. The categories are "food secure," "food insecure without hunger," and "food insecure with hunger."

Based on the most recent food security survey data available, nearly 9 out of 10 U.S. households were food secure throughout the entire year, while 10.7% of the households were food insecure at some time during the year (Figure 8). Most food-insecure households obtained enough food to avoid hunger, but 3.3% of U.S. households were food insecure to the extent that one or more household members reported being hungry at least some time during the year because they could not afford enough food.

Food insecurity and hunger are not usually chronic conditions for those U.S. households that are affected by them. The U.S. food security measure classified households as food insecure, or food insecure with hunger, even if the condition occurred only for a brief period during the year. Thus, the rates of food insecurity and hunger on any given day are far below the measured annual rates. For example, the prevalence of hunger on a typical day in 2001 was estimated to be less than one fifth the annual rate, or about 0.5% of households.

In 2001, rates of food insecurity and hunger were low for households with elderly members and for married-couple families with children (Figure 9). In contrast, rates of food insecurity were higher than the national average for the following household types:

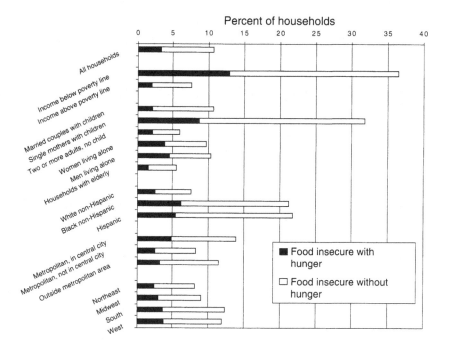

Figure 9. Prevalence of food insecurity by subgroup, 2001. *Source:* Nord et al. (2002).

- Households with incomes below the official poverty line
- Households with children, headed by a single woman
- Households headed by black or Hispanic persons

Also, food insecurity is a greater problem for households located in central cities and nonmetropolitan areas and in the southern and western regions of the country.

5. Adapting the U.S. measure for other countries

Methods based on similar approaches to those applied in the United States have been developed in other countries. In some cases, these have been based on translating questions in the U.S. module. Other applications have been based on additional research, including focus groups and cognitive testing of proposed questions and statistical analysis of survey data (Chung et al., 1997; Gittlesohn et al., 1998; Maxwell et al., 1999; Webb et al., 2002; Wolfe and Frongillo, 2001). Food security modules have been adapted for

three low-income populations: Orissa, India; Kampala, Uganda; and Bangladesh.

To achieve acceptable results the U.S. measure must be adapted to a new setting that is culturally, linguistically, and economically distinct from the United States. For use in low-income settings, additional attention may need to be given to incorporating the dimensions of frequency and duration of food deprivation into the measure (Hoddinott, 1999; Maxwell et al., 1999). In many poorer societies, a majority of the population faces food stress at times. The most important differences among households may be in how often this occurs and over how much of the year. Additionally, countries that face frequent acute shocks, droughts for example, are likely to differ from countries where exogenous shocks are rare.

6. Transfers and safety nets

Transfers and safety nets are created for moral, economic, and political reasons. The moral or humanitarian justification is "the removal or reduction of deprivation or vulnerability: food insecurity and hunger." The stated aim of many redistribution schemes is to reduce inequality and relative poverty. Economic efficiency justifications for transfers and safety net interventions rely on market failures most often related to the connection between health, education, and productivity. Selecting an appropriate intervention depends on the objectives of the intervention and the capacity of the country to implement it.

Over the past 30 years economic assistance and food assistance programs have helped protect households' food security when the market economy had failed to do so. These programs are intended to reduce vulnerability to food insecurity during economic downturns in the business cycle. Individuals with longer-term needs resulting from chronic illness, disability, or old age also rely on these assistance programs to maintain food security. Each program has its own objectives, its own eligibility criteria, its own benefit structure, and its own legislative oversight.

7. Economic and food assistance programs in the United States

Money income is an incomplete measure of a family's potential ability to fulfill basic needs. It often omits many goods and services such as housework and child care provided for within the household rather than purchased. In addition, it neglects any in-kind benefits that families receive from the government in the form of goods or services rather than cash. The largest of these programs are Food Stamps, Medicare, Medicaid, various housing subsidy programs, and aid to education.

Federally sponsored economic security programs in the United States were first enacted in response to the depressed economic situation in the 1930s. The Social Security Act of 1935 established two social insurance programs on a national scale to help prevent deprivation associated with old age and unemployment: a federal system of old age benefits for retired workers who had been employed in industry and commerce, and a federal–state system of unemployment insurance. The Social Security Act also provided federal grants to states for means-tested programs for the aged, blind, and disabled to supplement the incomes of persons who were either ineligible for Social Security or whose benefits could not provide a basic living. In 1972, the federally administered Supplemental Security Income (SSI) program replaced these grants (Table 1).

The original Social Security Act also provided for grants to enable states to extend and strengthen maternal and child health and welfare services. This provision evolved into the Aid to Families with Dependent Children program, which was replaced in 1996 with a new grant program to states for Temporary Assistance for Needy Families. U.S. workers with dependent

Table 1
Government transfer payments to individuals by type: 1990–2000 (million $)

Item	1990	1995	2000
Total	561,399	841,041	1,013,424
Retirement and disability insurance benefits payments	263,854	350,027	425,333
Medical payments	189,099	337,532	423,180
Income maintenance benefit payments	63,481	100,444	106,421
Supplemental Security Income (SSI)	16,670	27,637	31,675
Family assistance[a]	19,187	22,637	18,277
Food stamps	14,741	22,447	14,939
Other income maintenance[b]	12,883	27,634	41,530
Unemployment insurance benefit payments	18,208	21,864	20,707
Veterans benefit payments	17,687	20,545	24,939
Federal education and training assistance payments	7,300	9,007	10,729
Other payments to individuals	1,770	1,622	2,115

[a] Through 1995, consists of emergency assistance and aid to families with dependent children. Beginning with 1998, consists of benefits—known as Temporary Assistance for Needy Families (TANF).

[b] Consists largely of general assistance, expenditures for food under the supplemental program for women, infants, and children; refugee assistance; foster home care and adoption assistance; earned income tax credits; and energy assistance.

Source: U.S. Bureau of the Census, Statistical Abstract of the United States.

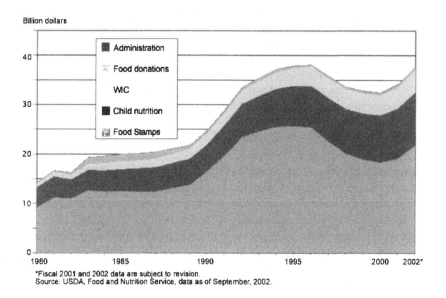

Figure 10. USDA expenditures on food assistance programs, fiscal 1980–2002*. Spending on food assistance programs increased in fiscal 2002.

children are given deductions in the computation of their federal income tax liability. In addition, since the enactment of the Earned Income Tax Credit in 1975, the working poor receive an additional reduction in their tax liability and, in some cases, a wage supplement.

U.S. agriculture and nutrition policy includes a number of farm program and food assistance and nutrition programs that lower food prices and also contribute to the social safety net. The core food assistance programs, managed by the U.S. Department of Agriculture, include the Food Stamp Program, the school meals programs, the Special Supplemental Nutrition Program for Women, Infants, and Children (WIC), and commodity distribution programs. These programs serve one in every six Americans at some point during the year. The federal government relies on state and local, public, and private agencies to administer, and in some cases contribute to the funding of, its food assistance efforts.

The Food Stamp Program is the foundation of the food assistance safety net. It provides benefits to qualifying families and supports markets for agricultural products. With program costs of $17.8 billion in fiscal 2001, it is the country's largest food assistance program. Using normal retail marketing channels, the Food Stamp Program provides qualified households with increased food purchasing power to acquire food. It offers the only form of assistance available nation-

wide to all households on the basis only of financial need, irrespective of family type, age, or disability. For many low-income households, the Food Stamp Program represents a major share of their household resources. For a typical low-income family with children, food stamps provide 25% of the family's total purchasing power (Figure 10).

The National School Lunch Program provides lunches free or at low cost to more than 27 million children each school day. In 1998, the program was expanded to offer snacks to children in after-school programs. Since 1972, the School Breakfast Program has also supported provision of breakfasts at schools. School districts and independent schools that choose to participate in one or more of the school meals programs receive cash subsidies from the federal government for each meal they serve. In return, they must serve meals that meet federal nutritional requirements, and they must offer free or reduced-price meals to low-income children.

Established in 1972 as a pilot program, WIC has grown rapidly and matured into a core component of the U.S. nutrition safety net. The program targets low-income women, infants, and children (up to the age of 5 years) who are at nutritional risk. WIC achieves this objective by providing (1) nutritious foods to supplement diets; (2) information on healthy eating; and (3) referrals to health care. It seeks to provide early intervention during critical times of growth and development that can help prevent future medical and developmental problems. In fiscal 2001, the program served an average of 7.3 million participants per month. Almost half of all infants and about one quarter of all children aged 1–4 years in the United States participate. Federal program costs totaled $4.2 billion in fiscal 2001, making WIC the country's third largest food assistance program, behind the Food Stamp Program and the school meals programs ($7.9 billion).

8. A food safety net

Social safety nets can be viewed as income insurance to help people through temporary livelihood shocks and stresses, such as those caused by drought, illness, unemployment, or displacement. Redistributive income transfers to chronically poor groups (e.g., the old or disabled) can be separated from systems of income insurance or safety nets for people who are acutely vulnerable to adverse events. A safety net targets two distinct sets of people: those unable to participate in the growth process and those who may be temporarily aversely affected when events take an unfavorable turn.

To provide an economic buffer, payments or transfers should rise during periods of economic downturn and contract during economic expansion. During both the recession of the early 1980s and the downturn of the 1990s estimates suggest that U.S. government transfers were responsible for significantly reducing the number of poor.

Recent increases in food stamp outlays highlight the role of food stamp programs as an important component of the social safety net both for the persistently poor and the working poor. Food stamps remain the sole federal entitlement program and, therefore, will likely be the primary personal income buffer operating during economic downturns, particularly for households that may have stronger ties to the workforce and move in an out of poverty.

The primary channel through which general economic conditions influence a household's income is earnings from employment. During an economic downturn unemployment rises. For the households whose members lose their job, their income falls. If these households have little unearned income and few savings they are likely to become eligible for food stamps, increasing the program's caseload. Unemployment is the primary channel through which an economic downturn affects the Food Stamp Program caseload for those attached to the labor force. There are other households that have little labor force attachment, and for them the economic conditions in the labor market are not as significant a determinant of their participation decision.

Recent evidence (Jolliffe et al., 2003) suggests the depth and severity of child poverty and poverty overall are significantly reduced by the Food Stamp Program. From 1988 to 2000, a time period capturing a recession and recovery, the Food Stamp Program evidenced large increases and then a subsequent decline in participation. Interestingly, adding Food Stamp Program benefits to income results in a small decrease in the incidence of poverty because the benefit structure is constructed so that as household income increases, food stamp benefits decrease. In general, for every dollar increase in income, food stamp benefits decrease by 30 cents. In addition, participation rates among poor households at the upper end of the poverty continuum are lower. A headcount measure of the effects on poverty status suggests little change in the poverty rate (Table 2).

By way of contrast, supplementing income by the value of food stamps has the effect of reducing the depth and severity of poverty, as measured by a poverty gap index, by 16% to 17%.[5] The results for child poverty are more striking yet. Supplementing income with food stamp benefits results in a mild reduction in the incidence of child poverty (from 4% to 7%). The severity and depth of child poverty ranges from 14% to 23% (Figure 11).

9. Sustainable reductions in poverty and food insecurity

In the lexicon of entitlements (Dreze and Sen, 1989), social safety nets are "entitlement protection" measures. The objective of these measures is to prevent or offset an acute decline in living standards following an economic shock. This contrasts with measures that

---

[5] The poverty gap index can be interpreted as the product of the headcount index and the income gap, where the income gap is the average shortfall of the poor as a fraction of the poverty line.

Table 2
Percentage reduction in poverty from food stamps, 1988–2000

Year	Head count index			Poverty gap index		
	Income only	Income + food stamps	Decline (%)	Income only	Income + food stamps	Decline (%)
1988	13.0	12.6	3.5	5.7	4.8	15.2
1989	12.8	12.3	4.4	5.5	4.6	15.6
1990	13.5	12.9	4.4	5.8	4.8	16.4
1991	14.2	13.4	5.4	6.2	5.2	17.2
1992	14.5	13.7	5.6	6.5	5.4	16.7
1993	15.1	14.3	5.4	6.8	5.7	16.6
1994	14.5	13.7	6.1	6.5	5.4	17.1
1995	13.8	13.0	6.2	6.1	5.1	16.4
1996	13.7	13.0	5.1	6.0	5.1	15.7
1997	13.3	12.6	4.8	6.0	5.2	13.4
1998	12.7	12.1	4.7	5.8	5.1	11.6
1999	11.8	11.3	4.3	5.3	4.8	10.6
2000	11.3	10.9	3.3	5.1	4.6	9.4

Source: Jolliffe et al. (2003).

enhance living standards to reduce chronic poverty and economic insecurity in the long term. A policy intervention such as food aid or a cash grant will allow the beneficiary to bridge their consumption deficit for as long as the transfer program continues. This welfare transfer will not reduce poverty sustainably, because it has no impact on productivity, it is not a livelihood-enhancing intervention.

Relying on undifferentiated economic growth is likely to prove increasingly inefficient for permanently or sustainably mitigating U.S. poverty and food insecurity. Macroeconomic progress was a powerful engine for reducing poverty in the United States during the 1940s and 1950s. It has been less successful in reducing poverty since the mid-1970s and early 1980s. Obvious possible explanations—poor aggregate economic performance during the 1970s and 1980s and a changing distribution of the gains from growth even with improved growth during the 1990s—apparently hold. Figure 12 plots the percentage of the population with incomes less than 125% of the official poverty rate against real personal income per capita for 1959 through 2001. Extending the measure of poverty to include people who are at 125% of the official poverty

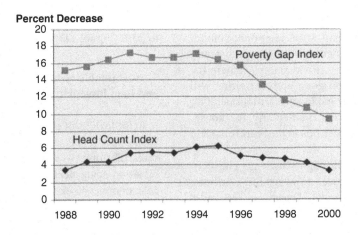

Figure 11. Percentage reduction in poverty from food stamps (1988–2000). *Source:* Jolliffe et al. (2003).

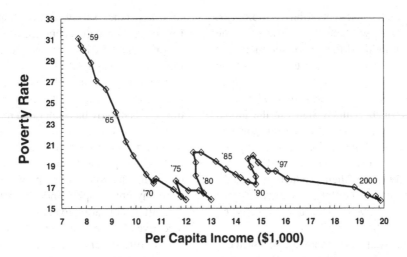

Figure 12. Relationship between poverty and income (1959–2001).

threshold captures a larger number of the working poor, low-income people whose poverty status is more sensitive to changes in the macroeconomy.

From 1959 until the late 1970s, fluctuations in the poverty rate paralleled changes in the performance of the macroeconomy. Business cycle upswings significantly reduced poverty while business cycle troughs increased poverty. Since the late 1970s the relationship between economic performance and poverty is less clear. From 1959 to 1989 per capita growth averaged a fairly constant 2.7%. From 1959 to 1969 the poverty rate declined dramatically but only modestly during the 1980s. During the high-growth period of the 1990s, where real per capita GDP increased by 34%, the U.S. poverty rate declined by only 13%. This contrasts with the 1960s where a 35% increase in per capita income was associated with a 43% decrease in poverty.

Although many reasons for the divergence between the historical poverty rate and economic growth have been advanced, only three or four are compelling. Some leading candidates include changing institutional wage-setting mechanisms with the decline of labor unions, a changing labor cohort, globalization of production and markets, shifting relative expenditures for goods and services, technological change associated with the digital revolution, and increased earnings instability (Levy and Murnane, 1992). No single cause is likely large enough to account for the divergence between economic growth and poverty, but labor market changes fostered by technological change and economic restructuring and shifts in relative wages are important.

Increasing wages for high-skilled individuals and deteriorating wage and career opportunities for many less-skilled workers over the past 15 years have resulted in increased earnings inequality. Earnings inequality has taken the form of polarization and the apparent hollowing-out of the income distribution, where the middle of the distribution has declined and the upper and lower tails have increased. The observed increase in earnings inequality has been driven by increased wage variation rather than changing hours of work. Polarization combined with nearly stagnant growth in average earnings has meant the proportion of men with earnings below $20,0000 and over $40,000 per year has both increased.

Research conducted during the early 1990s suggested low wages and not less work was responsible for lower earnings. For the working poor, low numbers of work hours appeared not to be a major factor in the incidence of poverty among households headed by either men or women. Instead, evidence suggested that, based on the characteristics of the heads of poor households, the expected average wage rates were low (Levy and Murnane, 1992).

Over the last two to three decades the demand-side forces of industrial restructuring, technological change, and shifting relative expenditures on services have reinforced the effects of the entry into the workforce of a well-educated generation of employees. The combination of these changes has increased the relative demand for highly skilled labor and increased earning inequality. Industrial restructuring cannot be separated from technological change. They are dual manifestations of the same phenomenon. The comparative advantage of the United States in the global market place is its highly skilled and well-educated workforce. The increased internationalization of product markets does not mean only industries in the United States that primarily export their production are subject to increased competition. Rather, workers in general have seen their relative compensation affected by profits and wages in other countries.

There are several studies that indicate recent technological change has favored the more skilled over the less skilled (Welch, 1970; Davis and Haltiwnager, 1991; Katz and Murphy, 1992; Sachs and Shatz, 1994). Faster technological change linked with greater spending on research and development has been associated with increasing pay differentials between less-educated and highly educated workers (Bartel and Lichtenberg, 1987). More recent studies indicate workers who use computers in the execution of their job received higher wages than those that did not (Krueger, 1993). These studies suggest the shift in demand away from unskilled and toward skilled labor in U.S. manufacturing is explained by the adoption of labor-saving technological change and a reallocation of production away from industries with a high labor production component (Berman et al., 1994).

10. Conclusions

Although food assistance programs are a critical component of the U.S. food security and income safety net, renewed economic growth will be critical for improving the sustainable level of food security of U.S. households. Targeted policies and programs that improve employment and earnings opportunities for the types of households that are most vulnerable to food insecurity—especially those with less-skilled or less-educated workers and those headed by single women with children—can also contribute to improving food security. Achievement of the targeted reductions in food insecurity and hunger will also require continued federal, state, and private commitments to the country's food assistance safety net. Innovative and principled improvements in the economic and nutrition safety net programs can further improve the likelihood of reaching food security goals.

The key to sustainably reducing poverty and food insecurity is to improve the returns to labor. Although the U.S. economy has generated many jobs, wages and benefits are often not sufficient to lift a family out of poverty. Improvements in education and job training are the primary venue for improving wages for the poor. The connections between education and the economy are complex and interdependent. The economy affects the attractiveness of education by creating incentives for the poor to continue schooling. In addition, family structure and stress and access to learning resources and experiences have important implications for educational success. We cannot focus solely on educational institutions to increase the quality of workers. Attention and, more importantly, financial resources must be focused on the lives of children in and out of school if we are to improve the knowledge, abilities, and attitudes of the future work force.

Looking toward the future, the long-term prospects for improving food security are likely to be driven by the same general forces shaping the U.S. economy— globalization of markets and cultures; advances in information and technology; and fundamental changes in the workforce. In the end, however, it is clear that the persistence of food insecurity in wealthy societies cannot be fully understood if attention is confined only to income. Food insecurity in the United States is associated with many causal factors of which low income is only one, although an important one. The social environment, the provision of medical care, the pattern of family life, and a variety of other factors affect food insecurity.

References

Bartel, A., and F. Lichtenberg, "The Comparative Advantage of Educated Workers in Implementing New Technology," *Review of Economics and Statistics* 69, no. 1 (1987), 1–11.

Berg, A., *The Nutrition Factor* (Brookings Institute: Washington, DC, 1973).

Berman, E., J. Bound, and Z. Griliches, "Changes in the Demand for Skilled Labor within U.S. Manufacturing: Evidence from the Annual Survey of Manufactures," *Quarterly Journal of Economics* 109, no. 2 (1994), 367–397.

Blank, R., *It Takes a Nation: A New Agenda for Fighting Poverty* (Princeton University Press: Princeton, NJ, 1997).

Chambers, R., "Editorial Introduction: Vulnerability, Coping, and Policy," *IDS Bulletin* 20 (1989), 1–7.

Chung, K., L. Haddad, J. Ramakrishna, and F. Riely, *Alternative Approaches to Locating the Food Insecure: Qualitative and Quantitative Evidence from South India*, Discussion Paper No. 22 (International Food Policy Research Institute: Washington, DC, 1997).

Davis, S., and J. Haltiwanger, "Wage Dispersion between and within U.S. Manufacturing Plants, 1963–1986," *Brookings Papers on Economic Activity* (1991), 115–200.

Dreze, J., and A. Sen, *Hunger and Public Action* (Clarendon Press: Oxford, 1989).

Duncan, G., *Year of Poverty, Years of Plenty: The Changing Economic Fortunes of American Workers and Families* (Institute for Social Research, University of Michigan, 1984).

European Commission, *Food Security Policy: Examination of Recent Experiences in Sub-Saharan Africa*, Commission Staff Paper, SEC (1988).

Food and Agriculture Organization of the United Nations, *The State of Food Insecurity in the World, 2002* (2002a).

International Scientific Symposium on Measurement and Assessment of Food Deprivation and Undernutrition: Executive Summary, United Nations Food and Agriculture Organization, FIVIMS (Food and Agriculture Organization of the United Nations: Rome, Italy, June 2002b).

Gittlesohn, J., S. Mookherji, and G. Pelto, "Operationalizing Household Food Security in Rural Nepal," *Food and Nutrition Bulletin* 19, no. 3 (1998), 210–222.

Hoddinott, J., *Choosing Outcome Indicators of Household Food Security*, Technical Guide No. 7 (International Food Policy Research Institute: Washington, DC, 1999).

Jolliffe, D., C. Gundersen, L. Tiehen, and J. Winicki, *Food Stamp Benefits and Child Poverty in the 1990s* (Economic Research Service, U.S. Department of Agriculture, 2003).

Joy, L., "Food and Nutrition and Planning," *Journal of Agricultural Economics* 24, no. 1 (January 1973), 196–197.

Kabeer, N., *Monitoring Poverty as if Gender Mattered: A Methodology for Rural Bangladesh*, Discussion Paper No. 255, Institute of Department Studies, University of Sussex, Brighton (1998).

Katz, L., and K. Murphy, "Changes in Relative Wages, 1963–1987: Supply and Demand Factors," *Quarterly Journal of Economics* 107 (1992), 1–34.

Kielman, A., C. Taylor, R. Farugee, C. DeSweemer, D. Chernichovshy, I. Uberoi, N. Masih, R. Parker, W. Reinke, D. Kakar, and R. Sacma, *Child and Maternal Health Services in Rural India: The Narangwal Experiment*, 2 vols (A World Bank Research Publication, Johns Hopkins University Press: Baltimore, 1983).

Krueger, A., "How Computers Changed the Wage Structure: Evidence from Microdata, 1984–1989," *Quarterly Journal of Economics* 108, no. 1 (1993), 33–36.

Levinson, F. M., *An Economic Analysis of Malnutrition among Young Children in Rural India*, Cornell-MIT International, Nutrition Policy Series (1974).

Levy, F., and R. Murnane, "U.S. Earnings Levels and Earnings Inequality: A Review of Recent Trends and Proposed Explanations," *Journal of Economic Literature* 29, no. 4 (1992), 1333–1381.

Maxwell, D., C. Ahiadeke, C. Levin, M. Armar-Klemesu, S. Zakariah, and G. M. Lamptey, "Alternative Food-Security Indicators: Revisiting the Frequency and Severity of 'Coping Strategies'," *Food Policy* 24 (1999), 411–429.

Maxwell, S., "The Evolution of Thinking about Food Security," in S. Maxwell and S. Devereux, eds., *Food Security in Sub-Saharan Africa* (ITDG Publishing: London, 2001).

Nord, M., M. Andrews, and S. Carlson, *Household Food Security in the United States, 2001* (Economic Research Service, U.S. Department of Agriculture, October 2002).

Payne, P., "Measuring Malnutrition," *IDS Bulletin* 21 (July 1990).

Reardon, T., and P. Matlon, "Seasonal Food Insecurity and Vulnerability in Drought Affected Regions of Burkina-Faso," in D. E. Sahn, ed., *Seasonal Variability in Third World Agriculture: The Consequences for Food Security* (Johns Hopkins University Press: Baltimore, 1989).

Sachs, J., and H. Shatz, "Trade and Jobs in U.S. Manufacturing," *Brookings Papers on Economic Activity* (1994), 1–84.

Sen, A., *Poverty and Famines: An Essay on Entitlement and Deprivation* (Clarendon Press: Oxford, 1981).

Siamwalla, A., and A. Valdes, "Food Insecurity in Developing Countries," *Food Policy* 5, no. 4 (November 1980), 250–272.

Staaz, J., "Food Security and Agricultural Policy: Summary," *Proceedings of the Agriculture-Nutrition Linkage Workshop* (1990).

Townsend, P., "Poverty as Relative Deprivation: Resources and Styles of Living," in D. Wedderburn, ed., *Poverty, Inequality and Class Structure* (Cambridge University Press: Cambridge, UK, 1974).

Webb, P., J. Coates, and R. Houser, "Challenges in Developing a 'Direct Measure' of Hunger and Food Insecurity for Bangladesh: Preliminary Findings from Ongoing Field Research," Paper prepared for the International Scientific Symposium on Measurement and Assessment of Food Deprivation and Under-Nutrition Sponsored by the Food and Agriculture Organization, Rome, Italy, June 26–28, 2002.

United States Bureau of the Census, *Poverty in the United States, Current Population Reports* (Government Printing Office: Washington, DC, Various Years).

U.S. Bureau of the Census, *Statistical Abstract of the United States* (Government Printing Office: Washington, DC, Various Years).

Welch, F., "Education in Production," *Journal of Political Economy* 78, no. 1 (1970), 35–59.

Wolfe, W., and E. Frongillo, "Building Household Food—Security Measurement Tools from the Ground Up," *Food and Nutrition Bulletin* 22, no. 1 (2001), 5–12.

Food safety, the media, and the information market

Johan F. M. Swinnen*, Jill McCluskey**, and Nathalie Francken*

Abstract

Availability of information has increased rapidly over the past decades. Yet, information on food safety is still considered problematic. Economists have extensively researched the effects of imperfect information. However, little attention has been paid to the institutional organization of the supply of information and the incentive schemes in the information market. This paper analyzes how and when information is supplied by media organizations, and what the implications are. We first develop a theoretical framework and afterwards provide empirical evidence from media coverage of two recent food safety crises in Europe.

JEL classification: L82, P16, Q18

Keywords: food safety; media; information market; political economy

1. Introduction

The . . . press appears to me to have passions and instincts of its own. . . . In America as in France it constitutes a singular power, so strangely composed of mingled good and evil that liberty could not live without it, and public order can hardly be maintained against it.

—Alexis de Tocqueville, 1853

Although households are flooded with information through dozens of TV channels, plenty of newspapers, journals, and radio, the public is said to be poorly informed when consumption declines dramatically following media reports on food contamination, or when European consumers oppose the introduction of genetically modified organisms (GMOs) in their food, despite claims by scientists and official institutions that those products are safe.

The main (implicit) assumption in the extensive literature on the impact of imperfect information (see

e.g., Akerlof, 1970; Stiglitz, 1993) is that such problems may be solved by improving provision of information. The problem is that information provision is assumed to be neutral. However, in reality, most information is not provided by institutions whose objective is to foster the public good, but by organizations that have an internal incentive to select certain information items and certain forms of information over others in their information distribution activities. Information is provided either by private sources with their own profit-maximizing objectives or by public sources that may formally be charged with providing objective information, but administrators and governments may have incentives to bias the information.

Increasingly, the most important ally of governments is no longer the police or the military; it is the media, as is illustrated in the following quote:

The surest thing that Tony Blair was kick-starting an election campaign wasn't his traditional trip to Buckingham Palace to tell the queen. More important was Rupert Murdoch's early-May call on 10 Downing Street. Through two decades and three prime ministers, the News Corp.'s chairman has reigned as Britain's political kingmaker.

—McGuire, 2001, p. 21

* Department of Economics and LICOS—Centre for Transition Economics, Katholieke Universiteit Leuven, Leuven, Belgium.
** Department of Agricultural Economics, Washington State University, Pullman, WA, USA.

A prime target of political organization is control over the media, as witnessed by recent maneuvering by, for example, Silvio Berlusconi in Italy and President Putin in Russia. The first target of a military coup or a popular uprising is no longer the police station, but the TV station, as is well illustrated by the following quote:

> The most important single pillar of Slobodan Milosevic's power [in Serbia] throughout the 1990s was state-run television, with its insidious diet of nationalist lies and regime propaganda.... The crucial moments of the Serbian revolution to overthrow Mr. Milosevic were also televisual. First the storming of the Belgrade Parliament, seen via satellite on CNN, BBC, and Sky and on embattled semi-independent provincial channels in Serbia itself. Second, the storming of the headquarters of state television, revealingly known as TV Bastille, and the subsequent appearance of the new president, Vojislav Kostunica, on that channel. Those were the two moments that told everybody it was over, even though all the traditional organs of power were still formally (and many so also very practically) in the hands of Mr. Milosevic. It had happened on television, so it had happened. Television made it true.

—Ash, 2000

This issue is particularly relevant for food safety and consumers' associated risk perceptions. Over 90% of consumers receive information about food and biotechnology primarily through the popular press and television (Hoban and Kendall, 1993). Extensive media coverage of an event can contribute to heightened perception of risk and amplified impacts (Burns et al., 1990). For example, Johnson (1988) shows how media coverage of product contamination by the pesticide ethylene dibromide (EDB) resulted in important disruptions in the market for grain products. Other recent examples are the media coverage of the Alar scare in apples in the United States and bovine spongiform encephalopathy (BSE), commonly known as "mad-cow disease," in Europe, Japan, and now Canada.

Alar, the trade name for daminozide, which regulates fruit set, size, coloring, and ripening, is used primarily on apples. In February 1989, the U.S. television news program *60 Minutes* aired a story on the Natural Resources Defense Council's (NRDC) findings that Alar is a cancer risk to children. In the following months, a number of media organizations featured the Alar story, resulting in a panic. School systems removed apples from their cafeterias, and supermarkets took apples off their shelves. U.S. apple growers lost millions and announced a voluntary ban effective fall 1989. In hindsight, analysts argue that the media confused a long-term cumulative effect with imminent danger, resulting in unnecessary panic and losses (Negin, 1996).

The BSE outbreak and its effects on the livestock industry, beef demand, and consumers' food safety perceptions have been studied mainly in Europe where a large number of countries have been affected. Verbeke et al. (2000) find that television coverage on meat safety had a negative effect on demand for red meat after the BSE outbreak in Belgium. Younger people, and households with young children, were the most susceptible to such negative media coverage. Verbeke and Ward (2001) find that advertising had only a minor impact on meat demand compared to negative media coverage in Belgium after the BSE discoveries.

It is remarkable how little attention has been paid so far to how the industrial organization of the media industry and the structure of the information market affect the quantity and quality of information supply, although some recent studies have emerged on this issue, analyzing the impact of media structures and ownership on information distribution and economic welfare (Besley and Burgess, 2001; Besley and Pratt, 2002; Djankov et al., 2001). In this regard, the objective of this paper is to contribute to this understanding by developing a theoretical framework of the information market and comparing it with the empirical observations on media reporting on two recent food safety crises. The paper starts with a theoretical analysis of the organization and incentive structure of the media industry, looking at both supply and demand factors, and discusses a set of general hypotheses on the characteristics of information provided by the media industry. This section draws from a formal model of the information market developed in Swinnen and McCluskey (2003). The second part of the paper compares the hypotheses with the results from an empirical study of media reporting on two recent food safety crises in Western Europe: the 1999 dioxin crisis that originated in Belgium, and the 2001 foot-and-mouth disease (FMD), which originated in the United Kingdom (Swinnen and Francken, 2002).

2. A positive theory of media and information

Newspapers, TV, radio, and other media simultaneously decide *what* (which issues) to report, *how*, that is, in which format (pictures, interviews, text, etc.), and which aspects (positive vs. negative aspects; business vs. environment effects; etc.). Therefore, define a "story" as a unit of media coverage. Each story is characterized by the issue on which it contains information, and by its attributes. Define $m(\theta)$ as the number of stories on a specific issue, with θ the set of attributes of the story. The attributes are a variety of characteristics, such as ideology, attitude (e.g., negative vs. positive), format, and regional coverage. For each of these attributes, one can assume a single-dimensional space between two extremes. The supply of stories and their attributes will depend on the structure of the media industry and on the demand for information on the issue.

2.1. Supply side of the market

Private commercial sources of information are increasingly important. While in the United States news coverage has always been largely in the hands of commercial companies, the emergence of private companies as the dominant source of information is a relatively new phenomenon in Europe. Until recently European TV and radio broadcasting were largely in the hands of state broadcasting companies, and companies publishing daily newspapers and popular journals were often closely aligned with political parties. All that has changed dramatically. Commercial TV and radio stations have emerged in Europe and are now the dominant channels. The written press has gradually devolved itself of the patronage of the political parties and is driven more by commercial than political objectives.

Nevertheless, many media organizations, either because of the preferences of their owners or because of the preferences of journalists who have sufficient autonomy to influence decision making of the media organization, do have their own attribute preferences, for example, on the ideological perspective of the stories. We therefore assume that the media organization is driven by the need for profits and its own attribute preferences. Each media company j has the following objective function for producing a story located at θ in attribute space:

$$U_j^m(\theta, \theta_j^m) = -\alpha(\theta - \theta_j^m)^2 + \beta\pi(\theta), \qquad (1)$$

where $\pi(\theta)$ is the profit function for the media company. Further, θ^m is the media company's preferred location in attribute space, and the parameters α and β reflect the relative importance of the profit objective in the company's objective function. For example, in case of the ideology attribute, if ideological propaganda is very important, α is large compared to β. An extreme example of this weighting would be the *Pravda*, the former Soviet News Agency. On the other hand, for commercial media companies, α is smaller, and can be zero if the company only cares about profits. Except for the two extreme cases, the company will trade-off between ideology and profits. The same holds for the other attributes. We define profits as $\pi(\theta) = pm(\theta) - F$, where p is the exogenous price of stories and F is the fixed cost of production.

We will first consider the case of a monopolist. The firm chooses the location to maximize its objective function

$$\max_{\theta} \ -\alpha(\theta - \theta^m)^2 + \beta pm(\theta) - F. \qquad (2)$$

The first-order condition is the following:

$$-2\alpha(\theta^* - \theta^m) + \beta pm'(\theta^*) = 0. \qquad (3)$$

The first-order condition exhibits the firm's trade-off between ideology and profits. The supply is determined by the ideological choice, θ, which solves the firm's first-order condition (3). The ideological choice and exogenous price determine the number of stories supplied:

$$m^* = m^*(\alpha, \beta, p, \theta^*, \theta^m). \qquad (4)$$

The equilibrium number of stories can be found by setting supply equal to demand.

Now we consider adding additional firms. In general, each media company will choose an ideological location, θ_j, to maximize its objective function relative to the location of the competition. When firms care about both location and profit, we obtain

$$\theta_j = \theta(\alpha, \beta, p, \theta_j^m, \theta_1, \ldots, \theta_{j-1}, \theta_{j+1}, \ldots, \theta_n). \qquad (5)$$

Adding additional media firms affects the ideological choice of media firm j because it now wants to differentiate its product relative to the competition. The decision making of the company will obviously be affected by the structure of the media industry and by the preferences of consumers. In a media environment characterized by competition and free entry, companies whose editorial policies are mainly concerned with ideology may find themselves either facing losses or a small part of the market unless the ideological preferences of the population fit perfectly with that of the media organization.

2.2. Demand side of the market

The reader uses the information provided by the newspaper to reduce the variance of his/her estimate of truth, which could relate to many things, including the financial health of a firm (as suggested by Mullainathan and Shleifer, 2002) or which politician will be elected. For simplicity, we just assume here that the consumption of media products will positively affect his/her income through an increase in the reader's knowledge on certain issues.[1] A consumer further obtains positive utility from consuming leisure and all other goods and obtains negative utility from consuming media stories that diverge from his or her attribute preference. Specifically, the utility a consumer obtains from m stories located at θ in attribute space is

$$U^c(\theta, \theta^c, L^l, x) = g(L^l, x) - cm(\theta - \theta^c)^2, \qquad (6)$$

where θ^c is the location in attribute space of the consumer's favorite type of attributes, c is a scaling parameter, m is the number of media stories purchased, L^l is time spent on leisure, $g(\cdot)$ is a concave function, and x is all other goods. The squared difference between the location of the media products and the consumer's favorite type of media negatively affects utility. That is, consumers get disutility from reading stories that diverge from their favorite types. For now, we assume that all consumers are identical except for ideological preferences. We assume that there are n consumers

whose favorite ideological locations, $\theta^{ci}, I = 1, \ldots, n$, are uniformly distributed in the ideological space.

The consumer maximizes utility subject to a budget constraint,

$$x + pm = f(L^w; k(mL^m)), \qquad (7)$$

and a time constraint,

$$1 = L^l + L^w + L^m, \qquad (8)$$

where p is the exogenous price of news stories, $f(\cdot)$ a concave production function, L^w is time spent working, L^m is time-processing media stories, and k is the level of knowledge or informational content gained from consuming media stories. The price of all other goods, x, is normalized to 1. Note that knowledge is gained from both story purchases and the time taken to process stories. Both are necessary for obvious reasons. Media consumption increases utility by increasing income as knowledge increases.

The first-order conditions yield the usual result that the net marginal benefit of time must be equal across the uses of processing media stories, work, and leisure. This model yields a demand equation for consumer i for news stories:

$$m_i = m_i(\theta, \theta_i^c, p, x), \qquad (9)$$

where demand is a function of the attributes of the stories, the consumer's favored attributes, price, and consumption of other goods.

3. Rationally ignorant consumers and negative news coverage

3.1. The rationally ignorant consumer hypothesis

A first result from this model is that it is rational for consumers to be imperfectly informed. There are three reasons why it is rational for most consumers not to inform themselves fully on an issue. First, most obviously, if the price (p) of news stories is high compared to the marginal benefits of information, it will cause consumers to stop purchasing information. With decreasing returns to information we see that it is rational for individuals not to be fully informed. Consumers will prefer to inform themselves only up to a

---

[1] The model in Swinnen and McCluskey (2003) is less restrictive and formally includes dynamic aspects and more details on how consumption of newspaper stories affects information variance and income.

point where further acquisition of information will be too costly, either because (with decreasing returns to information) the increase in income from more information becomes less than the cost (price p).

Second, reducing the price of stories will increase consumer information, but only up to a point. Even when stories are free, consumers will stop acquiring more information when the opportunity costs of processing the information become larger than the growth in income, or both. Opportunity costs play an important role, especially when considering trade-offs involved in information accumulation on many issues and problems.

A third reason why consumers may choose to be less than fully informed has to do with the attributes of the stories presenting the information. As explained above, consuming a story may have a negative impact on consumer welfare because of the story's ideological bias. From the first-order conditions, the marginal disutility from consuming media with a divergent ideology must be equal to the marginal net benefit from the increase in income due to knowledge. With decreasing returns to information, there is a point where consumers prefer not to inform themselves any further if consumer attitude preferences differ from those offered by the media.

3.2. Bad news hypothesis

A second result is that the generally recognized tendency of the popular media to publish mostly negative aspects of news items is driven by the demand of their audience, rather than by inherent preferences of the media itself. To understand this, consider for a moment that there are two types of stories: positive stories or "good news" ($\theta^A = G$) and negative stories or "bad news" ($\theta^A = B$). Think of good news as stories about happy endings, in which people made the right choices. Bad news stories are about unhappy endings, in which people made the wrong choices. When consumers read good news stories, they can make similar choices to increase their incomes. When they read bad news stories, they can choose to avoid bad outcomes and the resulting income losses.

Define the $m^B = m(\theta^A = B)$ as the demand for bad news stories, and $m^G = m(\theta^A = G)$ as the demand for good news stories. Assuming that the costs involved in purchasing and processing good news and bad news

stories is identical, it follows that individuals are more interested in bad news, m^B, than in good news, m^G. The expected value of additional information is higher when it concerns an issue with negative welfare effects than with positive welfare effects. Since $g(\cdot)$ is concave, the marginal loss in utility from not consuming the first bad news story is greater than the marginal gain in utility from consuming the first good news story. Consumers will choose story types until the marginal utility across story types is equal. By concavity, consumers will choose to consume more bad news stories than good news stories $m^{B*} > m^{G*}$.

4. Tabloids and the elite press

So far we have not considered variations in consumer characteristics and preferences. If consumers are heterogeneous, a variety of media products will emerge in an environment that allows entry of new media products and organizations.

To illustrate this, consider the simple case of a population with "high-skilled" and "low-skilled" people. Skills will affect preferences for both the types and the format of stories. First, skills affect consumers' ability to process the information contained in the stories, and therefore high-skilled consumers will prefer stories that contain more information, even if that makes the story less easily digestible ("more words and less pictures").[2] Second, skills will affect the relevance of the stories for the consumers. High-skilled people typically are more mobile, both professionally between sectors and geographically. This means that high-skilled consumers are more interested in stories that contain information that goes beyond local issues than are low-skilled consumers.

This will lead to the emergence of two differentiated media products. The first, which we refer to as the "elite press," targets the high-skilled part of the population bringing news with a wider geographical focus and in a format that is more difficult to process but contains more information. The second, which we refer to as the "tabloids," has a more local selection of news items, and is presented in an easily accessible format.

---

[2] There is some trade-off here since high-skilled people may have higher opportunity costs.

5. The dynamics of a story

An important extension of this simple model is to include the dynamics of the choice, i.e., not only *whether*, but *when* to publish a story. Collection of information requires time, effort, and other costs. Full information on an issue, in particular when it is a story about a "novel" item, emerges only gradually. Typically, early on only part of the information is available. Publishing a story based on part of the information has the risk of giving a "biased" perspective on the issue—and for the consumer it carries the risk of drawing incorrect conclusions, and thus of having possibly negative welfare effects. The risk for the media is that reporting a biased story may hurt its reputation, and thereby future profits.

However, bringing a story early, even based on limited information, and therefore probably biased, has potential benefits for both consumers and the media. A media organization which brings the story early may capture a larger market share and profits if consumers want whatever information they can get on a new issue. Consumers also face a trade-off. If the issue is something very important they know nothing about and which is potentially very important for their welfare (say, the first information about Anthrax letters sent to U.S. government offices), they may well be willing to take the risk of getting very biased stories in exchange for getting whatever information available ("any news is better than no news").

However, in other cases consumers may not accept very biased news, and may well react to this by adjusting their future story consumption (i.e., buying other newspapers if they feel that the information provided is too biased, even if it informs them earlier).

More generally, the trade-off between current profits and future profits will depend on the structure of the media market, on the characteristics of the company (tabloids vs. elite press), and on the demand side (consumers).

In this perspective there is an interesting secondary, but important, dynamic component: once one media company (newspaper, radio, or TV) starts reporting on a story (no matter how biased), it is a free game: everybody will now use the first story as a base for their own reporting. The dynamics are well summarized by the following quote:

In practice . . . even apparently responsible papers like *The Times* actually contribute to building up the [food] scare. Then, when the scare has run its course, they will argue against it. But when the scare dynamic is up and running, *The Times* and other broadsheets will join with the throng and become more tabloid than the tabloids.

—North, 2000, p. 8

There are two reasons for this. First, the competition (and consumers) forces them to bring something on the issue, otherwise the consumers will ask: "why is this issue not addressed in my newspaper, or on this TV journal? I should look at other media!" The second reason is that by commenting on a story launched by another media company, it provides them cover in case things go wrong, i.e., when the early information turns out to be very biased. They can then hide behind the fact that they did not bring it and only commented and reflected on a story launched by another media company. In terms of our model, the first factor reduces the current losses in waiting too long with the story, and the second factor reduces future losses of reputation by not waiting long enough until more information is available.

These dynamic effects have important implications for the role of the media in the distribution of information and its impact. This is illustrated by the following quote from North (2000) on the "1678 Papist Plot":

The central figure was a character called Titus Oates, who put about a series of lies to the effect that the Jesuits were planning to murder the king and invite in foreign powers to recover England for the Pope and Catholicism. In the ensuing panic, 35 entirely innocent people were executed. It was not until 1684 that the frenzy had completely subsided.

Interestingly, this "scare," indistinguishable in principle from modern scares, existed before even daily newspapers—much less television and radio—had been invented. That a scare dynamic can run without the media suggests that the media are not primarily responsible for the phenomenon. However, the Papist Plot ran for about 6 years, which suggests that the media intensifies the scale of the scare, while perhaps bringing it to a conclusion faster.

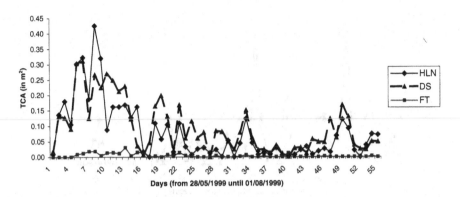

Figure 1. Total coverage of the dioxin crisis per day.

6. Empirical analysis of media reporting on food safety crises

This section summarizes the conclusions from an analysis of reporting by three newspapers on two recent food safety crises in Western Europe: the 1999 dioxin crisis and the 2001 FMD crisis. Both the newspapers and the crises were selected to analyze how different characteristics of the subjects and the media organizations affect the reporting.

6.1. Food safety crises

The dioxin crisis originated in Belgium. In May 1999, the media started reporting on dioxin contamination of food in Belgium. The resulting crisis caused strong consumer reactions, significant effects on food export markets, and had important political implications. Initially two government ministers were forced to resign, and it is widely believed that the dioxin crisis contributed significantly to the heavy electoral losses of the governing parties a few months later.

In February 2001, the FMD was discovered in the U.K. countryside. Despite drastic measures by the U.K. government, FMD was discovered on the European continent in the following weeks. The ensuing slaughter of millions of animals, blockades of local communities, and the associated effects, as well as the potential spread to other countries, were extensively reported in the West European press.

6.2. Newspapers

In order to draw conclusions about how reporting is affected by the characteristics of the event and by the nature of the media organization, reporting by three different newspapers on these two food crises that originated in different countries is analyzed. The newspapers include two Belgian newspapers that differ in audience and media strategy—a so-called "popular" newspaper (*Het Laatste Nieuws* [HLN]) and a so-called "quality" newspaper (*De Standaard* [DS])—and one international newspaper (*Financial Times*, International Edition [FT]). Figures 1 and 2 present typical covers of both newspapers to illustrate the differences between the "popular" and "quality" press.

Both Belgian newspapers are in Dutch, which is spoken by approximately 60% of the population. HLN is the most popular newspaper in the Dutch-speaking part of Belgium with a market share of 27%. Based on a ranking of the Belgian population according to income, profession, and education level of the household head from 1 to 8, its readership consists mainly of people from classes 3 to 5, with a large share of pensioners and both skilled and unskilled workers (CIM, 2001). In contrast, 56% of the readers of DS come from classes 1 and 2 of this ranking. DS has a total market share of only 8% but is considered the leading "quality" newspaper. More than half of its readers have a higher education degree (CIM, 2001).

The FT is a world business newspaper, based in London, with almost 5,000,000 issues sold per day and a global readership estimated at over 1.3 million

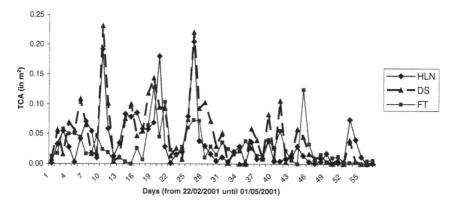

Figure 2. Total coverage of the FMD crisis per day.

people in more than 140 countries. Its audience consists mainly of business leaders, government employees, international entrepreneurs, bankers, investors, teachers, and students (Pearson, 2003).

6.3. Empirical methodology

The coverage is analyzed over a period of 2 months after the news "broke": May 28–August 1, 1999 for the dioxin crisis and February 22–May 1, 2001 for the FMD crisis. The attention paid by the newspaper to the crises is measured by constructing indicators of area coverage by the articles. The global article surface is considered the most valuable unit (Berger, 1998). More specifically, the coverage area (CA_{ij}) of food crisis i in newspaper j is calculated as

$$CA_{ij} = \sum_{p=1}^{P} \gamma_p CA_{ij}^N, \qquad (10)$$

where N is the total number of pages, γ_p is the weight of page number p, and CA_{ij}^N is the area covered by an article on food crisis i on page p. The weight of the pages in consumer perception (attention) or in the editors' choice is not known. To ensure that the choice of weights does not affect the results, a set of sensitivity analyses was run with different sets of weights. The conclusions were generally robust to the choice of weights. For each crisis and newspaper three different indicator variables were calculated per day: the "total coverage area" (TCA), which is the combination of the "illustrated coverage area" (ICA) and the "text coverage area" (XCA).

In addition, the reporting on the FMD crisis by the Belgian newspapers was categorized under either a "domestic" (Belgian) or "international" focus, following a simple approach to content analysis (see e.g., Budd et al., 1967; Marks et al., 2000a, 2000b). All articles with a focus on Belgium, for example, precautionary measures taken by the Belgian government, potential Belgian cases of the disease, national, and regional consequences of the outbreak etc., were categorized under "domestic" news. "International" news was regarded as news with an international (or "non-Belgian") focus.

6.4. Empirical results

The results of the analyses are summarized in the tables and figures. The first conclusion is the remarkable similarity between the Belgian newspapers in terms of total attention to the food safety crises (see Figures 1–4). The total coverage of the dioxin crisis and, to a lesser extent the FMD crisis, follows roughly the same pattern, with the bulk of the coverage in the first 20 days of the dioxin crisis and between days 10–30 of the FMD crisis. The correlation of daily total coverage (TCA) is 81% for the dioxin crisis and 76% for the FMD crisis (see Table 1).

The coverage by the FT is different: there is very little reporting on the dioxin crisis. As an illustration: the highest coverage of the dioxin crisis by the FT is on day 13, the day before the weekend of the elections in Belgium, when the FT speculates about the possible consequences of the crisis on the election outcome. The reporting on FMD is also less and more equally

Table 1
Correlation of the coverage dynamics

			DS & HLN	HLN & FT	DS & FT
DIOXIN		TCA	0.81	0.55	0.54
		XCA	0.81	0.49	0.52
		ICA	0.68	0.38	0.43
FMD	Total	TCA	0.76	0.21	0.46
		XCA	0.74	0.18	0.47
		ICA	0.60	0.19	0.27
	Domestic	TCA	0.81	n.a.	n.a.
		XCA	0.81	n.a.	n.a.
		ICA	0.77	n.a.	n.a.
	International	TCA	0.45	n.a.	n.a.
		XCA	0.37	n.a.	n.a.
		ICA	0.31	n.a.	n.a.

TCA = total coverage area; XCA = text coverage area; ICA = illustration coverage area; n.a. = not applicable. All these areas are measured in m^2.

spread over the day 10–50 period. Still, comparing FT reporting with HLN and DS shows a closer similarity between FT and DS, the two quality newspapers, with an average correlation of 50%, than between FT and HLN with an average correlation of 38%.

Despite the general similarities between HLN and DS in aggregate coverage, there are important differences in the format and focus of the coverage. There is a clear *difference in* terms of *format*: the average share of illustrations in total coverage is 35% in HLN, considerably more than in DS (26%) and in the FT (20%), although the illustration share in the FT is significantly higher for the FMD case, at 26% similar to that of DS (see Table 2).

There are remarkable *differences in regional coverage* between HLN and DS. DS covers more international aspects of food crises than HLN (see Figures 3 and 4). The difference in total coverage between DS

Table 2
Illustration share (in %)

			HLN	DS	FT
DIOXIN		ICA/TCA	34	27	13
FMD	Total	ICA/TCA	35	25	26
	Domestic	ICA/TCA	32	24	n.a.
	International	ICA/TCA	43	25	n.a.

TCA = total coverage area; ICA = illustration coverage area; n.a. = not applicable. All these areas are measured in m^2.

and HLN in the FMD crisis is almost entirely due to differences in coverage of international aspects of the crisis. Table 1 shows how the correlation in daily total coverage of HLN and DS is almost identical for domestic aspects of the FMD crisis (81%, which is identical to the correlation for the dioxin crisis, which was—from a Belgian perspective—almost uniquely a domestic issue), and much larger than the 45% correlation coefficient for the international aspects.

This is also illustrated by the following cases. All three newspapers published their first article on FMD on the same day, but the focus was different. The title of the DS article was "Europe stops the import of British cloven-hoofed," focusing on Europe and not on Belgium in particular, whereas the HLN article was titled "Belgium prohibited the import of British cattle because of foot-and-mouth," an article with a complete national emphasis. Further, DS has the highest "international" coverage on day 19 (see Figure 4), when the first case of FMD was found on the European continent, specifically France. The epidemic was now crossing the borders. In contrast, HLN has the highest "international" coverage on day 53, when it reports that humans had been infected by FMD in the United Kingdom.

Further, while the *dynamics* of total coverage appear rather similar at first sight, there are important differences. The tabloid press, HLN, is earlier both in its initial coverage of the crises and in reducing attention. This is well illustrated in Figures 5 and 6, which compare the share of aggregated total coverage over time. It is clear from these figures that initial coverage is more intense in the popular press, but coverage also reduces faster in the popular press. This observation is consistent with the notion that the same forces of competition for an audience in the media leads to an intensification of the media attention early on, but also to rapid decline in attention afterward.

An interesting difference between HLN and DS can be found in the *"nature" of the coverage*. This can best be illustrated with some case studies. A first illustrative case is the "Klemskerke case," which is summarized in Box 1. The Klemskerke case was a dream for the popular media: it combined intense personal drama with high-level politics, the central role played by an organically raised lamb, yielding images of high emotional value, and with bargaining and uncertainty (!) on the final outcome, which lasted for several days. In

Johan F. M. Swinnen, Jill McCluskey, and Nathalie Francken

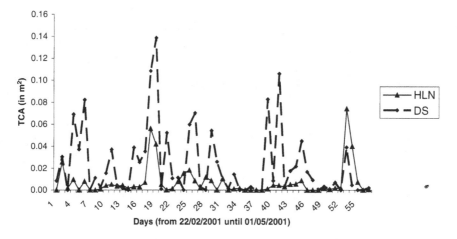

Figure 3. Total coverage of international news on FMD per day.

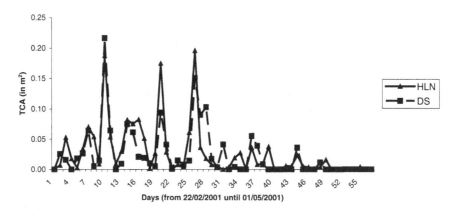

Figure 4. Total coverage of domestic news on FMD per day.

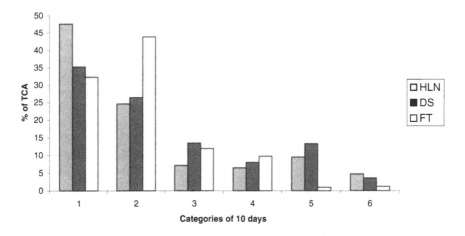

Figure 5. Total coverage (in % of TCA) of the dioxin crisis by the three newspapers.

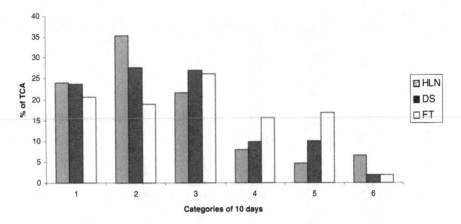

Figure 6. Total coverage (in % of TCA) of the FMD crisis by the three newspapers.

Klemskerke is the name of a village in Belgium. An organic farmer in Klemskerke imported 12 goats from Wales (Great Britain) before the FMD outbreak. After the FMD outbreak, the Belgian government decided that all the livestock of all farmers who had imported live animals from the UK had to be slaughtered. This included the organic farmer's stock of 270, mostly home reared and specially selected, organically raised goats. The government's decision resulted in wide protests from the local and organic community, attempting to block the implementation through political pressure, blockades, and legal action. While the actions proved ultimately unsuccessful and the government's decision was implemented, the story lasted for several days and strongly captured the attention of the media.

Box. 1. The Case of Klemskerke.

HLN more than 20% of total domestic coverage and 18% of total coverage of the entire FMD crisis over 3 months went to this single case (see Table 3). While DS paid less attention to this case, it is telling that, despite its "quality image," it still devoted more than 12% of aggregate total domestic coverage on the FMD crisis to this issue.

Another illustrative case is HLN reporting on FMD. On day 53, HLN reports that in the U.K. symptoms of FMD had been found in a human being. There are two interesting observations on this case. First, this is the most extensive coverage of international issues of the entire FMD crisis. Second, when a few days later it was announced that the person was not affected by FMD and that the symptoms were due to another cause, this was presented in a small article on page 17. Hence,

this case provides worrying evidence on the selectivity of reporting, and of information distribution through the media. It seems to illustrate that "(only) those who shout first, get heard."

Table 3
Coverage of the Klemskerke case by the Belgian newspapers

		HLN	DS
Klemskerke	TCA	0.37	0.18
	XCA	0.23	0.12
	ICA	0.14	0.06
	% of TCA FMD domestic	23	13
	% of TCA FMD	18	7

TCA = total coverage area; XCA = text coverage area; ICA = illustration coverage area. All these areas are measured in m^2.

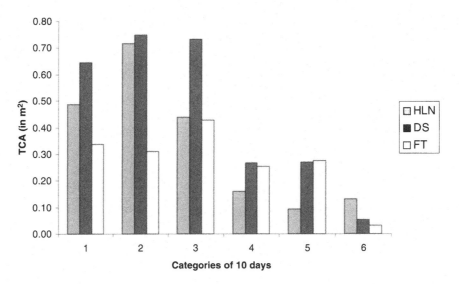

Figure 7. Total coverage of FMD after the first FMD case.

Finally, here is some evidence on "the bad news hypothesis." While it is hard to systematically collect evidence on good versus bad news, it appears that there is indeed a bias in favor of negative news reporting. In a way the entire reporting in the first months of the FMD and dioxin crises as essentially "bad news." Quite tellingly, in the months after the crises were under control, few articles appeared in the press. Figures 7 and 8 illustrate how in the 100 days after the last FMD case was diagnosed, every day that went by without a new case was "good news," hardly any articles appeared in the press. At the end of the 100 days, DS carried a small article announcing that the FMD crisis was now officially over. Nothing was reported in HLN. In the FT the announcement of the formal end of the FMD crisis was hidden inside a page 6 article on the costs of the FMD crisis—hence even the good news was buried in an essentially negative news story.

7. Conclusions

The availability of information has increased rapidly over the past decades. Yet, public information on

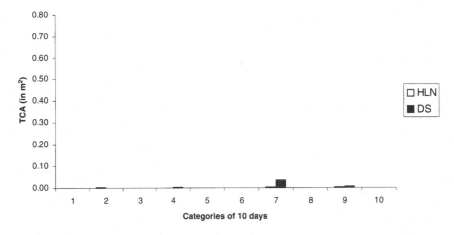

Figure 8. Total coverage of FMD after the last FMD case.

food safety issues is, rightly or wrongly, still considered problematic. While economists have extensively researched the effects of imperfect information, relatively little attention has been paid to the structure of the information market and its implications.

This paper makes a contribution to this emerging research field by studying how information is likely to be supplied by commercial media organizations, and how this is affected by the nature of the issue and by the characteristics of the population and the media industry structure.

More specifically, the analysis shows that consumers are likely to remain imperfectly informed on most issues because of the costs of information, primarily the opportunity costs of information processing. Second, a series of characteristics (related to novelty, emotional value, credibility, and uncertainty) that affect the demand for information from specific sources are identified, and an explanation is provided on how negative news coverage is likely to dominate positive news stories because of the welfare effects. Finally, the ways in which tabloids and elite presses are likely to emerge endogenously with sufficiently heterogeneous populations are discussed. In a dynamic framework, tabloids are likely to launch stories, but that if the issue is important enough, competitive forces will induce elite presses to follow, even before being able to fully verify the story.

The empirical evidence from an analysis of Belgian newspapers shows that there is a similarity between them in terms of total attention to the food safety crises, while there are important differences in the format and focus of the coverage. In terms of dynamics, the popular press is both earlier and more intense in its initial coverage of the crises, but loses interest more quickly. This observation is consistent with the notion that the same forces of competition for an audience in the media leads to an intensification of media attention early on, but also to rapid decline in attention afterward.

There is also evidence that early claims, even when false, are reported much more extensively than eventual corrections. This selectivity of reporting, and of information distribution through the media, seems to illustrate that "(only) those who shout first, get heard." Finally, while it is hard to prove formally, it appears that there is indeed a bias in favor of negative news.

Based on previous studies, media coverage of a food safety crisis increases the perceived risk of consuming the associated product. In turn, increased perceived risk affects demand for the product. This provides a strong additional incentive for firms to avoid causing food safety hazards. If an incident does occur, firms have an additional incentive to employ corrective measures as soon as possible to limit damage. The problem of false claims is serious and results in tarnished firm reputations. Because the tabloid press is more likely to get the story out fast, the probability of errors is increased. Liability issues should be considered in this area.

References

Akerlof, G., "The Market for 'Lemons': Qualitative Uncertainty and the Market Mechanism," *Quarterly Journal of Economics* 84 (1970), 488–500.

Ash, T. G., "Winner in a Long-Running Soap Opera: The Television Camera," *International Herald Tribune* (November 29, 2000).

Berger, A. A., *Media Analysis Techniques*, 2nd edition (Sage: Thousand Oaks, CA, 1998), p. 240.

Besley, T., and R. Burgess, "Political Agency, Government Responsiveness and the Role of the Media," *European Economic Review* 45 (2001), 629–640.

Besley, T., and A. Pratt, "Handcuffs for the Grabbing Hand? Media Capture and Government Accountability," CEPR working paper no. 3132 (2002), p. 39.

Budd, R. W., R. K. Thorp, and L. Donohew, *Content Analysis of Communications* (Macmillan: New York, 1967).

Burns, W., R. Kasperson, J. Kasperson, O. Renn, S. Emani, and P. Slovic, "Social Amplification of Risk: An Empirical Study," Unpublished report available from Decision Research, Eugene, Oregon (1990).

Centrum voor Informatie over de Media (CIM). Available at http://www.cim.be.

De Tocqueville, A., "Democracy in America," Reprinted in Vintage Classics Edition (Vol. 1) 1990 (1853), p. 185.

Djankov, S., C. McLiesh, T. Nenova, and A. Shleifer, "Who owns the media?" NBER working paper no. w8288 (2001), p. 29.

Hoban, T. J., and P. A. Kendall, *Consumer Attitudes about Food Biotechnology* (North Carolina Cooperative Extension Service: Raleigh, NC, 1993).

Johnson, F. R., "Economic Costs of Misinforming about Risk: The EDB Scare and the Media," *Risk Analysis* 8, no. 2 (1988), 261–270.

Marks, L. A., N. G. Kalaitzandonakes, K. Allison, and L. Zakharova, "Time Series Analysis of Risk Frames in Media Communication of Agrobiotechnology," in: Paper Presented at the Fourth International Conference of the Economics of Agricultural Biotechnology, Ravello, Italy, August 24–28, 2000 (2000a).

Marks, L., S. Mooney, and N. Kalaitzandonakes, "Quantifying Scientific Risk Communications of Agrobiotechnology," in: Paper

Presented at the Fourth International Conference of the Economics of Agricultural Biotechnology, Ravello, Italy, August 24–28, 2000 (2000b).

Mullainathan, S., and A. Shleifer, *Media Bias*. Harvard Institute Research working paper no. 1981; MIT Department of Economics working paper no. 02-33 (2002).

McGuire, S., "Blair vs. The Press," *Newsweek* (May 21, 2001), pp. 18–22.

Negin, E., "The Alar "Scare" Was for Real: And So Is That 'Veggie Hate-Crime' Movement," *Columbia Journal of Review* 35, no. 3 (1996), 13–15.

North, R., "Scared to Death: The Anatomy of the Food Scare Phenomenon," in Paper Presented at the Symposium "Gezond Voedsel: Genese en Dynamiek van een Crisis," Katholieke Universiteit Leuven, March 31, 2000 (2000).

Pearson. Available at http://www.pearson.com.

Stiglitz, J. E., *Information and Economic Analysis* (Oxford University Press: Oxford, 1993).

Swinnen, J., and N. Francken. "Food Crises and the Political Economy of the Media," PRG working paper (2002). Also available at http://www.econ.kuleuven.ac.be/prg.

Swinnen, J., and J. McCluskey, "The Rationally Ignorant Consumer, Negative News, and Other Theories of the Information Market," PRG working paper (2003).

Van Dooren, P., "Report of workshop on Greenpeace and the Agricultural Industry," *Ethical Perspectives* 7, no. 2–3 (2000), 172.

Verbeke, W., and R. W. Ward, "A Fresh Meat Almost Ideal Demand System Incorporating Negative TV Press and Advertising Impact," *Agricultural Economics* 25 (2001), 359–374.

Verbeke, W., R. W. Ward, and J. Viaene, "Probit Analysis of Fresh Meat Consumption in Belgium: Exploring BSE and Television Communication Impact," *Agribusiness* 16 (2000), 215–234.

Agricultural biotechnology: implications for food security

Vittorio Santaniello*

Abstract

In 2015 under nourishment and famine will still be at higher levels than the targets set by the World Food Conference. Agricultural biotechnology is the major technological innovation to be made available to farmers after the end of the green revolution. The research activities of the biotech community, to provide solutions to the agricultural production problems, is intense and the results might be far reaching. The development of those technologies has been at times controversial but economic analysis of their impact have shown that producers and consumers, especially in developing countries, can benefit substantially. Although agricultural biotechnology is not a silver bullet to solve food insecurity problems, it can provide a significant help. Those technologies however need to be linked to the real needs of farmers and consumers.

JEL classification:

Keywords: agricultural biotechnology; crop varieties; research and development in agriculture; food security; gene revolution

1. Introduction

It is already generally accepted that the 2015 target—set by the World Food Conference convened by FAO—of reducing by 50% the number of undernourished, from 800 to 400 million, will not be met. The most likely figure is that by that time there will still be 600 million undernourished people in the world.

In the second half of the past century the world has experienced an unprecedented food production increase; in particular in some areas of Asia where food scarcity has traditionally been a source of special concern. Significant results have been obtained in East Asia thanks in large part to the progress made in China. Success stories are also present in some areas of sub-Saharan Africa, especially where the ecology and a favorable policy environment have made this possible. Eight of the best and six of the worst-performing countries in this battle are located in sub-Saharan Africa (Meyers, 2001).

Progress has also been made in the quality of diets, as is evident in the increase in life expectancy at birth and by a decrease in infant mortality in most areas of the underdeveloped world.

In the fight against hunger the best-performing countries have experienced growth in GDP as well as in agricultural production.

This increased aggregate production that has resulted from the first Green Revolution has made possible a substantial decrease in food prices, and has had favorable effects on the food security of those who buy part or their entire food basket. The former are prevalent in the urban areas but are present as well in the rural areas.

A new revolution, a gene revolution, is now on the way. There are, however, fears that the opportunities offered by the gene revolution could worsen the relative position of small farmers, especially those in the developing countries, making them even more threatened by market forces while losing ties with their traditional knowledge.

Advocates of these new technologies, while not completely ignoring the potential adverse effects that they may have on income distribution, maintain that these new seed-based technologies are scale neutral and accessible to farmers even without complementary inputs.

* University of Rome "Tor Vergata," Italy.

From a global point of view it is evident that aggregate world food supply could easily satisfy global food demand. This, however, is not a solution to the moral, economic, and political problems posed by the food shortage faced by hundreds of millions of food-insecure people. In some cases, the surplus generated in the well-endowed regions of the world, can even become part of the problem. Poverty in fact is an important component of the food security issue. Food surplus producers in developed areas are obviously often able, more than willing, and ready to supply areas where food shortages are more severe. However, while this may ease acute short-run food security problems, it could delay the implementation of lasting solutions in a more distant future. The objective of sound policy in this area should be to promote production, and efficiency in production, where food is needed the most, i.e., among the poor farmers in disadvantaged areas.

What can biotechnology do to favor this process?

2. The technology and its development

Biotechnology is a basket of tools that have in common the use of DNA manipulation procedures to obtain products or define new processes with desired characteristics. The cadre of processes and products is rather large, although those that have recently received the greatest attention are the genetically modified organisms or varieties.

These genetically engineered varieties are the result of two separate stages in the production process. The first leads to the creation of a receptor variety that has an adequate expression of a character of economic value. The second encompasses the production, starting from the receptor, of marketable varieties of the same species.

The former can be defined more properly as a biotech process, which requires heavy investment and advanced research capabilities, while the latter requires the use of more traditional breeding techniques (Traxler et al., 1999). Those who are involved in the first stage gain a comparative advantage that allows them to be actively present in the second. With a receptor variety on hand the production of a marketable variety can be obtained with a relatively minor effort. This contributes to explaining the importance allocated by biotech companies to intellectual property rights.

It has been suggested that the presence of this two-stage process could offer an opportunity for cooperation between the private and the public sectors, where the former include those private companies that are the leaders in this area of activity, and the latter could be the National Agricultural Research Services (NARS) which, being closer to the needs of local farmers are better apt to select the mix of traits that are most in demand by farmers.

In contrast to the first Green Revolution, innovations in the biotech era are mainly in private hands, within a context where public research, extension, and public seed companies are in a perilous state or have been largely dismantled. Moreover, however, while cooperation between the public and the private sectors is often advocated, it is seldom put into practice.

3. What is in the pipeline?

It is not possible to provide a comprehensive overview of the present efforts of the scientific biotech community to address the producer's problems in the developing world, or to show what agricultural biotechnology can do to solve them. This paper in fact has a far more modest and limited scope. Nevertheless, a few examples will be provided with the aim of showing how different and far reaching are the efforts now underway in biotech laboratories and in field experiments.

New fields of agricultural biotechnology research are promising, and are gaining increased support. It might be interesting to note that in November 2004 the Vatican released a document showing open support for agricultural biotechnology in general, and for the developing world specifically. The document will call upon industrialized countries to help the Third World to develop and implement these new technologies to effectively fight food insecurity and poverty.

The Rice Genome projects and the proteomics research that has followed are well known, and need not be discussed here. In addition, however, a large range of other research is being conducted that may not enjoy the same prominence, but that nevertheless will surely produce significant results in the not too distant future. Some examples include:

• CYMMIT and IITA, with the financial support of the Rockefeller Foundation, are mapping the gene resistance to Stringa that infests 21 million hectares of maize in Africa. This map should be available soon, and will allow marker-assisted backcrossing into diverse locally adapted varieties (Mannyong et al., 2003).

- Rice Yellow Mottle Virus (RYMV) is an important disease in African rice production. Endemic in African traditional agriculture, it is also increasingly present in modern irrigated schemes and in experimental fields among Asian exotic varieties. For lowland cultivars RYMV can reduce yields by up to 97%, while with more tolerant upland cultivars damage is more limited, but can be as high as 54%. Control of RYMV is difficult because the virus is highly infectious, and the epidemiology and the role of vectors is not well understood. Natural resistance exists in African rice varieties, but introduction into new varieties is not possible because of fertility barriers, poligenicity of the resistance trait, and its recessive nature. Biotechnologists have apparently been successful in "vaccinating" new varieties by introducing in their genomes fragments of RYMV genome and generating in those transgenic varieties pathogenic-derived resistance (Pinto et al.,1999).

- *Bt* (*Bacillus thurigensis*) technology is presently the most widely employed mean to introduce pest resistance into crop varieties. A considerable amount of research, however, is underway to identify alternative routes, like identification of genes that confer natural plant resistance or enzymes and other inhibitors with pesticide resistance. Virus resistance is already amply used and more is under investigation. (Hilder and Hamilton, 1994; Khush and Brar, 1998; Flasinski et al., 2000).

- *Bt* rice is already available and attempts are now under way to produce *Bt* hybrid rice. Hybrid rice (HR) was introduced commercially for the first time in China in 1976 where it now covers 13 million hectares. Rice heterosys is an important technology that is used in several Asian countries, including India, Thailand, Vietnam, Indonesia, the Philippines, and Malaysia. HR has a yield advantage over inbred varieties of up to 20%, and allowed China to produce an additional 300 million tons of paddy between 1976 and 1994. The increased production of rice obtained in the other Asian countries would have required 6 million hectares of land without the HR varieties. Besides being more productive HR is more responsive to fertilizer and more adaptable to different environments than other varieties. On the negative side, however, HR is more vulnerable to pests and diseases, especially to stem borers. Chemical control of these insects is difficult because their larvae remain in the open for only a short time before penetrating the stem. Until now transgenic HR produced with *nptII* and *bar* genes has indicated poor field performance and therefore limited potential commercial value. New *Bt* HR varieties that are produced with a different set of genes, i.e., the *cry* genes, and that are highly resistant to the larvae of leaf folder and yellow stem borer have been tested. Although introduced transgenes have some effects on yield components (number of panicles, number and weights of filled grains) this does not seem to affect rice yields negatively. In field tests the yield of *Bt* HR have in fact been 29% higher then the non-*Bt* HR control varieties (Tu et al., 2000).

- Efforts to produce transgenic varieties that are effectively resistant to fungal attack have been less successful in controlling crop losses. Moreover, at least until now, transgenic varieties expressing antifungal proteins have had disappointing results in field tests although trials and research are still intensively conducted. A gene, for example, has been isolated in the seed of alfalfa (*Medicago Sativa*) and transplanted into potato plants that produces and antifungal peptide designated as *alfAFP*, which inhibits elongation of pregerminated spores of pathogen such as *Alternaria solani* and *Fusarium culmorum* (Ai Giao Gao et al., 2000). Progress in the area of fungal protection will be highly beneficial for enhancing food security, especially in humid tropical and subtropical agriculture. In those regions of the world ecological conditions are particularly conducive to pre- and post-harvest losses caused by fungal agents.

Finally, mention should be made of the efforts made to produce drought, salinity, and acid resistance or nutrient-enriched varieties with vitamins, iron, and amino acids.

4. Benefits and beneficiaries of agricultural biotechnologies

Only a relatively small number of developing countries have introduced genetically modified crops.

Actual open field experience and data on the performance of these crops in developing countries are, therefore, limited and often refer to trial plots. Several studies on the effects of biotech in developing countries are, therefore, of an *ex ante* nature and tend to estimate what would be the effects of those crops if they were introduced.

Biotechnology often has been a controversial development, and this has limited its diffusion in the developing world. Some developing countries fear that its adoption would act as a further barrier to access to lucrative markets in Europe and Japan. This attitude can, however, backfire and make the situation unmanageable, especially where such countries are not able to police a ban. The introduction of genetically modified (GM) varieties can in fact take place in an uncontrolled fashion, as has been the case in Brazil, where it is estimated that 3 million hectares of soybean are GM herbicide resistant (Sampaio, 2002).

There is a substantial difference between the effect of new traits embodied in biotech varieties for farmers in developed and developing countries. In a developed country a *Bt* variety decreases the number of treatments that a farmer has to apply to a crop. In a developing country a *Bt* crop, in a low input agriculture, can make the difference between a reasonable harvest and a significantly reduced harvest. Most of the available data on the effects of biotech varieties in developed countries in fact tend to show that the yield effects are negligible and in some cases even negative. On the contrary, a different scenario may instead prevail in developing countries where the ecology and the techniques presently in use often cause severe yield losses.

As will be seen later, this could result in dualism even within developing countries themselves, between large and modern farmers on the one hand and poor and backward small holders on the other.

From a more general point of view the sharing of the benefits of any technological improvement in agriculture depends upon several factors. De Janvry et al. (1999) showed that sharing of benefits between rural and urban poor depends critically on the type of crop and on the nature of the technological improvement. Notwithstanding, the rural poor are always the greater beneficiaries. However, their share is still larger if the crop is not traded and/or the technological improvement is scale neutral. In the case of a cash crop

technological improvements tend to lower the market price and therefore the value of the farmer's marketable surplus. In the latter case the largest benefits for the rural poor come from an increase in family consumption.

Qaim and Zilberman, (2003) report the results of a 2001 study on *Bt* cotton, conducted in seven Indian States. This experiment, which was carried out on experimental stations and that tried to duplicate the technology adopted by farmers, demonstrated a reduction of losses of 80–87% of produce. It should be considered, however, that the 2001 crop season was exceptionally severe for bollworm in India. Those results are in line with other information gathered from entomologists that indicate that losses of 50–60% are to be considered common.

An *ex ante* evaluation of the introduction of the combined resistance to three tuber-borne diseases, which affect the potato in Mexico and often cause an estimated 25% yield loss, reports that the resistance has been genetically engineered into two modern and into one traditional variety (Qaim, 1999). Consumer and producer both benefit from the new technology. However, benefits for farmers are more limited because the combinations of price and quantity variations are such that profit increases but slightly. More significant, instead, are benefits to consumers who are able to buy this produce at a lower price. Yield increases will be higher for small traditional farmers than for larger farmers. The latter in fact already have access to modern varieties and production techniques. Yield increases for traditional, medium, and large farmers were estimated to be in the range of 45%, 28%, and 15%, respectively, while per unit cost was reduced by 32%, 22%, and 13%. However, the overall benefits to small farmers were adversely influenced by the constraints in the distribution network of the GM traditional variety. The introduction of the new technology *ceteris paribus*— will, therefore, improve the condition of those small farmers who are adopters, but worsen the income distribution within the sector.

Pachico et al. (2002) have analyzed the economic and employment implications of the adoption of one of three technological innovations aimed at improving cassava production in Colombia. In their exercise they simulated the release and adoption of a transgenic herbicide-resistant variety (HRV), a traditional breeding high-yielding variety (HYV), and the mechanization of the planting and harvesting of this crop.

Estimates of production changes were based upon the judgment of cassava scientists and the analysis was conducted for six regions in northern Colombia, where cassava producers are among the poorest people, and are usually located in the most disadvantaged areas. Any improvement in their economic condition will, therefore, surely affect their capacity to access a more convenient source of nutrition. The HRV ranked first in all six regions in terms of reduction in production costs, thanks mainly to the reduced requirement for labor for weeding. The position of the other technologies varied depending upon the prevailing ecological conditions. The HYV performed better in the more favored regions, while in the less favored regions mechanization offered a better potential opportunity.

In addition to this cost analysis the authors estimated the effects on economic surplus brought about by the supply-shifting effects of the new technologies, and therefore the effects upon consumer and producer welfare. In all cases HRV produced the largest economic surplus effects, while HYV and mechanization ranked second and third, respectively. In all cases the simulation allocated 40% of the increased surpluses to consumers and 60% to producers. However, all the technologies led to a lower level of employment. This is true also for the introduction of the HYV, regardless of the fact that its adoption would require more labor per hectare. Given an inelastic demand for cassava, the increased production per hectare lowers the area needed to satisfy the quantity demanded and therefore the labor requirements at the aggregated level.

The authors do not speculate on how all these changes would be influenced by the introduction of a technology fee. However, it is clear that in a situation were weeding is performed with the help of landless labor, as is often the case; these new technologies are bound to worsen the economic condition of this social group. The effect will be more pronounced in the case of the HRV, as its introduction has the greater effect on the demand for labor.

One of the few studies available on the effects of the introduction of the *Bt* cotton in a smallholder area in a developing country has been conducted by Ismael et al. (2000) in the Makhatini Flats in the KwaZulu-Natal province in South Africa. This is one of the lesser well-endowed cotton-producing areas in South Africa, and producers face uncertain tenure. Moreover, the area is characterized by labor shortages as many men migrate

to town. *Bt* cotton was introduced for the first time in this area in 1998, and since then the percentage of small holders growing *Bt* cotton has been growing constantly. Preliminary data indicate that for 2002, i.e., after 4 years, 90% of the 3,000 smallholders that grow cotton were employing the *Bt* variety. A panel made up of a random sample of 100 farmers, and including both *Bt* adopters and nonadopters, has been studied to identify the yield effects, the factors that have influenced the adoption and the economic impact on the adopters. A fraction of the farmers in the panel were already *Bt* adopters in the first year, and there were more in the second year. None of the adopters in the first year dropped the *Bt* variety in the second year. This is a first indication of the farmer's acceptance of the new variety.

Farmers in the panel belonged to an association that gave them the opportunity to exchange experiences and information in organized meetings. A private company provided credit, seeds, and other farm inputs, and bought the cotton from farmers. In the first year the yields of the *Bt* adopters were 39% higher than those achieved by the nonadopters, even though the adopters used 20% less seed due to the higher cost of the seed. The *Bt* variety also performed better in the wet crop year, producing yields that were 33% higher, because the rains wash off the chemicals used to fight pests and makes treatment more difficult to apply. The *Bt* variety was still accountable for increased production even when due consideration is given to the fact that first adopters usually are the better-off farmers who would have better results even with traditional technology.

In a later paper Thirtle et al. (2003) revisited the data, employing a stochastic efficiency frontier and the DEA model to better account for differences in farm size and labor use. This second analysis allowed the authors to estimate total, technical, and scale efficiency separately, and confirmed that the *Bt* variety was responsible for the better performance of the cotton crop in the smallholders' fields.

A sample of 299 Argentinean farmers was analyzed by Qaim and de Janvry (2003) to determine the income and environmental effects of the introduction of *Bt* cotton in the major cotton-growing regions of Chaco and Santiago del Estero. Farmers were either *Bt* adopters or nonadopters, and were divided into small (less than 90 ha) and large (more than 90 ha) categories. It was found that *Bt* had a positive effect on yields, leading to an average increase of 32%, while chemical use declined by 50%. However, small farmers usually cultivated cotton with low inputs, while large farmers employed a more sophisticated technology. In both groups of farmers the introduction of *Bt* varieties resulted in a yield increase and a reduction in the use of chemicals, but not uniformly. Yield increases were more pronounced for small farmers, who had lower yields to start with, and who hardly used chemicals. In contrast, the benefits for the larger and more advanced farmers were mostly in the form of less intense applications of chemicals.

Finally Traxler et al. (2003) report the successful introduction of *Bt* cotton in Camarca Lagunera in Mexico. Here benefits, as has also been found elsewhere, depended upon the level of lepidoptera infestation.

5. Some institutional aspects of the introduction of biotechnology in developing countries

Intellectual property rights, appropriate biosafety regulations together with the capacity of effectively implement them, and the promotion of local research capacity and cooperation between the private and the public sectors are some of the institutional issues that are most often raised and considered as essential preconditions for the effective promotion of agricultural biotechnology in developing countries.

Experience with the Green Revolution shows how the lack of local quality traits or traits adapted for local climate conditions are critical to farmers' acceptance of new varieties, and therefore to their level of adoption (Santaniello, 2002*)*. In Thailand, for example, the rate of adoption of MV rice has been constrained because breeders have been not been able to produce varieties that are comparable to Jasmine rice from a quality perspective. HYVs that last more than 110 days and are intolerant to drought cannot be grown in eastern India, where the rainy season is short and the monsoon erratic. In regions of deep flooding (basins of Bangladesh and Cambodia, part of Uttar Pradesh) semi-dwarf varieties cannot be grown because of the risk of submergence. In Brazil only 25% of rice is under modern varieties because of the lack of suitable drought-tolerant varieties for the uplands (Hossain et al. (2003).

A national agricultural research capacity, a sound science base, and experience in traditional breeding are necessary preconditions for the development of a national biotech capacity. Evenson and Gollin (1997) in a recent, well-documented book on the Green Revolution, show that four decades of agricultural innovations have made evident the importance of the National Agricultural Research Systems (NARS) and the pivotal role that the International Agricultural Research System (IARS) has played in stimulating their growth and feeding their research activities.

For the still large number of countries where NARS have not yet developed or grown to maturity, the role of the IARS is still vital. As in the Green Revolution, the role of the IARS can be of pivotal importance in promoting agricultural biotechnology in the developing world, especially if they can adapt their policy and structure to the needs of the gene revolution. Here cooperation between private and public sectors needs to be seriously explored.

The number of "orphan" countries where neither the public nor the private sectors are active is large, probably close to 50. Here the public sector is inactive because of a lack of funds, and the private sector because the size of the market is too small and IPR protection is probably not effective. As food insecurity problems are most acute in these countries, innovative ways of promoting this partnership are urgently required.

Kremer (2003) and Master (2003) address the problem of market failure in the African research market and propose two alternative systems of public/private cooperation to stimulate scientific innovation and adoption among farmers. These schemes aim at promoting the acquisition by the public sectors of innovations, produced by the private sector, at a price based upon their estimated social value.

The Mexican experience (Traxler et al., 2003) provides clear evidence that the presence of a large national agricultural research system, the size of the agricultural area, the capacity of a university-based research establishment, and the availability of credit and technical assistance to smallholders have all been crucial to the successful introduction of *Bt* cotton in that country. In the specific case of the *Comarca Lagunera*, the capacity of the seed company to recoup the benefits of its research investments was made easier by the control that the marketing structure could exercise over producers. In another situation, however, where producers are less vertically integrated such control by the seed companies would be more problematic and therefore those companied less willing to make improved seed available to farmers.

The relevance of a well-functioning IPR system to local production, and the transfer and the adaptation of technological innovations have been questioned. Here it may be interesting to note that biotechnology innovators have ways of enforcing appropriability that are unknown in other more traditional fields such as mechanics, physics, and chemistry. These means could lower the cost of enforcing the ownership of proprietary knowledge and open *de facto* markets that, due to limited size or institutional factors, are at present neglected by the private sector.

It is well known that hybrid seeds are protected because the benefits of heterosys are limited only to the first generation. Breeders that produce hybrids can enforce their proprietary rights by limiting control to the parental lines; hence the enforcement of IPR is limited to the parental lines.

Maize is the best known case of a hybrid crop whose seeds are widely used in developing countries, although with large variation between regions. Three quarters of the total maize area in developing countries is seeded with modern varieties. Some 90% of the more than half a million tons of maize seeds that are sold annually by the seed industry in the developing countries are hybrids developed and sold by the private sector (Morris et al., 2003).

Genetic Use Restriction Technologies (GURTS) are biotechnology-based techniques that prevent unauthorized use of genetically improved varieties could fulfill the same role in this regard. GURT technologies can operate through a mechanism that makes the second-generation seed sterile (V-GURT) or which requires the application of a chemical inducer for the expression of the desired trait (T-GURT). In the latter case farmers can still use their own duplicated seed, but need to buy the inducer, which is patent protected, to activate the specific trait (FAO, 2002). The V-GURT technology operates through the activation or deactivation of a gene by an inducer to impede the embryo formation in a second-generation seed, or to block the growth of a vegetatively reproduced plant. In all cases the crop varieties do not necessarily need to be protected by patents or other IPR. The protection operates at the level of techniques needed to generate the V-GURT varieties or upon the molecule of the T-GURT inducer.

The GURT technologies can therefore be used to protect biologically improved varieties produced either transgenically or with traditional breeding, as they are, in fact, not limited to transgenic varieties, although several technical problems have yet to be solved. It is, therefore, not infeasible that the V-GURT technologies may supplant other types of protections in the future, notwithstanding negative public perceptions. The marketing of GURT varieties will not, therefore, require the elaborate institutional framework needed to implement other types of IPR for crops. However, the use of these technologies still lies in the future. Five to ten years are required for the V-GURT varieties to come to market, while the T-GURT varieties are much closer to the market.

There have been concerns (Swanson and Goeschl, 2002) that the GURT varieties might disrupt the seed improvement activities of farmers that operate in low-input farming systems, although it can be assumed that the effect on farmers that are already using modern varieties would be minimal. The detrimental effects on low-input farmers are, however, a possible side effect of the spreading of those technologies. The mere fact that those farming systems have limited links with input markets, however, makes this possibility rather unlikely. The success of these techniques might cause a widening of the gap between those who have access to them and those who do not, contributing to the worsening of the income distribution that is usually the negative side effect of many technological innovations.

GURT technologies could provide an incentive to private research to enter in the markets for inbred crops or for species where hybrid technologies have not been successful, and in seed markets where it is not now financially attractive to invest due to their limited size. GURT technologies will lower the transaction costs of IPR protection, but will increase the production costs of those varieties. The overall effect will depend, among others, on the compensatory effects of those cost variations. The International Union for the Protection of New Varieties of Plants (UPOV) system will be most influenced because the farmers' privilege and the existing research exemption will be directly affected, unless reverse engineering becomes acceptable. If the GURT technologies limit the flow of elite lines between developing countries they may also negatively influence the rate of growth of crop genetic improvement.

On the other hand, the rate of growth of crop improvements might improve because these technologies will offer innovators a better chance of recouping a share of the benefits generated by their work, especially because the time coverage of the GURT is not limited, and this is unlike patents whose protection has a limited timespan. Moreover in the case of the latter the economic life of a protected variety is often shorter than the legal protection, because of the introduction of improved and more effective competing varieties.

A well-functioning seed market is considered essential to the adoption of the new improved varieties. This is certainly true. Dalton and Guei (2003) note that the low adoption rate of new upland rice varieties in Nigeria is often attributed to weakness in the extension service and in the seed distribution service. However, this explanation must be seen against the fact that the varieties adapted to the local needs of farmers have been widely adopted in this case. These farmers, have, instead, resisted the introduction of modern varieties that were not able to outperform traditional seeds. The role of the informal seed market should not be undervalued, even in situations where public-produced varieties play an important role.

6. Summary and conclusions

This paper presents some indications of how agricultural biotechnology can help in assuring a better chance of obtaining a higher food security for the rural poor. A significant research effort is now underway to help solve production problems in developing countries. Contrary to many technological innovations of the past, biotech innovations are scale neutral, and therefore well able to help small farmers. These benefits will, however, will reach poor farmers only if the innovations are closely linked to the farmers' needs and if a national research system is properly developed.

Agricultural biotechnology can increase yields, improve the environment by decreasing the use of chemical inputs, cooperate in reducing soil erosion, and decrease the need for new land to respond to the increase in food demand. However, it is not wise to raise unreasonable expectations about what agricultural biotechnology can do. Agricultural biotechnologies are not the magic silver bullet that will eliminate food insecurity and poverty. The spread of a message that is unrealistically optimistic would be ethically wrong, economically erroneous, and politically counterproductive. Nevertheless, agricultural biotechnology can support to the efforts aimed at increasing the life expectancy and quality of many in the developing world in much the same way that the high-yielding varieties did in the Green Revolution.

References

Ai Giao, Gao, et al. "Fungal Pathogen Protection in Potato by expression of a Plant Defensive Peptide," *Nature Biotechnology* 18, no. 12 (2000), 1307–1310.

Dalton, T. J., and R. G. Guei, "Ecological Diversity and Rice Varietal Improvement in West Africa," in R. E. Evenson and D. Gollin (Eds.), *Crop Variety Improvement and Its Effect on Productivity: The Impact of International Agricultural Research* (CABI: Delémont, Switzerland, 2003).

De Janvry, A., E. Sadoulet, G. Graff, and D. Zilberman, "Agricultural Biotechnology and Poverty: Can the Potential Be Made a Reality?" ICABR Conference on The Shape of the Coming Biotechnology Transformation: Strategic Investment and Policy Approach from an Economic Perspective, Rome and Ravello, Italy, June 17–19, 1999.

Evenson, R. E., and D. Gollin, "Genetic Resources, International Organizations and Improvement in Rice Varieties," *Economic Development and Cultural Change* 45, no. 3 (1997), 441–500.

FAO, "Potential Effects of Genetic Use Restriction Technologies (GURTs) on Agricultural Biodiversity and Agricultural Production Systems," Commission on Genetic Resources for Food and Agriculture, CGRFA/WG-PGR -1/10/March (2002).

Flasinski, S., V. C. M. Aquino, R. A. Hautea, W. K. Kaniewski, N. D. Lam, C. A. Ong, V. Pillai, and K. Ronyanon, "Value of Engineered Resistance in Crop Plant and Technology Cooperation with Developing Countries," in R. Evenson, V. Santaniello, and D. Zilbermann (Eds.), *Economic and Social Issues in Agricultural Biotechnology* (CABI: Delémont, Switzerland, 2000).

Hilder V., and W. Hamilton, "Biotechnology and Prospects for Improving Crop Resistance," in C. Blak and Swettmore (Eds.), *Crop Protection in the Developing World* (British Crop Protection Council: Farnham, 1994).

Hossain, M., D. Gollin, V. Gabanilla, E. Carrera, N. Johsion, G. S. Khush, and G. McLaren, "International Research and Genetic Improvement in Rice: Evidence from Asia and Latin America," in R. E. Evenson and D. Gollin (Eds.), *Crop Variety Improvement and Its Effect on Productivity: The Impact of International Agricultural Research* (CABI: Delémont, Switzerland, 2003).

Ismael, Y., L. Beyers, C. Thirtle, and J. Piesse, "Efficiency Effects of Bt Cotton Adoption by Small Holders in Makhatini Flats, KwaZulu-Natal, South Africa," in R. Evenson, V. Santaniello, and D. Zilbermann (Eds.), *Economic and Social Issues in Agricultural Biotechnology* (CABI: Delémont, Switzerland, 2000).

Khush, G. S., and D. S. Brar, "The Application of Biotechnology to Rice," in C. L. Ives, and B. M. Bedford (Eds.), *Agricultural Biotechnology in Agricultural Development* (CABI: Delémont, Switzerland, 1998).

Kremer, M., "Encouraging Private Sector for Tropical Agriculture," 7th ICABR International Conference on Public Goods and Public Policy for Agricultural Biotechnology. Ravello, Italy (July, 2003).

Mannyong, V. M., J. G. Kling, K. O. Makinde, S. O. Ajala, and A. Menkir, "Impact of IITA Germplasm Improvements on Maize Production in West and Central Africa," in R. E. Evenson and D. Gollin (Eds.), *Crop Variety Improvement and Its Effect on Productivity: The Impact of International Agricultural Research* (CABI: Delémont, Switzerland, 2003).

Master, W. A., "Research Prices: a Mechanism for Innovation in African Agriculture," 7th ICABR International Conference on Public Goods and Public Policy for Agricultural Biotechnology. Ravello, Italy (July, 2003).

Meyers, W. T., *Sustainable Food Security for All by 2020* International Conference, Bonn (September, 4–6, 2001).

Morris, M., M. Mecuria, and R. Gerpacio, "Impact of CYMMIT Maize and Breeding Research," in R. E. Evenson and D. Gollin (Eds.), *Crop Variety Improvement and Its Effect on Productivity: The Impact of International Agricultural Research* (CABI: Delémont, Switzerland, 2003).

Pachico, D., Z. Escobar, L. Rivas, V. Graffet, and S. Perez, "Income and Employment Effects of Transgenic Herbicide Resistant Cassavain Colombia: A Preliminary Simulation," in R. E. Evenson, V. Santaniello, and D. Zilberman (Eds.), *Economic and Social Issues in Agricultural Biotech* (CABI: Delémont, Switzerland, 2002).

Pinto, Y. M., R. A. Kok, and D. C. Baulcombe, "Resistance to Rice Yellow Mottle Virus (RYMV) in Cultivated African Rice Varieties Containing RYMV Trangenes," *Nature Biotechnology* 17 (1999), 702–707.

Matin, Q., "Modern Biotechnology for Small Scale Farmers in Developing Countries: Contradictions or Promising Options," Paper presented at the International Consortium on Agricultural Biotechnology Research (ICABR) Conference: The Shape of the Coming Biotechnology Transformation: Strategic Investment and Policy Approaches from an Economic Perspective, University of Rome Tor Vergata (17–19 June, 1999).

Matin, Q., and A. de Janvry, "Bt Cotton, Pesticide Use and Resistance Development in Argentina," 7th ICABR International Conference on Public Goods and Public Policy for Agricultural Biotechnology, Ravello, Italy (July, 2003).

Matin, Q., and D. Zilberman, "Yield Effects of Genetically Modified Crops in Developing Countries," *Science* 299 (2003), 900–902.

Santaniello, V., "Biotechnology and Traditional Breeding in Sub-Saharan Africa," in T. M. Swanson (Eds.), *Biotechnology, Agriculture and the Developing World: The Distributional Implications of Technology Change* (Edward Elgar: Northampton, UK, 2002).

Sampaio, M. A. A., "Technology Offices. Intellectual Property and Much More," *FAO – Tor Vergata Expert's Workshop on Public Agricultural Research: The Impact of IPRs on Biotechnology in Developing Countries*, University of Rome "Tor Vergata" (2002).

Swanson, T., and T. Goeschl, "The Impact of GURTs: Agricultural R&D and Appropriation Mechanism," in T. M. Swanson (Ed.), *Biotechnology, Agriculture and the Developing World: The Distributional Implications of Technological Change* (Edward Elgar: Northampton, 2002).

Thirtle, C., L. Beyers, D. Hadley, and J. Piesse, "Measuring the Benefits of GM Using Cost DEA," 7th ICABR International Conference on Public Goods and Public Policy for Agricultural Biotechnology Ravello, Italy (July, 2003).

Traxler, G., J. B. Falck-Zepeda, and G. Sam, "Genes, Germplasm and Developing Country Access to Genetically Modified Crops Varieties," Paper presented at the International Consortium on Agricultural Biotechnology Research (ICABR) Conference: The Shape of the Coming Biotechnology Transformation: Strategic Investment and Policy Approaches from an Economic Perspective, University of Rome Tor Vergata (17–19 June, 1999).

Traxler, G., S. Godoy-Avila, J. Falck-Zepeda, and J. De Jesus Espinoza Arelland, "Transgenic Cotton in Mexico: A Case Study of Comarca Lagunera," in N. G. Kaitzandonakes (Ed.), *The Economic and Environmental Impact of Agbiotech: A Global Perspective* (Kluwer Academic/Plenum: New York, 2003).

Tu, J., G. Zhang, K. Datta, C. Xu, Y. He, Q. Zhang, G. S. Khush, and S. K. Datta, "Field Performance of Transgenic Elite Commercial Hybrid Rice Expressing Baccillus Turigiensis Delta—Endotoxin," *Nature Biotechnology* 18 (2000), 1101–1104.

Plenary 4

The poverty of sustainability: rescuing economics from platitudes

Daniel W. Bromley*

Abstract

The idea of sustainability has become confused and incoherent. If sustainability is to regain a plausible pertinence to economic policy it must be understood to encompass two realms: (1) human interaction with nature; and (2) human interaction with others with respect to their interaction with nature. The on-going redefinition of the *purposes of nature* requires that institutions—norms, rules-property regimes—undergo constant evolution so that human action conduces to nondestructive action. Caution in the social realm is the greatest risk to environmental sustainability.

JEL classification: O13

Keywords: sustainability; institutions; evolution; purposes of nature

1. Introduction

The idea of sustainability has had a curious life history. It started out as an interesting idea, then became tied to the economics literature in growth theory, soon became a rallying cry for those opposed to globalization, and now finds itself part of the official title of offices, divisions, bureaus, and directorates in many of the world's most visible organizations. From the title of this article the reader may be led to suppose that the idea of sustainability has become a mere platitude. Sustainability is, in fact, a platitude precisely because the term conveys nothing of substance.

The original idea of sustainability concerned consumption levels that would meet the "needs" of current people without compromising the "needs" of future generations (Dixon and Fallon, 1989; WCED, 1987). What exactly constitutes the "needs" of present and future people is an empirical challenge of unsurpassed difficulty. Indeed, it is so difficult that the quest to give this idea empirical content induces yet another round of platitudes. Note that the original idea was one of constraining current consumption such that those of us now living—who necessarily stand as dictators over the endowments and consumption opportunities available to future people—would not so foul the natural world that those who are to come after us would inherit a vast wasteland. A perusal of the current literature shows that the idea of sustainability has now become transformed into a conversation about consumption "entitlements"—clean water, health, housing, nutrition, education, employment, and income (Parris, 2003). Of course, there remain discussions about maintaining life-support systems, and here the emphasis concerns climate change, atmospheric ozone, the oceans, biodiversity, chemical contamination, deforestation, and land-use issues.

While this session of the conference edition concerns environmental stewardship, it is essential to indulge the emergent trajectory of the general discourse on sustainability by extending the discussion to encompass the nexus between humans and nature. That is, nature must be connected to people, and people to nature. In more specific terms, the central role of the *constructed domain within which people interact with each other as they go about interacting with their physical surroundings will be emphasized*. The three other plenary papers in this session on environmental stewardship follow a similar pattern—two focusing on agriculture and the environment (López, and Rola and Coxhead), and the other

---

* Department of Agricultural and Applied Economics, University of Wisconsin-Madison, Madison, WI, USA.

focusing on resource degradation and poverty (Ehui and Pender).

A major part of the difficulty with the current idea of sustainability among economists is its focus on capital—both natural and man-made—rather than on the institutional arrangements (norms, working rules, and property regimes) that give economic value to particular actions and not to others. There is a long history in economics concerning the precise meaning of the idea of "capital." But the real problem is that capital as an economic concept is incoherent and incomplete without reference to the institutional arrangements that indicate the ways in which that physical object called "capital" can or cannot be used in an economy. Examples of these issues include to whom does the capital belong? Who may control its use? Who may and may not receive its income stream? What are the social parameters of acceptable use of that object? Who may use it to obtain credit and to settle debt? The issue here concerns the social and economic content of what are called *natural* and *man-made capital*. And that social and economic content is determined precisely by the institutional structure of an economy. Rendering the idea of sustainability useful requires moving beyond the traditional focus on natural and man-made capital for the simple reason that the idea of capital is entirely dependent on the socially constructed rules that relate individuals and groups to physical objects—whether naturally occurring or humanly constructed.

This implies that while the precise meaning of sustainable development is open to debate, there can be no doubt that the ecological dimension of sustainability cannot be considered and understood apart from the social dimension. This necessarily follows from the fact that the social dimension concerns how and why humans interact with their physical surroundings as they do. Are tropical forests being cleared at a rate that concerns ecologists and atmospheric scientists? Why is this happening? Is soil erosion in agricultural areas threatening future agricultural production and river ecosystems? If so, what are the plausible reasons for this unwanted outcome? Do industrial and agricultural chemicals pose a threat to living organisms? And if so, why? Are unique habitats—repositories of rare genetic resources—being savaged in the name of "progress?" If so, whose idea of "progress" is driving these outcomes? Each of these physical eventualities represents the possible outcome of human interaction with the environment. More important, these physical manifestations of human behavior are also manifestations of human interaction in a social and economic domain. If we are to understand sustainability we must be concerned with the ways in which humans relate to each other—and why those particular interactions produce particular implications for the natural environment.

The research challenge here is to understand human behaviors *not* at the point where individuals interact with nature.[1] Rather, we must understand human behaviors from the point where individuals are driven to act not out of choice but out of necessity. A government heavily indebted to foreign creditors is a government without many choices. Being landless is to be without compelling choices. Farmers, who cultivate steep hillsides, thus giving rise to soil erosion, can be said to exercise choice in only a very limited sense of that word. Clear thinking about sustainability is not advanced if analyses start from the notion that most of the participants in the systems being studied act on the basis of free choice. Choosing between the slums and the remote hills may look like choice to some, but it is a categorical mistake to call such behavior the result of "choice." When necessity forces actions there is little scope for choice. If you cannot move, you are not *choosing* to stand still.

The problem, therefore, is to understand the conditions in which individuals and groups find themselves acting—not choosing as an expression of free will, but responding as a manifestation of necessity. Everyone is embedded in a structure of economic and social relations that are not of their choosing. Humans are born into such a structure and, depending on the luck of that birth, they stand a reasonably good chance—or no chance at all—of influencing that structure in the future. Regardless of their capacity to alter that structure, everyone faces differential opportunities to move fluidly within that structure, or to be thwarted by it at most every turn. And this raises an interesting issue in the matter of sustainability.

To talk of the sustainability of social and economic arrangements—the working rules and property

---

[1] This seems to be the tradition in much of this work where economists seek to understand tropical deforestation by constructing econometric models with "explanatory" variables such as miles of road, "weak" property rights, rates of in-migration, etc. For a critical methodological look at this genre of work see Bromley (1999).

regimes—is to raise an awkward question. Is the focus on the sustainability of the arrangements regardless of the social and ecological consequences that flow from them? Or is it the maintenance of a process of gradually searching for—and evolving into—new institutional arrangements that will assure both ecological integrity and the general ennobling of human life over the long run? That is, is the concern to maintain (sustain) a specific *structure* or a particular *process*? This question reminds us that traditional labels and approaches can be problematic. Notice that cautious approaches to environmental behaviors may be precisely what are needed to avoid serious ecological disasters. Humans must be careful with the forest, careful with genetic resources, cautious with endangered species, and indeed circumspect about the arrogance of human domination of nature. Conservative principles serve well in the realm of protecting the environment against the onslaught of human exploitation. Sustainable development is, in a sense, a cautious and precautionary approach to how humans will interact with nature.

But caution in the social and economic realm may well be the enemy of sustainability. This paradox arises because solutions to existing destructive uses of nature may indeed entail quite drastic changes in the working rules and their correlated organizational manifestations that now constitute plausible reasons for destructive behaviors toward the environment. If steep hillsides and other fragile lands are overrun with migrants desperate for food and livelihood then one must ask why the fragile hillsides represent the only option for those seeking a better life. What if there are large expanses of quite good agricultural land that might be made available for these landless people, yet which are currently protected by a set of social and economic relations that lead to conditions of great income disparities and landlessness? Those individuals well served by prevailing institutional arrangements from which massive landlessness springs may not be eager for this attention and thus a conservative approach to social and economic relations may turn out to be the enemy of ecological and social sustainability.

If timber concessionaires are destroying forests then the question must be asked: "Why is this behavior allowed?" (Ascher, 1999; Bromley, 1999). These forest practices constitute serious threats to nature and if the prevailing institutional arrangements are seen as the reasons for the results (the plausible explanation of the behaviors) then those institutional arrangements are immediately suspect. To the extent that certain segments of society are well served by those working rules—and if they were not well served by them it might be impossible to explain the existence of such rules—altering current behaviors and practices inimical to ecological sustainability threatens the presumed goodness (instrumentality) of the existing working rules. And once there is talk of the need to alter existing working rules and practices, particular vested interests—well served by those rules—can be expected to mobilize against the proposed changes.

The challenge, therefore, in understanding sustainability, is to search for an understanding—an explanation—of the reasons for prevailing rules. Many of the working rules and property regimes that mediate human action toward the environment are products of the traditional idea that conquering nature was a plausible means of inducing economic development. Nature has traditionally been seen as a storehouse of raw materials whose proper purpose was to serve human extraction and use. That is, nature existed to be subjugated to the human will, and her bounty— timber, minerals, fish, water, kinetic energy for hydroelectric generation, coal, oil, natural gas, solar energy—was there to serve human desires. In addition to this provision of raw materials, the purpose of nature was also to provide a stream of resource services—carrying away human and industrial waste. Accordingly, the institutional arrangements pertinent to human–nature interactions throughout much of human history have been predicated on this view of the purposes of nature.

But when the purpose of nature is itself contested— as it surely has been for some time now—then caution in the social and economic realm, where caution means a rigid and aggressive defense of the prevailing institutional setup, instead of enhancing ecological sustainability will almost certainly undermine it. This threat from a cautious strategy arises because the existing institutions and organizations were crafted and refined during an era when there was a different purpose of nature than that which is now emerging. With new and evolving ideas about the purposes of nature it follows that there must be reconsideration of the institutions that mediate human interaction in the social and economic realm, but also in interacting with nature. If the new purpose of nature is not reflected in modified institutional arrangements then nature will continue to suffer, and eventually it will be impossible to maintain existing social and economic relations. It is for this reason that caution in the social realm might very well lead to serious threats to nature.

This brings about an interesting twist, in the sense that sustainability in the social and economic realm depends on *constant change in social and economic institutions, and not in their preservation.* Social and economic stasis is the enemy of environmental sustainability. There must be means whereby the institutions of nation-states can be continually modified in accord with the inevitable evolution in the imagined purposes of nature. It may seem odd that sustainability implies change and evolution rather than caution and stasis, but this essential evolution is driven by the fact that the purposes of nature are changing. If institutional arrangements fail to adjust accordingly, social processes will be threatened and out of that threat will emerge a profound danger of accelerated harm to nature.

The correlated point here is that the standard policy prescription to flow from much of the economics literature is that property rights must be secure in order to protect nature. Indeed, if there is one aspect of the Washington Consensus that pertains to environmental policy it is this constant harangue about the manifold wonders of secure property rights. Unfortunately, this prescription is flawed on two grounds. The first flaw is a theoretical one. Those economists who pronounce with great conviction on this subject reveal their ignorance of the *iron law of the discount rate* (Page, 1977). The obvious implication of the iron law of the discount rate was made clear for fisheries over 30 years ago by Colin Clark who found that "depending on certain easily stated biological and economic conditions, extermination of the entire population may appear as the most attractive policy, *even to an individual resource owner*" (Clark, 1973, p. 950) (emphasis added). Clark's analysis shows that private ownership is consistent with resource destruction (Pearce and Turner, 1990). The iron law of the discount rate dispenses with the notion that private property is *sufficient* to ensure wise resource management.

The second flaw comes in the idea that private ownership is still *necessary* for stewardship. Those who insist that secure property rights are necessary for the protection of nature confuse the general proposition about property rights with the specific proposition

(Becker, 1977). That is, the advocacy of *clear property rights* has been distorted into the idea that only *individual property rights* will do the work of protecting nature. Since the iron law of the discount rate defeats the sufficiency claim, and since many nations have effective regimes of both common property and state property, we see that private property is neither necessary nor sufficient to protect valuable aspects of nature. What is essential is that *some* property regime is in place so that the natural resource is not an open access resource (Bromley, 1991). The decisive issue here is that any property regime—to be worthy of that name—requires the presence of an enforcement (compliance) structure.

Property regimes are not some divine intervention revealing to mere humans the "truth" about human interactions with nature. Rather, property regimes at any moment simply reflect the collective determination of which settings and circumstances seem worthy of extraordinary protection. Settings and circumstances are not protected with a rights regime because they are "property." Rather, those settings and circumstances deemed of extraordinary importance come to acquire the protection that we associate with property rights (Bromley, 1991)—an important point when institutions and the continual evolution in the purposes of nature are considered. Recall that each generation has inherited its values, its institutional arrangements, and its governance structures from those who came before. The law in general and property law in particular, at any moment reflects that heritage.

Public policy is best understood as collective action in restraint and liberation of individual action. Since collective action results in new institutions (new working rules) these new working rules differentially restrain and liberate particular individuals in their actions. These working rules also expand individual action in the sense that new working rules augment the capacities of certain members of a particular society to have their interests given protection. When the nation-state grants rights to individuals—but especially property rights—the state is thereby expanding the reach of the individual. This follows from the fact that to have a right is to have the capacity to *compel* the state to act to protect your interests.

New public policy is simply the application of new collective action that will simultaneously restrain and liberate the field of action—the choice domain—of individuals. If firms are no longer able to discharge their wastes into nearby rivers then their field of action has been restrained, and the field of action of those who prefer clean rivers has been enlarged (liberated). If land reform expands the choice domain of the landless then it simultaneously constrains the choice domain of those who previously imposed their will on landless peasants. If timber concessionaires are restrained from aggressive harvesting of trees in fragile ecosystems then those who suffered at the hands of deforestation have been liberated from this imposition.

These issues in sustainability then emphasize the importance of the processes whereby institutional arrangements change over time, i.e., of the need for an evolutionary environmental economics.

2. An evolutionary environmental economics[2]

The idea of sustainability can only come to have coherence and operational content if it is understood to relate to a process whereby the working rules and entitlements that mediate individual choice sets are continually modified ("worked out") in response to *new emergent ideas about the purposes of nature*. The work of Thorstein Veblen provides a good starting point in the development of this line of thought. It is both ironic and unfortunate that his popular book *The Theory of the Leisure Class* is well known to most economists (perhaps because of its catchy metaphors (conspicuous consumption, snob effect, conspicuous waste) while his much more profound and substantive article in the *Quarterly Journal of Economics* entitled "Why Is Economics Not an Evolutionary Science?" has been ignored. This is ironic because the *Quarterly Journal of Economics* was, at the time (1898), the most prestigious outlet for economists. And it is unfortunate because Veblen's perceptive evolutionary insights were soon to be surpassed and overwhelmed by the static marginalist equilibrium economics of Robbins, Edgeworth, Hicks, Kaldor, and Samuelson in what has come to be called "the ordinalist revolution" (Cooter and Rappoport, 1984).[3]

The flaw in ordinalist economics is that it skirts the issue of value (embedding it in the ultimate relativist triumph of the indifference curve). When combined with the pernicious idea of equilibrium, the individual is thereby emasculated from having any role to play other than performing the right calculations in order to achieve some alleged optimum. Notice that the concept of equilibrium celebrates and ratifies the notion of arrival—of attaining something that henceforth will be automatically maintained, at least until the next exogenous shock perturbs the system. That so few economists are troubled by the centrality of equilibrium models and metaphors in economics says more about our fascination with physics and its machines than with the ongoing and evolving process of people getting a living from their interaction with each other—and with nature (Mirowski, 1989). To suppose that the concept of equilibrium is useful in this pursuit of understanding and explaining human action remains one of the more enduring puzzles in contemporary economics (Brock and Colander, 2000). Indeed, the concept of equilibrium, with its message of stationarity, stands as one of the paramount hurdles to clear thinking about sustainability broadly defined.

When economists undertake economic analysis and economic advice, the standard approach invariably entails thinking about some desired state of efficiency running off into the future that will serve us well until some perturbation upsets this happy state. We are not sure what will change, but we are sure that when it changes the economy will adjust to some magical new equilibrium pathway. Of course, increasing or decreasing returns may complicate matters. And of course, externalities can make this attainment difficult. But once these minor inconveniences in market-produced outcomes have been rectified, all will be efficient once again. This smoothly running machine remains the dominant mental (and analytical) model of much of contemporary economics. But the simplicity of the machine is precisely its abiding weakness. To assume that the human condition is correctly described and modeled as a tractable and monocausal mechanism is to do serious violence to reality. Consider the following quote from Veblen:

The economic life history of the individual is a cumulative process of adaptation of means and ends that cumulatively change as the process goes on,

---

[2] See Norgaard (1981) for a prescient account of coevolution in social and ecological systems.

[3] Another of Veblen's profound works pertinent to this theme is "The Limitations of Marginal Utility," 1909.

both the agent and his environment being at any point the outcomes of the last (the previous) process. His methods of life to-day are enforced upon him by his habits of life carried over from yesterday and by the circumstances left as the mechanical residue of the life of yesterday . . . What is true of the individual in this respect is true of the group in which he lives. All economic change is a change in the economic community . . . . The change is always . . . a change in habits of thought . . . but . . . there remains the generic fact that their (an individual's) life is an unfolding activity of a teleological kind . . . . The economic life history of any community . . . is shaped by men's interest in the material means of life . . . . Primarily and most obviously, it has guided the formation, the cumulative growth, of . . . economic institutions' (Veblen, 1990, pp. 74–76).

The essential point here is that successive generations are the necessary *creators* of the structures and functions of the local environment within which they are embedded. That is, individuals often make and re-make their economic settings and circumstances. Of equal importance, from the outset, individuals are *constituted* by the settings and circumstances in which they have been shaped and find themselves embedded. That is, the constructed social and economic settings and circumstances come, to a certain extent, to *form* individuals and to predispose them to certain "habits of mind." John R. Commons referred to this as the "instituted personality." It is this perpetual interaction between individuals and their constructed surroundings that led Commons to refer to the process of "artificial selection." That is, while biological evolution may be "natural," social evolution is constructed ("artificial") (Commons, 1931, 1968, 1990). It is constructed precisely because individuals are capable of receiving feedback from actions taken, processing the lessons from that feedback, and re-constructing the norms, working rules, and entitlements (property regimes) that stand as the plausible explanation for the outcomes now realized to be in need of modification. Notice that these ideas of both Veblen and Commons provide the basis for thinking of economics in evolutionary terms.

This evolutionary approach is impossible in the equilibrium models and metaphors of contemporary economics. In the currently accepted view of human action, the individual is—as Veblen put it—nothing but a "lightning calculator of pleasure and pain." Veblen pointed out that this hedonistic formulation forces us to assume that the individual has neither antecedent nor consequent. More specifically, "He is an isolated, definitive human datum, in stable equilibrium except for the buffets of the impinging forces that displace him in one direction or another . . . the hedonistic man is not a prime mover. He is not the seat of a process of living, except in the sense that he is subject to a series of permutations enforced upon him by circumstances external and alien to him" (Veblen, 1990, pp. 73–74).

The essence of an evolutionary economics is seen in the fact that when existing institutional arrangements are found to be the plausible reason for behaviors that lead to unacceptable environmental outcomes, there will soon be citizen pressure on these institutional arrangements. In the early stages of this process those seeking change will be small in number though possibly loud in voice. Their efforts will be resisted and dismissed as the special pleadings of a particular minority. This has certainly been the case for environmental activists the world over. The practice of politics and of policy reform is the process of bringing others along to one's perspective. As the vocal minority mobilizes arguments in its behalf, soon others will join in. When their numbers, and the volume of their collective voice, reach a critical threshold they will be noticed. Suddenly, it will be realized that there is a "policy problem" that may no longer be safely ignored. It is at this point that the resiliency of a nation's institutional arrangements will come under scrutiny. If these arrangements are rigid and resistant to change, and if the groundswell for change gains momentum, it will not be long until these two forces will collide.

One way to think of evolution in the institutional arrangements of society is to understand the syllogism of practical reason. Practical reason brings together two kinds of premises. The first we call the *volitional premise*. This premise can be thought of as outcome in the future for the sake of which a particular event (or action) must be undertaken today. If there is a new felt need to protect fragile hillsides then particular actions are required now. If there is a new felt need to protect unique ecosystems from destruction, then particular actions must be taken now. The policy question is: if we wish particular future states then what must be done now to realize those states? Beyond the volitional premise, practical reason requires an *epistemic premise*. The epistemic premise mobilizes current knowledge—both "scientific" and traditional—to offer a plausible guide for what is necessary that the volitional premise might be realized. If it is intended that fragile hillsides are protected in the future from both migrants and the timber companies then the epistemic premise indicates those actions that offer plausible means whereby those intentions might be realized.

Thus, new public policy is the conjunction of new collective intentions, new working rules (new institutions) that are entailed by the epistemic premise, and the presumption of compliance. That is, the policy process always starts with a consideration of particular desired outcomes in the future (the volitional premise). The question becomes, how clean do we want our water to be? Or, the question becomes, what sort of natural environments ought to be bequeathed to future persons? Or the question may concern the appearance of the countryside. From the answers to those questions an emerging consensus will ultimately prevail—and it might take a very long time—that advocates new parameters for water quality, or new rules for habitat preservation, or new rules about deforestation.

An evolutionary perspective on the topic of sustainability suggests a need to understand the reasons for actions as running from the future back to the present. Recall that when new policies are contemplated, the essential question is: what outcome in the future would justify a particular course of action today? Another way to put this is to say that a particular event in the future is the *reason* (or the explanation) for the action taken today. Or, what purpose in the future did today's action serve?

When policy is understood in this way it is possible to understand that particular aspects of the natural environment are preserved today not because it is suddenly economically efficient to do so, but because of a collective commitment regarding how the future ought to be constituted and how it ought to unfold. Thinking of sustainability in this way shows that deforestation in the developing world continues not because of weak property rights, or not because of roads, but because it serves the purposes of the current government to allow it to happen. It helps us to see that biodiversity is allowed to be destroyed because doing so serves the interests of those in control of the machinery of state.

In contrast to this evolutionary approach, traditional policy analysis seeks to explain and justify future economic circumstance in terms of the present. When economists calculate the present-valued benefits and costs of possible actions to protect nature, this serves as an example of letting the future fall victim to a decision approach that considers the future in terms of how well it serves present interests rather than considering the present in terms of how well it serves the interests of the future (and those who will live then). The human will in action—*prospective volition*—assesses the present in terms of the future. Reasoning "backward" is precisely the act of understanding the present in terms of the future, and deciding how we wish the future to unfold for us. Prospective volition is the human will in action, informed and motivated by the plausible purposes of the future with respect to governance structures and processes. Are governance structures secure in serving the future if they permit devastation of the forests? Are governance structures secure if they ignore the relentless poverty of the majority of their citizens? Are governance structures secure if they serve only a tiny fraction of the population?

Sustainability requires not the precautionary principle but the *prudence principle*. Prudence entails an understanding of the need to modify existing institutional arrangements in recognition of the *evolving purposes of nature*. Environmental policy must be seen as a process whereby volitional premises are transformed into meaningful operational strategies and programs that will render the goals attainable. The "collective action" component of this definition tells us that new institutions—new policies—are the product of legislatures and courts whose job it is to translate nascent political sentiments into new rules which, with luck and careful analysis, will lead to new behaviors that are less destructive of biological resources. This serves to remind us that the problem of biological destruction is first addressed by understanding that the existing rules and customs constitute particular perverse incentives and sanctions for local people and thus constitute the plausible explanation for destructive use patterns of biological resources. New policy goals thus represent a conscious change of course. When the leaders of a number of nations declare that henceforth it will be their individual and collective policy to protect the world's biodiversity from future threats, the first necessary ingredient is in place. But good intentions are not enough—such goals must be matched by new institutions yielding a new constellation of incentives and sanctions that will lead to desired outcomes. These new institutions will entail new property relations among those with varying interests in the maintenance of biodiversity. Finally, any structure of new institutions must be accompanied by a correlated structure of compliance provisions that will assure new behaviors in keeping with the intentions of the new policy.

It is helpful to recall that any new policy is both a prescription and a prediction. Policies *prescribe* because they tell what changes in the rules are necessary to bring about new behaviors with respect to biological resources. Policies *predict* because they tell that if particular changes in the working rules or property regimes are implemented then new behaviors are likely to result. But of course problems are often misdiagnosed, and therefore it is to be expected that some prescriptions and some predictions will be mistaken. There must be mechanisms and procedures in place to assess those new ecosystem outcomes against the declared purposes of conservation policy, and to allow correction and modification when discrepancies arise.

This suggests that the new institutions emanating from the policy process will likely hold implications for perceptions of rights and duties among those who have been traditional ecosystem inhabitants. As with biodiversity conservation, the policy problem is to design a resource management regime—a new institutional setup—that would give those currently unhappy with the status quo a new and more satisfactory institutional setup, yet at the same time leave those whose behavior must change (the "losers") no worse off than they are at present. That process of searching for Pareto safety entails *the asking for and the giving of reasons* (Brandom, 1994, 2000). Successful policy implementation entails sharing constructed accounts—called *created imaginings* by G. L. S. Shackle—in order that those who think they will gain and those who think they will lose can gradually come to grips with this evolving playing out of their own very particular settings and circumstances. And, of course, individuals will create quite different imaginings about possible outcomes. This should not surprise us. We have different imaginings because the *available actions* are novel events in our lives. We have *not done that before*, so why should it be supposed that each of us could have definitive data and similar imaginings concerning

precisely what will transpire? As Shackle says, "An action which can still be chosen or rejected has no objective outcome" (Shackle, 1961, p. 143).

The usual economic response to this statement would be to agree and add that we will therefore assign probabilities to future states so that proper calculations might then proceed. But this response misses the point. Shackle means here that it is impossible to offer a plausible description (account, prediction) of these alternative future states since those states have not existed before. All we have in our mind about those future states is contending *thoughts and imaginings*. Assigning probabilities to necessarily imagined and constructed outcomes in the future is to impart a false sense of *precision* when, in fact, *accuracy* is the unavoidable issue here. And the matter of accuracy must remain unresolved since we will never know what the future holds until we "arrive there." We can discuss it, describe it, form quite firm convictions about it, but all of this discussion is nothing but a process of working out what the future *might* be—and it has little bearing on what the future *will* be.

Notice that evolutionary economics deals with this problem quite differently from what is found in conventional approaches to collective choice. In the standard story, the benefits and the costs are calculated by "experts" (that's us) and then communicated to the citizenry so that they can make a "rational" choice. In the evolutionary approach, those estimates of gains and losses are reckoned by the individuals affected by such policies. That process of assessing impacts is itself one that accords a singular importance to the working out of perceptions of new settings and circumstances. It is a process that the pragmatist philosopher Charles Sanders Peirce (1934, 1997) would call the *fixing of belief*. And as Peirce insisted, *a belief is that upon which we are prepared to act*. I follow Shackle (1961) in his criticism of the standard economic approach that the ends of action are fixed, and that the individual need only address alternative means to those predetermined ends. I am certainly not alone here. Many writers suggest that it is precisely here that the rational choice theory goes off the rails—for the simple reason that the concept of *choice* as it is used in economics becomes incoherent. Or, as Amartya Sen has observed, contemporary economics turns the idea of choice into a mere play on words (Sen, 1977). Notice that if ends are given, and all that remains is for the individual to compute the most efficacious means to achieve those ends, this is not *choice* but mere *calculation*. Individuals who can only calculate are not *choosing* among alternative actions—they are calculating to find the "best" means. Notice that this route leaves the individual, once the calculations have been made, with *no choices to make*. As long as the individual could not "rationally" have done other than what the calculations revealed to be the rational choice, the agent did not *exercise choice* (Lawson, 1997).

Indeed, Shackle has insisted that:

> Conventional economics is not about choice, but about acting according to necessity . . . . Choice in such a theory is empty, and conventional economics should abandon the word . . . . The escape from necessity . . . lies in the *creation of ends*, and this is possible because ends, so long as they remain available and liable to rejection or adoption, must inevitably be experiences by imagination or anticipation and not by external occurrence. Choice, inescapably, is choice amongst thoughts, and thoughts . . . are not given (Shackle, 1961, pp. 272–73).

3. Summary and implications

> . . . the fundamental premise of pragmatism's theory of action . . . does not conceive of action as the pursuit of ends that the contemplative subject establishes *a priori* and then resolves to accomplish; the world is not held to be mere material at the disposal of human intentionality. Quite to the contrary, pragmatism maintains that we find our ends in the world, and that prior to any setting of ends we are already, through our praxis, embedded in various situations.
>
> —Joas, 1993, p. 130

Sustainability can be rescued from platitudes and incoherence by rediscovering the evolutionary predecessors of the ordinalist revolution in economics, and by connecting that with the idea of Shackle's *created imaginings* about future outcomes. I used to believe that conversations about sustainability were conversations about what is worth saving for the future (Bromley, 1998). I no longer believe that. Nor is sustainability usefully thought of in terms of how much of something (some natural capital) ought to be saved

for the future. I now insist that sustainability is best thought of *as looking for those aspects of our natural and constructed settings and circumstances for which we can, at the moment, mobilize the best reasons to make sure that they are passed on to future persons.* This is not a process in which we seek to maximize time paths of consumption or welfare into the infinite future. It is, instead, a process in which we search for the best reasons to bequeath a particular endowment bundle to those who will follow. And that task is precisely the subject matter of a properly constituted evolutionary economics. Unfortunately, not much has changed in the 100 years since Veblen thought about the topic.

References

Ascher, W., *Why Governments Waste Natural Resources* (Johns Hopkins University Press: Baltimore, 1999).

Becker, L. C., *Property Rights* (Routledge and Kegan Paul: London, 1977).

Brandom, R. B., *Making it Explicit: Reasoning, Representing, and Discursive Commitment* (Harvard University Press: Cambridge, MA, 1994).

Brandom, R. B., *Articulating Reasons* (Harvard University Press: Cambridge, MA, 2000).

Brock, W. A., and D. Colander, "Complexity and Policy," in D. Colander, ed., *The Complexity Vision and the Teaching of Economics* (Elgar: Cheltenham, UK, 2000).

Bromley, D. W., *Environment and Economy: Property Rights and Public Policy* (Basil Blackwell: Oxford, 1991).

Bromley, D. W., "Searching for Sustainability: The Poverty of Spontaneous Order," *Ecological Economics* 24 (1998), 231–240.

Bromley, D. W., "Deforestation: Institutional Causes and Solutions," in M. Palo and J. Uusivuori, eds., *World Forests, Society and Environment* (Kluwer: Dordrecht, Chapter 9, 1999).

Clark, C., "Profit Maximization and the Extinction of Animal Species," *Journal of Political Economy* 81 (1973), 950–961.

Commons, J. R., "Institutional Economics," *American Economic Review* 21 (1931), 648–657.

Commons, J. R., *Legal Foundations of Capitalism* (University of Wisconsin Press: Madison, WI, 1968).

Commons, J. R., *Institutional Economics: Its Place in Political Economy* (Transaction Publishers: New Brunswick, NJ, 1990).

Cooter, R., and P. Rappoport, "Were the Ordinalists Wrong about Welfare Economics?" *Journal of Economic Literature* 22 (1984), 507–530.

Dixon, J. A., and L. A. Fallon, "The Concept of Sustainability: Origins, Extensions, and Usefulness for Policy," *Society and Natural Resources* 2 (1989), 73–84.

Joas, H., *Pragmatism and Social Theory* (University of Chicago Press: Chicago, 1993).

Lawson, T., *Economics and Reality* (Routledge: London, 1997).

Mirowski, P., *More Heat Than Light* (Cambridge University Press: Cambridge, UK, 1989).

Norgaard, R. B., "Sociosystem and Ecosystem Coevolution in the Amazon," *Journal of Environmental Economics and Management* 8 (1981), 238–254.

Page, T., *Conservation and Economic Efficiency* (Johns Hopkins University Press: Baltimore, 1977).

Parris, T. M., "Toward a Sustainability Transition," *Environment* (2003), 13–22.

Pearce, D. W., and R. K. Turner, *The Economics of Natural Resources and the Environment* (Harvester Wheatsheaf: London, 1990).

Peirce, C. S., *Collected Papers*, Vol 5 (Harvard University Press: Cambridge, MA, 1934).

Peirce, C. S., "How to Make Our Ideas Clear," in L. Menand, ed., *Pragmatism* (Vintage Books: New York, 1997).

Sen, A., "Rational Fools: A Critique of the Behavioral Foundations of Economic Theory," *Philosophy and Public Affairs* 6 (1977), 317–344.

Shackle, G. L. S., *Decision, Order, and Time in Human Affairs* (Cambridge University Press: Cambridge, UK, 1961).

Veblen, T., "Why Is Economics Not an Evolutionary Science?" in *The Place of Science in Modern Civilization* (Transaction Publishers: New Brunswick, NJ, 1990a), pp. 56–81 [originally published in *The Quarterly Journal of Economics* 12, no. 4 (1898), 373–397].

Veblen, T., "The Limitations of Marginal Utility," in *The Place of Science in Modern Civilization* (Transaction Publishers: New Brunswick, NJ, 1990b), pp. 231–251 [originally published in *The Journal of Political Economy* 17, no. 9 (1909), 620–636].

WCED, (World Commission on Environment and Development). *Our Common Future* (Oxford University Press: Oxford, 1987).

Under-investing in public goods: evidence, causes, and consequences for agricultural development, equity, and the environment

Ramón López*

Abstract

A common factor that explains why agriculture causes too much environmental degradation, grows too slowly, and has been ineffective in reducing rural poverty is the generalized tendency by governments to under invest in public goods despite the high rates of return to such investments. A large share of rural public expenditures is deviated to private goods (mostly subsidies to the wealthy), which generally have low or even negative rates of return. Behind such an obviously aberrant choice are political economy forces; a highly unequal political lobby market leads to government policies that are biased in favor of economic elites and detrimental for both the environment and rural development. Globalization may affect this important distortion on the allocation of government expenditures in various ways. One such way is by restricting the ability of governments to repress the political mobilization of the poor to counter the almost unchallenged power of the elites in the lobby market. This may contribute toward creating conditions that are more consistent with sustainable and socially equitable development.

JEL classification: Q18

Keywords: public expenditures; agriculture; environment; poverty; political economy; globalization

1. Introduction

The first and central question of this paper is why have the environmental effects of agriculture been so negative in most developing countries?[1] A useful conceptual framework to analyze this issue has to be much broader than the conventional externality approach that attributes environmental degradation to its most proximate cause, the unsolved externalities (if they could only be "internalized"). The environmental impact of agriculture cannot be separated from the performance of the sector, including growth rates, distributional impacts and, most importantly, the underlying political economy process that determines how public resources are allocated in rural areas. To further orient the analysis, two additional questions that are highly complementary to the first are postulated in this paper.

The second question needs some introduction. Recent studies have shown that in the relatively few cases when growth in agriculture and related rural industries has been respectable, large poverty-reducing effects, especially for that segment of the rural population working directly in the modern agricultural sector, have been documented. More importantly, fast agricultural growth has importantly contributed to diminishing poverty in urban sectors, particularly via the unskilled labor market (López and Anriquez, 2003). Subsistence and semi-subsistence farmers, certain rural ethnic groups and others, which in many poor countries constitute the majority of the rural population, have, by contrast, apparently received little benefits

* Department of Agricultural and Resource Economics, University of Maryland at College Park, College Park, MD, USA.

[1] Foster et al. (2002), for example, find that increases in agricultural productivity in India reduce forest areas more than proportionally. Several other studies confirm the negative impacts of agricultural expansion upon water resources and other environmental resources. See also Abdelgalil and Cohen (2001) and Dasgupta et al. (2001) among others, for empirical evidence on the environmental impacts of agriculture.

from agricultural growth.[2] This fact, in part, explains why rural poverty has remained high and intense even in middle-income countries (López and Valdés, 2000). It is thus paradoxical that rapid agricultural growth, whenever and wherever it has occurred, has been good for reducing poverty in non-rural areas but it has been less powerful in promoting higher incomes among the poorest segments of the rural population. The second question is now natural: Why has agricultural growth not benefited these groups?

The rate of growth of agricultural and other rural industries in many countries may have been just too slow to induce sufficient spillovers to benefit a broader segment of the rural population.[3] Until the early 1990s it was fashionable to ascribe this slow growth to large macroeconomic and trade policy distortions that kept agricultural commodity prices artificially low. Generous agriculture-specific policies only partially offset the effects of these macroeconomic policies (Krueger et al., 1991).

Several countries have, however, largely removed the macroeconomic distortions against agriculture while still keeping highly favorable sector-specific policies (World Bank, 2001). Input and credit subsidies, a favorable tax treatment of agricultural income, and large government expenditures in the sector have remained in place after the removal of the macroeconomic and trade distortions. Despite this, annual agricultural growth in the countries that adopted the reforms has rarely surpassed their historical 1–2.5% rates. Hence, the third question is the following: Why has such slow growth continued even in countries that have removed anti-agriculture macroeconomic biases? A complementary question is the following: Are there other remaining, perhaps more important, distortions that impair the ability of the sector to grow faster? Moreover, are these "other distortions" factors that explains the seemingly low effectiveness of even fast-growing agriculture to increase the income of important segments of the rural population that are not directly linked to modern agriculture?

---

[2] Binswanger and Deininger (1997) in their review of empirical studies found that, with the exception of a few Asian countries, small farmers typically experience few welfare improvements out of agricultural growth.

[3] According to the World Bank (2000), annual agricultural GDP growth over the last two decades has been about 2% in Latin America and Asia and negligible in Africa.

It is argued below that the poor performance of agriculture in many developing countries, the persistence of rural poverty, and a large part of the negative impacts of agriculture upon the rural environment are associated with a more fundamental distortion in the allocation of public expenditures that leads to a chronic undersupply of public goods. Investments in public goods get crowded out from government budgets by massive expenditures in subsidies to the wealthy and other expenditures in private goods that play no role in ameliorating market imperfections. In turn, the undersupply of public goods is at least in part related to political economy forces.

2. A conceptual framework

These three questions could, indeed, be interrelated. The conventional approach of assuming independent producers/consumers responding to and being affected by market price incentives is not useful, at least not in understanding some of the remaining, unsolved questions. The literature on market failure recognizes the existence of other interactions among individuals that tend to play a role in resource allocation and wealth distribution when markets do not exist or fail (de Janvry et al., 1991; Stiglitz, 1991).

Additionally, the political economy literature that emphasizes government policies and public expenditures for sale, which are directed to serve those that are able to pay for them, is another pillar of the ensuing analysis (Bernheim and Whinston, 1986; Grossman and Helpman, 1994). Ironically, in this view the allocation of public resources (through policies and expenditures) to pressure groups, in part arises out of the development of a "market": economic groups bid for public resources in the form of bribes, political contributions, etc., and the government allocates the prize to those willing to bid the most. The literature of collective action (Olson, 1965; Ostrom, 1990) is also central to the ensuing analysis. This literature emphasizes certain characteristics of social groups that may facilitate their ability to act collectively in search of common objectives, including rent extraction.

Some authors have argued that competitive lobbying by interest groups may lead to policies that increase growth and efficiency, at least under certain conditions (Becker, 1985). However, the implied conditions for this to happen are quite unlikely to be satisfied in developing countries. On the contrary, competition for government policies and expenditures is *unequal* and leads to distortions and losses of output.

The distortions that are focused on here are not the traditional price-related distortions so popular in the economics literature. Instead, the distortions caused by the crowding out of expenditures in public and semi-public goods due to excessive government expenditures in private goods that are motivated by political economy considerations are addressed. The combination of interactions arising out of market failure and political economy mechanisms provides a powerful tool for understanding why governments systematically under-invest in public goods. This, in turn, causes agricultural growth to be deleterious for the environment while at the same time causing it to be too slow and too biased to induce significant welfare gains for subsistence and semi-subsistence rural producers and other agrarian communities.

Three key actors are considered:

(1) The commercial operators (C), comprising large and medium-sized producers (including farmers, agro-processors, and other large producers in related rural industries) whose production is market oriented, relying mostly on hired workers.

(2) Other producers (P); that is subsistence and semi-subsistence producers, comprising independent producers as well as communal producers who share part of their resources in common property. These producers are only partially integrated into the commodity markets, and rely mostly on their family labor for their subsistence. Unlike C producers, P producers generally have no legal rights upon their land, forest, and water resources, or at least face tenure insecurity.[4]

(3) The government (G). Apart from setting policies, the most important role of government is to allocate public expenditures, and to regulate the use and appropriation of public resources including public lands, forests, water, and other resources that are not subject to well-defined property rights.

A fourth group, a non-rural C (e.g., industrialists, financial entrepreneurs, etc.) that may affect policies and allocations to rural areas, could also be included, but is left out to simplify the analysis and focus on the rural sector. Therefore, the following is considered: *given* a fixed volume of public resources that G has available for rural areas, how is it distributed between C and P and what are the environmental and social equity implications of such a distribution? That is, a two-stage process is assumed: in stage I the overall level of support policies is set for the rural sector, presumably on the basis of competition between rural and non-rural lobby groups. In stage II, the public resources available for the rural sector are distributed by G through political allocations. The focus here is on stage II.

2.1. "Buying" policies and public resources

G has as an objective function to maximize the welfare not of society but of government bureaucrats themselves. One way of increasing the welfare of G is to elicit bribes from producers in exchange for favors in the form of orienting public resources to the producers who pay such bribes. G also attains welfare gains by directing policies and public resources in such a way as to increase social welfare.[5] Thus, the objective function of the government is a weighted average of the welfare of those who can bribe or provide campaign contributions to G, and the welfare of the rest of society. A measure of the degree of corruption of a government is the difference between the weight of bribe contributors in its objective function and the share of this group in the total population.

While most competition is assumed to take place through economic means (campaign contributions, bribing, etc), it is also possible to allow for noneconomic forms of eliciting benefits and policies from the government. The use of political organization, strikes, civil unrest, etc. is sometimes a recourse available to both C and P as a means to pressure governments. For the rural poor, however, with few exceptions, this recourse has not always been available in the past, in large part because of geographic isolation, poor

---

[4] The World Bank (1997) has reported that even in a relatively prosperous middle-income country such as Chile, over 60% of the small-size farmers do not have legal land titles.

[5] If C is a small minority of the population, as is normally the case in poor countries, then social welfare is ruled mostly by the welfare of group P where the majority of the population belongs.

communications, and low education levels. In addition, governments have frequently been able to use their repressive apparatus with few constraints to suffocate such political mobilization of the poor. As argued later, though, for various reasons these noneconomic instruments of pressure may become more prominent in the future. (In many cases the noneconomic instruments do not even need to materialize to become effective. It is enough that governments know that they exist as a last recourse to restrain the impact and effectiveness of economic instruments of lobbying).

2.2. Imperfect capital markets and unequal competition in the lobby "market"

Only C can offer bribes to G because unlike P, group C is able to exploit all profitable "investment" opportunities, including bribing G, in return for special favors. A key reason why C and not P can bribe lies in the capital markets. Capital market imperfections have been well documented in the literature, which has consistently shown that these imperfections lead to tight rationing of credit to small enterprises, including peasants.[6] Hence, P is assumed to have little, if any, access to capital markets while C faces no credit constraints at all.[7] This implies that group C individuals are more able to invest in bribing than those in group P. That is, capital market imperfections spill over into the political lobby "market" inducing severe inequities in access to government favors. Moreover, the fact that group C comprises relatively few individuals while group P consists of so many and dispersed households, makes it easier for C than P to lobby G in a coordinated fashion. This, of course, follows directly from accepted and empirically corroborated postulates of the theory of collective action.

Thus, there exists a highly unbalanced lobbying system where C, because of its wealth and access to capital markets, has the means to optimally invest in bribing G, and, because of its small group size has the ability to act as an interest group in a coordinated and consistent way vis-à-vis G. In contrast to C, P has neither the financial means nor the adequate group size and homogeneity to bribe G in the same way. The fact that in most developing countries government programs and policies for agriculture are distributed in a highly biased way, with the commercial sector receiving most of the benefits, is certainly consistent with this view (Binswanger and Deininger, 1997).

It is here where the synergy between political economy–corruption and market failure is most important: conventional models of political economy and bribery, when considering competition among individuals or groups for government favors, assume a level playing field, i.e., competition among individuals who have an equal or similar ability to pay bribes. These models generally ignore unequal competition where a segment of the private sector can bribe at a much lower cost than others. In the notation used here, competition on a level playing field may take place *within* group C, as individuals belonging to C are assumed to have similar access to capital markets. However, the interest is to focus on the highly unequal competition *between* groups C and P.

Recent empirical evidence shows that, contrary to conventional political economy postulates that emphasize competition among commercial producers for government favors, product- or commodity-based interest groups are unimportant determinants of government credit allocations while farm size is the key determinant (Helfand, 2001). Since competition *within* group C is likely to be reflected in commodity-based competition, while competition *between* C and P producers is probably better reflected in land size (which may be considered a good proxy for wealth); this evidence is highly consistent with the emphasis on C–P (unequal) competition rather than competition within C as the driving force behind government policies.

2.3. The lobby market and efficiency: the conventional view

The highly unequal competition for government favors leads G to bias the allocation of public policies and expenditures in favor of C (and against P). This biased allocation not only has obvious distributional effects, but also negative efficiency effects. The conventional argument that political lobbying may not

---

[6] Helfand, 2001, for example, shows that credit availability in Brazil was heavily biased in favor of large-scale farmers; Baydas et al., 1994 show similar findings for Ecuador.

[7] A common way of modeling capital market imperfection is to assume that the borrowing capacity of households is equal to a fraction of the household's wealth. Since C has much greater wealth than P, it is natural to expect that C will have access to much more credit than P.

have deleterious efficiency effects assumes that individuals or groups lobby because they can obtain large benefits, while individuals who get less out of government expenditures or policies will lobby less intensively. The implication, of course, is that if lobbying takes place in an environment of perfect competition, it will cause an allocation of government resources to those that obtain the highest marginal value out of them. That is, the outcome of the political economy process would be efficient.

2.4. The lobby market and efficiency: the role of unequal competition

Of course, when lobbying is not exactly a competitive activity, but on the contrary is subject to dramatically unequal competition, there could be many uses of government resources that have a higher marginal value but are not funded simply because those who would benefit from them do not have enough capacity to lobby the government. More importantly, there is a tendency to demand from the government the provision of private goods instead of public or semi-public goods. In fact, C will have little incentive to spend on lobbying efforts for the sake of public goods that by definition cannot be privately appropriated. This causes crowding out of public expenditures within the limited public budget and, consequently, scarcity of public goods, which are important factors of production. Moreover, public goods are usually complements rather than substitutes of private investment.

Hence the efficiency losses are double: (1) A loss due to the wrong allocation of government expenditures to producers that may obtain a low marginal product out of those resources to the detriment of producers that could get higher marginal products; (2) A loss due to a misallocation caused by supplying too few public goods (see López, 2003 for an analysis of the economic growth effects of this). Apart from the efficiency effects, unequal lobbying may have severe social equity impacts. It is also likely to dramatically exacerbate environmental degradation caused by agriculture. These issues are further discussed in the next section.

2.5. Equity and environmental consequences of unequal political competition

Unequal competition in the political lobby market causes the allocation of public expenditures to be biased in favor of private goods benefiting C and against the provision of public and semi-public goods, many of which are vital to the welfare of P. Provision of public education and health care, both key public or semi-public goods, is important for P to enhance their human capital and, consequently, their ability to increase income. This is particularly important for poor households that are unable to access these services through the private sector. Even if the rate of return of human capital for these households is high, they are unable to invest in these assets unless the public sector provides them at low cost. Capital market failures generally prevent poor households from accessing credit to finance profitable investments in human capital and, hence, in the absence of government intervention, these households are unable to acquire human capital from the private sector. The under-supply of public goods leads to low investments in human capital by poor households and, consequently, adversely affects their income potential. Thus, unequal competition in the lobby market, which originates in unequal wealth distribution and capital market failures, further worsen social equity.

A component of the public goods menu is the provision of public protection of the environment through public investments in protection and rehabilitation of ecosystems, as well as in the creation of institutions that mitigate environmental externalities. A frequent manifestation of the crowding out of public investment by the provision of subsidies and other private goods is the minimal provision of environmental public goods and institutions. Thus, an obvious implication of the government's emphasis on supplying private rather than public goods is the lack of investment in the environment and the lack of monitoring and enforcement of environmental regulations. This makes environmental and natural resource degradation much more likely. Natural resource degradation also has second-round negative equity effects, as the poor are more dependent on such resources than the non-poor and, consequently, producers P have their income potential reduced.

2.6. Biased public expenditure allocation and private investment

The biased composition of public expenditures has two conflicting effects upon private investments. On the one hand, the high emphasis by government on subsidies and the supply of private goods may, under certain conditions, be an incentive to invest more. On the other hand, the low supply of public goods reduces the marginal returns to private investment over the long run. The profitability of private investment in the long run depends on an adequate supply of public goods, including human capital, infrastructure, and natural resources. In the long run the slow growth of human capital and the degradation of natural capital reduce the incentives for private investment as the marginal returns to private capital are not supported by an adequate growth of public factors of production. Private capital and public assets are, therefore, highly complementary factors of production (World Bank, 2000; López, 2003).

The net effect is, in principle, ambiguous. However, there are conceptual reasons to expect that the investment-promoting effect (the first effect) is likely to be weak (see below). This theoretical prediction is corroborated by empirical studies discussed later, which show that the strength of the first effect is indeed quantitatively small as most public subsidies in reality promote greater consumption by the wealthy instead of more investment. By contrast, emerging evidence regarding the investment-inhibiting effect of subsidies (the second effect) suggests that it is quite large (World Bank, 2000). Thus, the net effect of biasing the structure of public expenditures in favor of private goods and against public goods is not likely to promote growth.

2.7. The double crowding out

Unequal competition for government expenditures and policies leads to the crowding out of investments in public goods within the limited government budget. There is, however, a second type of crowding out: as a consequence of the subsidized provision of private goods by the government, group C may invest less, not more as superficial analysis would suggest. The reason is that the goods provided by the government are usually substitutes for private investment, and thus much of the support of government to agriculture may have little net effect on agricultural growth. In fact, much of the investments made by the government in response to lobbying by C would be implemented by the private sector itself if it were not for the knowledge that the government provides them at a much lower cost to them.

Consider what is often regarded as a "desirable" subsidy; the government offers to pay a portion the costs of a particular investment. Assume further the best possible circumstance in terms of the allocation of the subsidy. The subsidy is, of course, rationed as the funds are obviously less than the demand, but their allocation among producers is transparent, not subject to corruption. Consider an investor that is able to extract a profitable return out of an investment (even in the absence of the subsidy) that potentially may qualify for the subsidy. Suppose that in that year there was a large demand for the subsidy and that the investor was not lucky enough to get the subsidy. The producer may go ahead with the investment (and never get the subsidy) anyways since it is a profitable investment. Alternatively, she/he may opt to postpone the investment and try again next year in the hope of then getting the subsidy. If the expected value of the subsidy is sufficiently large to compensate the foregone profits in one year the producer may decide to delay the investment.[8] Thus, investments that are privately (and socially) profitable may be postponed as a consequence of the existence of the subsidy.

Among the investors that actually get the subsidy there are two types. Those that would have implemented the investment anyways and those that would have not (because they would not be able to get high-enough returns) but they, in fact, invest because they got the subsidy. For the former the subsidy was ineffective—the subsidy is likely to promote more consumption by them rather than more investment as intended. For the latter the subsidy was effective in causing them to invest but at a low social return. Thus, the subsidy scheme does two things, it increases consumption of some producers and it causes a reallocation of investment from producers that are able to obtain high (private and social) rates of return to the investment to producers that obtain a low social return. The net effect on total investment is ambiguous, but the efficiency impact is negative.

The above example is not just a curiosity. It illustrates a phenomenon that has received important empirical support in recent years. That subsidies, at least

---

[8] Suppose the subsidy is 50% of the value of the investment cost and the rate of return per annum of the investment is quite high, 20%. Assume further that the producer estimates that the probability of getting the subsidy in the next year is 0.5. If the producer is risk neutral he/she will decide to delay the investment by one year.

in the form in which they are usually allocated, do not generally promote investment or more R&D has been shown by several studies in various countries. Empirical studies using detailed firm-level data by Bregman et al. (1999) for Israel, Fakin (1995) for Poland, Lee (1996) for Korea, Estache and Gaspar (1995) for Brazil, and several others have shown that subsidies and corporate tax concessions are at best ineffective in promoting investment and technological adoption and, in some instances, even counterproductive. Crowding out of private investment as a consequence of the subsidies occurs.

3. How public expenditure allocation biases are manifested

There are two broad types of interventions that tend to have negative effects upon both the environment and the poor (and detrimental effects on overall growth of agriculture).

First are "development expenditures." In most cases these are fiscal incentives that only (or mostly) C can access, such as tax exemptions (available only to those who pay taxes, generally belonging to C) and credit subsidies (available only to those who can access credit). Outright financial grants to certain projects, publicly funded infrastructure such as dams, targeted mostly to increase the wealth of C with sometime negative impacts upon P, are also considered "development expenditures."[9] Examples of "development expenditures" are the massive fiscal incentives currently underway for the development of the Brazilian Amazon region and the promotion of tree crop production in the outer islands in Indonesia.[10]

"Development expenditures" basically constitute a give-away of public resources for the obvious benefit of C, with ambiguous indirect impacts upon P. These programs cause efficiency losses due to a significant misallocation of public resources: these fiscal resources are generally invested in *private* goods while its opportunity cost is the foregone investment in *public* goods.

---

[9] Three recent publications provide empirical support regarding the large size of public subsidies that are detrimental for development: Asher, 1999; Myers and Kent, 2001, and Van Beers and de Moor, 2001.

[10] See Calmon (2003) for evidence about the new type of fiscal incentives that encourages "development" and deforestation of the Amazon Region.

When the government decides to invest one dollar in private goods (directed at C), it is one dollar less to be invested in education, health, the environment, and other public and semi-public goods. It happens that the dividends of investments in true public or semi-public goods accrue more directly and in a much greater proportion to P than "development expenditures," which only indirectly (mostly via employment effects) *may* benefit P.

An important component of the "development expenditures" are transfers (to C of course) in the form of free access to natural resources that are owned by society as a whole, such as forests, water resources, and others (Binswanger and Deininger, 1997). The fact that these natural resources are potentially available to C at little cost promotes greater lobbying efforts by C to persuade G to open up access to more public natural resources. Knowledge that irrigation water, for example, can be obtained from public irrigation projects at little or no cost induces C to spend more effort to cause G to finance public irrigation projects. These projects often have low social rates of return but high private rates of return (to be appropriated by C). The main reason why the private rate of return is high is simply that C usually pays only a minor fraction of the cost of water. Similarly, the fact that forest lands can be accessed at about zero price is an incentive for C to spend greater efforts and money to "bribe" the government into building more infrastructure and services in forested areas. Again, these projects often have low social returns but high private rates of return for those that are able to appropriate their (usually short-term) economic benefits.

Thus, the almost universal tendency to give away state natural resources (lands, water, mines, etc.) at almost no cost has not only distributional impacts (usually regressive, as those who gain such free access are in group C and not in P), but also resource allocation effects and negative environmental impacts. This is due to the fact that the opening up of new lands for agriculture is *not* exogenous but is at least in part determined through the system of bribes, lobbying, and influence peddling. If these decisions can be bought, the perspective of zero cost for the use of natural resources greatly increases the lobbying efforts of C to promote such opening up of new frontiers.

Second are government omissions. Governments sometimes fail to prevent the usurpation of land and other resources by P. A classical example is enclosures, where land held by communities (usually in common access without formal legal titles) becomes valuable for commercial interests. Usurpation of the land and expulsion of the peasants without compensation happened in the early stages of the industrial revolution in Europe. It happened in the eighteenth and nineteenth centuries in the United States and Latin America with native lands and still happens today in poor countries whenever resources of P become valuable to C.[11] In most cases, governments do little to protect P or to enforce compensation for the lost resources. The implication is that the poor end up paying part of the costs of economic growth.

Government omissions are also ominous causes of environmental degradation. Failure to enforce existing environmental regulations is apparently a more important cause of environmental destruction than lack of or insufficient regulations. Lack of environmental enforcement is usually attributed to insufficient funding and institutional capacity. This explanation is, of course, superficial. Lack of funding and capacity for environmental enforcement corresponds to a lack of priority in governments who find it more appropriate to devote their efforts to other activities, including those described above.

The almost complete lack of enforcement of environmental regulations unambiguously allows for greater profits for C, rapid resource degradation, and greater environmental losses. For P, it has at best an ambiguous impact: it may cause greater employment of the poor by C and perhaps some short-term benefits as producers, but most of the long-run effects are clearly detrimental to P. Unlike C, the poor are more dependent on natural resources as a source of income and have more restricted opportunities outside the rural sector. The (short-run) dividends of degrading the rural environment are obtained largely by C, while most of the (long-run) costs are paid by P.

4. Under-investment in public goods: empirical evidence

Governments in developing countries systematically under-provide public goods as a result of the political

lobby that gives incentives to politicians to spend public resources in private goods instead. Yet there is empirical evidence showing that two important public goods, education and agricultural research, have extremely high rates of return while at the same time governments have reduced rather than increased investments in such goods.

The literature reports such high returns with an amazing degree of consensus for many countries around the world. Investments in formal education (especially in secondary education), agricultural research, agricultural extension, and investments in the management of certain natural resources is reported to have extremely high rates of return. The permanence of such high returns *per se* does not necessarily reflect under-investment, mainly given the existence of significant non-convexities. Non-convexities may imply that the marginal returns to these assets do not necessarily fall or that they decrease only very slowly with their accumulation. Thus, if this is the case, even a rapid accumulation of the assets would do little to reduce their rates of return. However, given such high returns, one would expect a great emphasis from governments on investment in such assets. Yet this is not the case. In fact, in the majority of the developing countries, investment in human and environmental assets has not even kept up with population growth. That is, per capita human and environmental wealth appears to be declining.

4.1. High rates of return to education

Two recent surveys have reviewed returns to education, one by Psacharopoulos (1994) and another, an update of that survey, by Psacharopoulos and Patinos (2002). They report findings of hundreds of studies around the world that have used a variety of methodologies and diverse data and over different time periods over the past three decades or so. Despite this variability in data, countries, and methodology, there is a high degree of homogeneity in the results for most countries. In fact, the calculated rates of return found in the majority of the countries analyzed are extremely high. The average private rate of return for investment in primary education is about 30%, while the average social rate of return was about 20%. For many countries the social rates of return reach levels in excess of 30%.[12]

---

[11] For many recent examples in Latin America and throughout the world see Kates and Haarman (1992).

[12] Examples of most recent studies include Brazil (35.6% for primary and 21% for higher education); Uganda (66% for primary

The returns to primary and secondary education are both below 15% in only a handful of countries. In addition, from the evidence for countries that have more than one study, it follows that in the majority the rates of return to education have not declined over time.

Many projects that are implemented in developing and developed countries have much lower *ex ante* rates of return. Despite these high rates of return, in many developing countries high school drop-out rates are substantial, especially at the late primary and high school levels in rural areas. Even in middle-income countries such as Chile, Brazil, and Mexico, high school drop-out rates reach 40% to 50% (World Bank, 2000).

4.2. High and increasing rates of return to agricultural research and extension

A survey by Alston et al. (2000) reviewed almost 300 studies that evaluated social rates of return to agriculture research and farm extension in about 95 countries. The methodologies and data used varied dramatically across the many studies. The simple mean (social) rate of return for agricultural research among all studies in developing countries was over 50%, while the mean rate of return for public expenditures in agricultural extension was even higher, of the order of 80%. In most countries these rates rarely fall below 30%, still obviously a fantastic pay-off. Exploiting the fact that there are many countries for which there is more than one comparable study available, the authors conclude that, as in the case of returns to education, there is no evidence to support the view that the rates of return have declined over time. Despite this massive social profitability, studies often report that, with few exceptions, countries are not expanding agricultural R&D and many have indeed cut them back drastically.[13]

---

and 28.6% for secondary education); Morocco (50% for primary and 10% for secondary education); Taiwan (27.7% for primary and 17.7% for higher education); and India (17.6% for primary and 18.2% for higher education). These are social rates of return, with the exception of India. Private rates are even higher.

[13] The case of Peru is illustrative. In the mid-1990s the government decided to privatize agricultural research. The government sold 21 agricultural experimental farms where most of the agricultural research in the country was performed. The result was that by the year 2000, 20 of the 21 experimental stations had been transformed into commercial farm operations, and agricultural research in Peru has practically become extinct.

4.3. Investment in human capital, R&D, and the environment lag behind population growth

The emerging literature on "genuine savings" provides a clearer picture of the real changes in wealth over time.[14] The World Bank has provided estimates of genuine investment for many countries by adding net investment in human and natural capital to the estimates of net investments in physical capital (Hamilton, 2003). Apart from extending the analysis to more than 110 countries, an important modification over the previous estimates of genuine savings made by the World Bank is that now measures of change of net wealth are expressed on a per capita basis. Per capita rather than total wealth change is an adequate and consistent measure of welfare change (Dasgupta and Maler, 2002). The measure of per capita genuine savings as defined by Hamilton in his country estimates equals net investment in manufactured or physical capital minus depletion of natural resources plus net investment in education, health, and R&D.

The estimates for 1997 show that out of 90 low- and middle-income countries in Asia, Africa, and Latin America, 71 (or about 80% of them) exhibit negative per capita changes in wealth. While these estimates cover a large sample of countries, the fact that they refer to only one year raises the question of how representative this year might be. An analysis using the same definition of wealth as Hamilton but that covered a 20-year period was performed by Dasgupta (2003). Five Asian countries (Bangladesh, India, China, Nepal, and Pakistan), and 20 countries in sub-Saharan Africa over the period 1973–1993 were considered. This analysis shows similar results to Hamilton's. Not only has sub-Saharan Africa experienced a decrease in per capita net wealth; four of the five Asian countries also showed negative per capita wealth changes. The only exception is China, which has managed to accumulate wealth faster than its population growth.

The majority of the countries considered by Hamilton (2003) and Dasgupta (2003) show positive per capita growth rates for physical capital, implying that the reason for the negative growth rates of total

---

[14] Genuine savings is a national accounting aggregate designed to measure the net change in total assets including human and natural capital, in addition to the standard national accounts measures of change in physical and financial capital.

wealth is that human, knowledge, and environmental assets are growing at a rate below that of population. By implication, therefore, some 80% of the countries considered are experiencing reductions in their per capita human and environmental wealth. Since at least some countries may be compensating the declines of human and environmental assets with positive per capita growth in physical assets, the number of countries experiencing declines in human and environmental assets may be even larger.

The high rates of return to public goods and the fact that, notwithstanding such high rates of return, government investment in public goods has not even kept pace with population growth is a clear indication that the supply of public goods is insufficient. At the same time, governments spend a large share of their budgets on subsidies and other private goods. According to Van Beers and de Moor (2001), developing-country governments spend in total more than 6% of their countries' GDP and more than 30% of government revenues on subsidies, many of them environmentally perverse. A few recent country studies reach similar conclusions. Based on estimates by Calmon (2003) for Brazil, for example, it is possible to calculate that almost 50% of all public expenditures in the rural sector by the federal government is indeed subsidies to mostly large commercial operators and speculators.

A few studies have provided an evaluation of the impact of such subsidies. Bregman et al. (1999), Estache and Gaspar (1995), Lee (1996), Oman (2000), and World Bank (2000), among many others provide empirical evidence from many countries showing that government subsidies not only mostly benefit the wealthy, but also that their effectiveness in promoting more investment and output is low. Government subsidies effectively increase consumption of the wealthy instead of promoting more investment. That is, unlike investments in public goods, the rates of return of government expenditures on subsidies are low. Governments spend little in goods that have large social rates of return (public goods) and instead, they spend a large share of their budgets in goods with dubious rates of return (private goods). The conclusion is obvious: The composition of public expenditures is socially inefficient. That is, income will increase if governments reallocate public expenditures from private to public goods. This is the key distortion that is emphasized in this article.

4.4. Subsidies and agricultural growth: new empirical evidence

A recent study by López (2004) has analyzed new detailed panel data on the allocation of rural public expenditures elaborated by FAO for ten countries in Latin America for the period 1985–2000.[15] López, using various econometric approaches, has shown that the allocation of public expenditures is a key element in explaining agriculture per capita agricultural GDP, rural poverty, and land expansion of agriculture into frontier (often forested) areas.

The key findings are: (i) The countries in the sample devote a large share of the total expenditures in the rural sector to nonsocial subsidies, on average about 50%; (ii) Subsidies have a negative and highly significant effect on per capita agriculture GDP. Even a modest reduction of the share of subsidies in total rural government expenditures may cause a major increase in agricultural per capita GDP: Reducing the share of subsidies from 50% to just 45% may cause a permanent increase of agriculture GDP of 2.3%; (iii) Subsidies dramatically contribute to worsening poverty and to increase in the reliance of agriculture growth on area expansion rather than on intensification.

5. Back to the original questions: new insights?

Agricultural growth is more environmentally destructive than it needs to be because G gives away the environment to C in exchange for bribes and political contributions. Growth of agriculture and other rural sectors (in C) creates jobs for part of the unskilled and in doing so it improves labor market conditions for the poor. However, for rural subsistence and semi-subsistence producers that are not absorbed into the labor force working for C, agricultural growth hardly creates any benefits. While income growth of group C is complementary with income improvements of hired workers, in large part it competes through nonmarket mechanisms with subsistence and semi-subsistence peasants and communities. That is, part of the expansion of the "modern" C sector is financed by the losses of P. In addition, due to the double crowding-out effects

[15] The ten countries are: Costa Rica, Dominican Republic, Ecuador, Honduras, Jamaica, Panama, Paraguay, Peru, Uruguay, and Venezuela.

induced by public policies, large inefficiencies prevail causing growth in C to be too slow to absorb a greater part of the rural labor force in that sector and causing stagnation for P.

Does this mean that agricultural growth is always detrimental to those unable to benefit from greater job opportunities in C? The answer is, of course, no. Our hypothesis is that the impact of agricultural growth upon poor self-employed or mostly self-employed rural households greatly depends on how growth in C originates. If the instruments used to promote economic growth are those discussed above (which unfortunately appear to be most pervasive across the developing world), growth is likely to be too slow to benefit many of the rural poor and is effectively partly financed on the backs of the poor. If, instead, growth were induced through greater investment in truly public or semi-public goods relying on more neutral policies, then both its rate (and its stability over time) and its poverty effects would to be more desirable.

Similarly, agricultural growth has so many negative environmental consequences because of the biased instruments used to promote growth of the preferred groups at all costs. If, instead, more neutral instruments were used and government public allocations emphasized public and semi-public good investment as an engine of growth, the environmental consequences of agricultural growth would be more benign. Thus, an important insight following from the analysis is that to understand the implications of growth the focus needs to be not only on the speed of growth but, more importantly, *on the sources and instruments used to promote such growth.*

6. How globalization affects nonmarket interactions

Understanding agricultural growth within a political economy-cum-nonmarket interaction framework provides some unexpected implications concerning the potential long-run consequences of globalization. A key implication of the analysis presented above is that many of the negative impacts of agriculture upon the environment and the poor arise out of a highly unequal system of accessing government resources by the economic elites vis-à-vis the poor. Some of the influences of globalization may in fact contribute to

soften such unequal political power while others may worsen it.

Globalization usually involves several things, three of which are considered here: (1) more openness to international trade in goods, services, and capital; (2) greater exposure to international norms and patterns of quality for internationally traded goods and services, mostly imposed by developed countries; (3) greater integration of civil society into international information, international networks, participative and democratic values.

6.1. Trade liberalization

Increased trade openness has often implied dramatic changes in relative prices that have caused significant changes in the economic power within the rural sector. Traditionally dominant groups within C have become less able to influence government policies while new power groups have emerged. Also, to the extent that a significant part of the agricultural sector increases its profitability, the stakes of the peddling game get bigger. One could expect increased lobbying efforts to attract an even greater share of the government-controlled resources by C as these resources now, with freer trade, have a higher rate of return. Hence this, *ceteris paribus*, may induce an even more biased allocation of public resources to the private goods accessible to C to the detriment of the provision of public goods. That is, the pure trade liberalization component of globalization may exacerbate the low effectiveness of agriculture to reduce certain forms of rural poverty and anti-environment consequences.

6.2. International norms and standards

Developed countries impose international standards and norms affecting exports from developing countries. These norms usually concern sanitary, environmental, and child labor use. Some of them satisfy genuine objectives while others are simply hidden ways of protecting the domestic industries in developed countries. Apart from formal official norms there exist informal certification procedures to which a segment of the importers in the industrialized countries adhere.

What are the consequences of integrating a developing country into the system of environmental international norms that effectively internationalizes enforcement? As indicated earlier, governments provide little enforcement of environmental laws as a consequence of their desire to benefit C for the sake of bribes or as a means of accelerating "growth." This failure to enforce norms not only makes growth less environmentally friendly but also affects the poor (especially the subsistence and semi-subsistence households and communities) who generally are most dependent on natural resources for their subsistence. External enforcement of environmental norms affecting export industries tends to be stricter than domestic enforcement. In some instances this leads to substantial improvements in the management of pest control and fertilization, often inducing lower doses, the use of less toxic products, and their application at more opportune times. Thus, increased integration into global markets may be a good substitute, under some circumstances, for a lack of domestic enforcement, making agricultural growth less environmentally taxing and, sometimes, less harmful to the rural poor.

6.3. Globalization of civil society

This is perhaps the most important impact of globalization expected within the conceptual framework used in this analysis. As discussed above, part of the power of the elites is manifested in their ability to dispossess the rural poor from part of their land and other resources as they become commercially valuable to them. Native communities have historically felt the impact of this process that has been either implicitly or explicitly supported by governments. Globalization makes this harder to occur as exposure of these events to international public opinion could generate accusations of human rights violations that make the government liable to international trade sanctions and boycotts. In general, dictatorships and human rights violations are increasingly less accepted at the world level. Moreover, because globalization causes greater international trade dependence, such violations can now be more easily punished through trade bans and the like.

Greater openness to democracy and more constraints upon government imposed by international attention usually imply more participation of civil society in public decisions and less leeway for political and military repression. (Would the Zapatista movement in Mexico have been tolerated 20 or 30 years ago, when

Mexico was not yet fully integrated into global markets? Would the Sem Terra movement in Brazil have been able to avoid heavy government repression 30 years ago?)

This greater political freedom has consequences for unequal competition in the political economy process discussed earlier: The poor can only counter the enormous advantage that the elites have in influencing the government through political organization and pressure that could go all the way from greater participation in elections and civic movements to strikes and riots. In the past, these direct political instruments have generally not been available, either because of a lack of institutional capacity among the poor and a lack of participative mechanisms or, more often, due to the threat of government repression. Globalization has led to the increased capacity of the poor to organize and participate in international networks that have greatly increased their knowledge and ability to organize through both financial and technical support from abroad (many indigenous organizations have, for example, emerged in Latin America in the 1990s, several of them with strong international links). This tends to partially overcome the first constraint that the poor face: lack of knowledge and lack of institutions to channel their pressure. In addition, the restricted ability of governments to repress tends to reduce the second constraint the poor face: They now have the ability to pressure governments through organized political responses when governments affect their interests.

Thus, an important effect of globalization can be to reduce the imbalances that exist between the poor and the elites in their capacity to lobby governments. This may lead to policies that are slightly less biased toward the latter, which is often translated into greater pressure to increase the supply of public goods. Given the current severe under-investment in public goods, an increase in its supply may be translated not only into faster growth but also into greater social equity and less environmental pressures associated with growth. That is, a broadening of the scope by which government lobbying may take place, may allow for more even competition between C and P. This, in turn, could make the outcome of the lobbying process more consistent with economic efficiency as predicted by Becker (1985) and his followers. Despite progress toward democratization, civil participation and greater tolerance of political action by the poor, however, it is questionable whether outcomes of economic efficiency, social equity, and environmental sustainability can be achieved solely by the potential balancing effect upon the political process that globalization may bring about.

7. Final remarks

Large historical inequities have led to a strong dichotomy where a small rural elite is able to bias the allocation of government expenditures and policies in their favor and to the detriment of the majority of poor and semi-poor farmers. This has not only distributive implications but also efficiency effects as a consequence of the biased structure of investment that it induces—too few public goods and too many government-provided private goods. Thus, there is a double crowding out: Crowding out of expenditures in public goods within the limited government expenditure budget, and the crowding out of private investment through government-subsidized provision of private goods. The result is under-investment in public goods such as the environment, education, health, and social security with negative consequences for growth, the environment, and the poor.

The key factor behind the above process is the highly unequal capacity of the poor vis-à-vis the elites to lobby governments. Globalization may contribute toward reducing such inequality, but without strong pressure from international organizations, explicitly targeting greater social participation, transparency, and democratization, real progress toward economic efficiency and toward policies that make growth environmentally sustainable and more pro-poor is likely to be slow. Similarly, there is a need to change the composition of the investment mix financed by international organizations toward a greater provision of public and semi-public goods and away from investments in private goods that usually reinforce the power dominance of the elites.

Acknowledgments

Research assistance was provided by Alex Lombardia (University of Maryland) and editorial assistance was given by Melanie Zimmermann (University of Bonn, ZEF). Comments by Stefanie Engel, Greg Galinato, and Bruce Gardner are appreciated.

References

Abdelgalil, A. E., and S. I. Cohen, "Policy Modelling of the Trade-off between Agricultural Development and Land Degradation—The Case of Sudan," *Journal of Policy Modelling* 23 (2001), 847–874.

Asher, W., *"Why Governments Waste Natural Resources—Policy Failures in Developing Countries"* (Johns Hopkins University Press: Baltimore, 1999).

Alston, J., M. Marra, P. Pardey, and P. Wyatt, "Research Return Redux: A Meta-Analysis and the Returns of R&D," *Australian Journal of Agricultural and Resource Economics* 44 (2000), 1364–1385.

Baydas, M., R. Meyer, and A. Aguilera, "Credit Rationing in Small-Scale Enterprises: Special Microenterprise Programmes in Ecuador," *Journal of Development Economics* 31 (1994), 279–309.

Becker, G., "A Theory of Competition among Pressure Groups for Political Influence," *Quarterly Journal of Economics* 98 (1985), 371–400.

Bernheim, B. D., and M. Whinston, "Menu Auctions, Resource Allocation and Economic Influence," *Quarterly Journal of Economics* 101 (1986), 1–31.

Binswanger, H., and K. Deininger, "Explaining Agricultural and Agrarian Policies in Developing Countries," *Journal of Economic Literature* 35 (1997), 1958–2005.

Bregman, A., M. Fuss, and H. Regev, "Effects of Capital Subsidization on Productivity in Israeli Industry," *Bank of Israel Economic Review* (1999), 77–101.

Calmon, P., "Capital Subsidies and the Quality of Growth in Brazil," Report prepared for the World Bank, Brazil Department (2003).

Dasgupta, P., "Sustainable Economic Development in the World of Today's Poor," in D. Simpson, M. Toman, and R. Ayres, eds., *Scarcity and Growth in the New Millennium* (Resources for the Future Inc., John Hopkins University Press: Baltimore, 2003), forthcoming.

Dasgupta, P., and K.-G. Maler, "Net National Product, *Wealth*, and Social Well-Being," *Environment and Development Economics* 5, no. 1–2 (2002), 69–93.

Dasgupta, S., N. Namingi, and C. Meisner, "Pesticide Use in Brazil in the Era of Agroindustrialization and Globalization," *Environment and Development Economics* 6 (2001), 459–482.

de Janvry, A., M. Fafchamps, and E. Sadoulet, "Peasant Household Behavior with Missing Markets: Some Paradoxes Explained," *Economic Journal* 101 (1991), 1400–1417.

Estache, A., and V. Gaspar, "Why Tax Incentives Do Not Promote Investment in Brazil," in A. Shah, ed., *Fiscal Incentives for Investment and Innovation* (Oxford University Press: Oxford, 1995).

Fakin, B., "Investment Subsidies during Transition," *Eastern European Economics* 33, no. 5 (1995), 62–74.

Foster, A., M. Rosenzweig, and J. Behrman, "Population Growth, Income Growth and Deforestation: Management of Village Common Land in India," Mimeo, Brown University (2002).

Grossman, G., and E. Helpman, "Protection for Sale," *American Economic Review* 85 (1994), 667–690.

224   *Ramón López*

Hamilton, K., "Sustaining Economic Welfare: Estimating Changes in Total and Per Capita Wealth," *Environment, Development and Sustainability* 5, no. 3–4 (2003), 419–436.

Helfand, S., "The Distribution of Subsidized Agricultural Credit in Brazil: Do Interest Groups Matter?" *Development and Change* 32 (2001), 465–490.

Kates, R., and V. Haarmann, "Where The Poor Live: Are the Assumptions Correct?" *Environment* 34 (1992), 4–28.

Krueger, A., M. Schiff, and A. Valdés, eds., *Political Economy of Agricultural Pricing Policy* (Johns Hopkins University Press: Baltimore, 1991).

Lee, J. W., "Government Interventions and Productivity Growth," *Journal of Economic Growth* 1, no. 3 (1996), 392–415.

López, R., "The Policy Roots of Socioeconomic Stagnation and Environmental Implosion: Latin America 1950–2000," *World Development* 31, no. 2 (2003), 259–280.

López, R., "The Structure of Public Expenditures, Agricultural Income and Rural Poverty: Evidence for Ten Latin American Countries," Unpublished, University of Maryland, College Park (2004).

López, R., and G. Anriquez, "Agricultural Growth and Poverty: The Case of Chile," FAO Report, Rome (2003).

López, R., and A. Valdés, *Rural Poverty in Latin America* (Macmillan Press: London, and Saint Martin's Press: New York, 2000).

Myers, N., and J. Kent, *Perverse Subsidies: How Tax Dollars Can Undercut the Environment and the Economy* (Island Press: London, 2001).

Olson, M., *The Logic of Collective Action: Public Goods and the Theory of Groups* (Harvard University Press: Cambridge, MA, 1965).

Oman, C., "Policy Competition for Foreign Investment" (OECD Development Centre: Paris, 2000).

Ostrom, E., *Governing the Commons: The Evolution of Institutions for Collective Action* (Cambridge University Press: Cambridge, UK, 1990).

Psacharopoulos, G., "Returns to Investment in Education: A Global Update," *World Development* 22 (1994), 1325–1343.

Psacharopoulos, G., and H. Patinos, "Returns to Investment in Education: A Further Update," World Bank Policy Research Working Paper 2881, Washington, DC (2002).

Stiglitz, J., "Rational Peasants, Efficient Institutions, and the Theory of Rural Organizations: Methodological Remarks for Development Economics," in P. Bardhan, ed., *The Theory of Agrarian Institutions* (Clarendon Press: Oxford, 1991).

Van Beers, C., and A. de Moor, *Public Subsidies and Policy Failures* (Edward Elgar: Northampton, MA, 2001).

The World Bank, "Chile: Poverty and Income Distribution in an High-Growth Economy, 1987–1995," Report No 16377-CH, Washington, DC (1997).

The World Bank, *The Quality of Growth* (Oxford University Press: Washington, DC, 2000).

The World Bank, *World Development Report 2002: Building Institutions for Markets* (Oxford University Press: Washington, DC, 2001).

Resource degradation, low agricultural productivity, and poverty in sub-Saharan Africa: pathways out of the spiral

Simeon Ehui*, John Pender**

Abstract

Sub-Saharan Africa (SSA) has the lowest agricultural productivity in the world, while almost half of the population lives below US$1 per day. The biggest development policy challenge is to find appropriate solutions to end hunger and poverty in the region. Building on several years of empirical research conducted in East Africa, this paper identifies potential strategies for sustainable development in this region. In general, the empirical evidence reviewed confirms that different strategies are needed in different development domains of SSA. Nevertheless, some elements will be common to all successful strategies, including assurance of peace and security, a stable macroeconomic environment, provision of incentives through markets where markets function, development of market institutions where they do not, and public and private investment in an appropriate mix of physical, human, natural, and social capital. The differences in strategies across these domains mainly reflect differences in the mix of those investments as influenced by different comparative advantages.

JEL classification: Q01, Q16, Q24, Q38

Keywords: resource degradation; agricultural productivity; policy options; development domains

1. Introduction

About two thirds of the 627 million people living in sub-Saharan Africa (SSA) depend on agriculture for their livelihoods. Almost half of them live on less than US$1 per day. Most of the poor live in rural areas, and it is estimated that there are 236 million agricultural poor representing 60% of the agricultural population and 80% of the total number of poor in the region (Dixon et al., 2001). Therefore, agriculture continues to remain important, and indicators of rural well-being are closely related with agricultural performance.

While agricultural output is growing, productivity is not. Food production per capita has declined by 17% in SSA from an already low level since 1970, the most of any major region of the world (Figure 1). Cereal yields have remained stagnant since the mid-1970s, while yields have doubled in other regions of the

developing world, and now average only one third of yields in other developing regions (Figure 2). Yields of other food crops and livestock have also declined since the 1970s (World Bank, 2000a). Beef yields have decreased by 10% since 1970 (Figure 3). Low productivity has eroded the competitiveness of African agriculture in the world market; as a result most countries in the region have become net importers of food commodities.

Many factors have been suggested as causes, with some emphasizing the Malthusian link between rapid population growth, low agricultural productivity, and resource degradation (World Commission on Environment and Development, 1987; Cleaver and Schreiber, 1994), and others market failures (Holden and Binswanger, 1998), or government and institutional failures (World Bank, 1994). Clearly, there is no shortage of reasons to explain the failure to initiate sustainable development. What are needed are effective strategies to reverse the downward spiral.

In this paper it is argued that no single strategy will work for SSA. The key is to identify and implement

* The World Bank, Washington, DC, USA.
** The International Food Policy Research Institute, Washington, DC, USA.

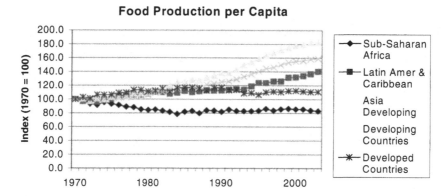

Figure 1. Food production per capita. *Source:* FAOSTAT 2004.

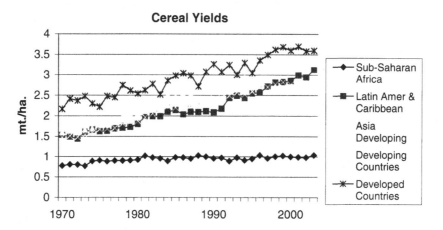

Figure 2. Cereal yields. *Source:* FAOSTAT 2004.

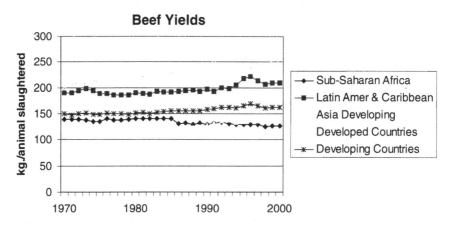

Figure 3. Beef yields. *Source:* FAOSTAT, 2004.

effective policies and strategies for different circumstances in the region. Essential to such strategies will be investment in an appropriate portfolio of physical, human, natural, and social capital, which will differ from one place to another depending on potential comparative advantage. The argument draws on the results of several years of empirical research conducted in East Africa where an attempt has been made to identify strategies for sustainable development based on comparative advantage. The lessons learned can serve as a basis for similar analyses in other regions of the subcontinent.

2. Explaining SSA's poor agricultural performance

Socioeconomic, policy, biophysical constraints, and unsustainable land management practices have been identified as major causes of low food productivity, soil fertility decline, and degradation of the agricultural land resource in SSA. The region has also been afflicted with poor resource endowments (including climate extremes, poor land quality, and endemic livestock and human diseases), and poor policies, including low public investments that have consistently undermined agriculture and the infrastructure and institutions that serve it (Binswanger and Townsend, 2000; Rosegrant et al., 2001).

2.1. Poor resource endowments

Adverse endowments and biophysical conditions have contributed to SSA's poor agricultural development. Although land is abundant, more than half of the region lies within the arid and semi-arid zones. High temperatures accelerate the degradation of organic matter, thus severely reducing the water-holding capacity of the soils and making them deficient in nitrogen and phosphorus. Only 6% of the land is of high agricultural potential, considering climate and soil constraints (Tegene and Wiebe, 2003). The soils in most of SSA are more marginal and less responsive to inputs that in Green Revolution areas of Asia (Voortman et al., 2000). Superimposed on these inherently fragile resources is the continuous removal through cropping of plant nutrients in quantities greater than those being returned by mineral or organic fertilizers. Average

rates of nutrient depletion during the past 30 years indicated losses of about 660 kg/ha of nitrogen, 75 kg/ha of phosphorous, and 450 kg/ha of potassium per year (Smaling et al., 1997). The situation is even more desperate in the arid zone, where the soils are shallow, calcerous, and saline.

The continuous decline in the stock of soil nutrients is linked to soil fertility management practices that are not properly aligned to continuous cultivation under increasing population pressure. Many Asian countries were able to achieve dramatic increases in crop yield through high rates of adoption of inorganic fertilizers on irrigated lands in conjunction with the high-yielding varieties of rice and wheat. Use of inorganic fertilizer in SSA averages less than 10 kg per cultivated hectare, less than 10% of the average intensity of fertilizer use in other developing regions of the world (Figure 4). Application of organic materials such as manure and compost, and use of nitrogen-fixing leguminous plants to restore soil fertility are also limited (Barrett et al., 2002; Nkonya et al., 2004; Place et al., 2003a).

Other natural factors that hinder agriculture include geography, the risk of drought, low-quality feed for livestock, and endemic livestock and human disease. Many countries are landlocked, and the resulting high transportation costs lead to high prices of imported inputs relative to local output prices (especially for export commodities). Uncertain rainfall is a major risk, especially in the dryland areas, also contributing to low adoption of fertilizer and other inputs in crop production. In the humid and sub-humid zones, where crop production is relatively less risky, trypanosomiasis reduces cattle and milk off-take by up to 30% and 40%, respectively, and reduces the work performance of draught animals by up to 30% (ILRI, 1998). Among the human diseases, malaria, tuberculosis and, more recently, HIV/AIDS have had a devastating impact on human lives and on agricultural performance. In 2000, about 70% of the people estimated to have AIDS worldwide were found in SSA, with 16 countries having more than 10% of their adult population affected (FAO, 2001).

2.2. Poor policy environments

Low producer prices for crops and livestock have discouraged farmers from investing in agriculture and

Fertilizer Use

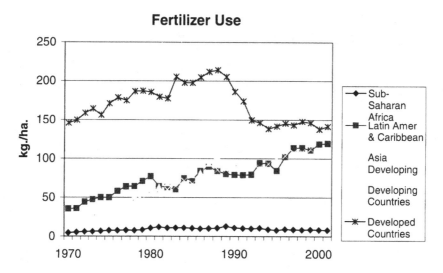

Figure 4. Use of inorganic fertilizer. *Source:* FAOSTAT 2004.

natural resource conservation measures. Disincentives were created by overvalued exchange rates in the 1970s and 1980s that resulted in loss of competitiveness in many countries. Agricultural marketing and input supply systems were dominated by the public sector, inhibiting the development of private traders and farmer cooperatives.

Despite high tax levels, there was little investment in rural public services and infrastructure. Urban bias favored investments that benefit more politically powerful urban elites (Lipton, 1993). Less than 10% of public spending has gone to agriculture (World Bank, 2000b). In addition, subsidies for fertilizer and credit usually benefited larger, export-oriented farmers (Binswanger and Townsend, 2000). While governments in East and South Asia encouraged the adoption of productivity-enhancing technologies through significant investments in the rural infrastructure necessary to enable the commercialization of smallholder agriculture, SSA's road and communication infrastructure remains undeveloped and inadequate (Rosegrant et al., 2001).

In part because of poor infrastructure, input markets have been poorly developed and inefficient, limiting long-term planning by farmers, while government policies have contributed to the poor performance of input markets. Fertilizer policy in SSA has generally been disastrous, with costly subsidies distorting markets during the 1970s and 1980s. Elimination of subsidies

and liberalization of exchange rates through structural adjustment programmes in the 1980s and 1990s resulted in higher fertilizer prices and diminishing use. Benin et al. (2003a, 2003b) argue that the low adoption and consumption of inorganic fertilizers is due mostly to high cost, lack of credit, delivery delays, lack of irrigation, low variable returns, and policy deficiencies. The current policy environment in SSA, which is generally opposed to efforts to stabilize commodity prices or subsidize fertilizer, other inputs, or credit, is less favorable to the adoption of Green Revolution technologies than the policies existing in countries where and when the Green Revolution occurred (Dorward et al., 2004). As a result of such limitations, organic sources of nutrients are often proposed as an alternative to inorganic fertilizers (Reijntjes et al., 1992). However, many organic methods are at best only a buffer to the system, do not redress nutrient depletion (Palm et al., 1997), may be limited in supply, and are difficult to use in sufficient quantities to restore soil nutrients due to the low concentration of nutrients in organic materials, especially phosphorus (Palm et al., 1997; Larson and Frisvold, 1996).

Output markets are also poorly developed and inefficient. High transport costs; high costs of vehicles, spare parts, and fuel; and poorly coordinated marketing chains depress farm prices. Lack of marketing institutions and infrastructure, such as appropriate grades and standards, price information systems, quality regulation, contract enforcement, storage facilities, and marketing credit all contribute further to high transaction costs (Gabre-Madhin and Amha, 2003). Heavy regulation of markets severely undermined the development of private markets in the 1970s and 1980s. Although markets have been liberalized, private investment is still often undermined by unpredictable food aid imports or other trade interventions, high taxes and fees on market activity, corruption, and other problems.

Complementary investment in irrigation systems has also been limited. Although irrigation potential is high, less than 4% of the total arable land has been irrigated since 1965. In 1965, there were more tractors per hectare of arable in SSA than in South and Southeast Asia, but by 1995, there were more than four times as many tractors per hectare of arable land in South and Southeast Asia. Public spending on agricultural research and development, despite their high returns (Alston et al., 2000), has also been poor, as they remain a low priority. Only 0.7% of agricultural GDP was spent on agricultural research in SSA in 1991, compared to 2.4% in developed countries (Alston et al., 2001).

3. Policies and strategies to reverse the downward spiral

The appropriate strategies will depend upon the potential comparative advantage of a particular location. Many factors combine to determine comparative advantage and the appropriate response to it, including agricultural potential, access to markets and infrastructure, and population density (Pender et al., 1999).

Agricultural potential determines the absolute advantage (technical efficiency) of producing agricultural commodities in different locations. Access to markets and infrastructure greatly influences which commodities have a comparative advantage in a particular location, given its agricultural potential. For example, a community may have absolute advantage in producing perishable vegetables, due to rainfall and soil conditions, but may not have a comparative advantage if it is located far from the market. Population pressure determines the labor intensity of farming (Boserup, 1965).

These factors interact with each other. For example, people tend to migrate to or stay in places with higher agricultural potential or market access, resulting in higher population densities in such areas. Conversely, population density can affect the development of markets and provision of infrastructure and rural services, or affect agricultural potential by affecting land degradation or improvement. Where population densities are high, per capita costs of investment in infrastructure and services are lower and transportation and transaction costs are generally lower, thus facilitating competitive markets for both outputs and inputs. In sparsely populated areas there is also little demand for credit and the supply of credit is limited by lack of diversification and absence of suitable collateral, as land has little value (Binswanger and Townsend, 2000). The resulting high transaction costs and absence of credit markets in sparsely populated areas inhibits technology adoption and sustainable management of the resource base.

The next section discusses as a case study the development strategies for the East African Highlands with a focus on Ethiopia, Kenya, and Uganda. The highlands, which are within the semi-arid, humid, and sub-humid zones and are primarily found in East Africa, cover only 5% of the total area, but contain about 14% of the population of the subcontinent due to the relatively favorable climate and soil conditions.

3.1. Development strategies for the East African highlands

The East African highlands—areas above 1,200 meters above sea level in Burundi, Ethiopia, Kenya, Rwanda, northern Tanzania, and Uganda—are home to 53% of the population of these countries (Place, 2000). Rural population densities are the highest in Africa; well above 500 people per square kilometer in some areas. Consequently, farm sizes are small, averaging around 1 hectare. Most of the highlands have favorable rainfall, averaging over 1,000 mm per year, and many sites have two growing seasons. There are a variety of soil types, though most are clayey with relatively good stability. Therefore, most of the area has high potential, although potential varies greatly and much farming takes place on steeply sloping land.

A wide range of agricultural crops is found, especially in the bimodal rainfall highlands, where high-value commodities such as tea, coffee, dairy, sugar cane, and fruits and other horticultural crops are common. The chief food crops are maize, banana, beans, teff, wheat, sorghum, cassava, and potatoes. Livestock are an important part of the farming system, providing draft power, manure for crops, food, and a source of cash income and wealth, especially in Ethiopia.

Soil erosion and soil fertility depletion are major problems. Crop production on steep slopes, often of annual crops with limited vegetative cover and limited use of soil and water conservation measures, contributes to high rates of erosion. Partly because of high erosion rates, soil nutrient depletion is higher than in other regions of SSA (Stoorvogel and Smaling, 1990). Low use of fertilizer and organic inputs also contributes to the nutrient depletion problem. Lack of cash and inefficient markets keep small farmers from using more fertilizer, except for the most profitable crops. As a result, researchers and technical assistance programs are working to identify inexpensive ways of improving nutrient management and organic nutrient production, such as improved fallow, biomass transfer, crop residue management, manure management, and composting.

Soil conservation efforts are needed in the lower-rainfall zones as well. There is potentially more scope for labor-intensive conservation measures if the work can be undertaken in the off-season and there is slack labor demand. In addition, because they conserve scarce water, soil conservation structures often are more profitable in lower-rainfall areas (Herweg, 1993). Vegetative approaches such as live barriers and leguminous cover crops may not be as suited to low-rainfall areas (Kaizzi et al., 2002).

3.1.1. High-potential areas with good market access

3.1.1.1. Central Kenya. In high-potential areas with good access to a large urban market, such as the central Kenya highlands close to Nairobi, there is potential for profitable production of high-value perishable commodities such as dairy products and horticultural crops. Dairy development in central Kenya is a major success story (Ngigi, 2004), stimulated by the favorable agro-climatic conditions, by access to the Nairobi market, and to livestock services and feed supplies, and by several decades of support through the Kenya Cooperative Creamery (KCC), a parastatal that until recently acted as a guaranteed buyer of milk output

(Place et al., 2003c; Staal et al., 2002; Ngigi, 2004). Although the KCC provided an assured market, its monopsony and monopoly power eventually inhibited dairy development. Liberalization of dairy marketing in the early 1990s led to a mushrooming of private firms, which have introduced new processed products and brands, contributing to the expanding demand for milk (Place et al., 2003c). There has been widespread adoption of high-yielding exotic and cross breeds since their introduction in the 1980s, accompanied by a reduction in the number of cows and a shift to zero-grazing feeding systems. Use of dairy goats is also increasing in lower altitude areas of the central highlands.

Vegetables, fruits, and nuts are also common in the central Kenyan highlands. Although farm sizes are small (averaging between 1 and 2 hectares), farmers have diversified (Place et al., 2003c; Owuor, 1999), growing six to seven different crops. Kenya is one of only a few countries in SSA that have developed a successful horticultural export sector, including significant participation of smallholders (Jaffee, 1995). This success was built upon the development of horticultural production for the domestic market and several technical, financial, infrastructural, and managerial advantages, in addition to favorable agro-climatic conditions, favorable location relative to the European market, and low labor costs. Even with these advantages, horticultural exports took more than ten years to develop, and only a small number of exporters have been able to achieve sustained growth (Jaffee, 1995). Thus, replicating this success is likely to be a slow and difficult process for other countries. In the near term, greater opportunities are likely in producing horticultural crops for the domestic market, especially in the large urban areas.

Place et al. (2003a) argue that commercialization and diversification into high-value commodities were stimulated by the production of traditional cash crops (coffee and tea) and associated institutions. Farmers developed a commercial orientation and were able to use the income from these crops, and access to credit from the tea and coffee cooperatives, to diversify as new opportunities arose. Increased global competition and declining world prices for traditional cash crops also provided a stimulus to diversification, as did declining farm sizes (Njuki and Verdeaux, 2001).

Commercial production of high-value commodities has been associated with the adoption of land-improving practices. Fertilizer use is higher in the central highlands than in the rest of the East African highlands, averaging over 100 kg/acre, and is much higher on higher-value crops (Owuor, 1999). More than half the farmers apply manure to all of their crops, except beans (Murithi, 1998). Intensive dairy production contributes substantially to the use of manure in crop production (Staal et al., 2002) well as contributing to planting of fodder trees, which are increasingly common (Place et al., 2003c). Not surprisingly, most farmers in the central highlands perceive that their land quality has improved since they acquired it (Place et al., 2003c).

Opportunities to earn an income from nonfarm activities are also relatively favorable in areas of good market access. Nonfarm activities accounted for 37% of household income on average in one study of the Kenya highlands, and household incomes are higher in the central Kenya highlands than in the East African highlands generally (Argwings-Kodhek et al., 1999). Farm and nonfarm income sources appear to be complementary, providing investment funds for each other and facilitating households' ability to take risks in new activities.

3.1.1.2. Central Uganda.

In central Uganda the use of fertilizers is negligible, except on a few large estates. Most smallholder farmers produce coffee for cash needs and bananas and/or maize for subsistence food needs, using traditional methods (Deininger and Okidi, 2001; Pender et al., 2001b). There is some production of horticultural crops, and dairy production is increasing, but these activities are much less advanced than in central Kenya.

The lower elevation of central Uganda poses a constraint to livestock and horticultural development, as pests and diseases are more of a threat at lower elevations in the humid tropics. Therefore, dairy and horticultural production are developing more in the drier areas of the southwest and at higher elevations (though still with good access to roads), than in the peri-urban areas around Kampala. Uganda has also not seen the strong and sustained support for dairy development that has existed in central Kenya, and the urban market is not as large in Kampala, while access to credit and technical assistance has also been more

limited. Although coffee production provides cash income that could be used to support investments in other commodities, coffee farmers' incomes were limited by the parastatal marketing system until the early 1990s. Liberalization, together with favorable world market prices, led to soaring returns to coffee in the mid-1990s, which encouraged investment in coffee but not in other commodities. Since world coffee prices have fallen recently, farmers now face greater incentives to diversify, though the response is still limited.

Nonfarm activities are increasing, and account for about 40% of household income in the bimodal high rainfall areas of Uganda (unpublished IFPRI data). Nonfarm activities contribute to farmers' ability to finance investments in farm activities, but may also reduce their interest in such investments, as greater opportunities for use of scarce labor and capital may be available in nonfarm activities. Recent empirical evidence shows that nonfarm activities contribute to higher crop production and less soil nutrient depletion by enabling greater use of fertilizers and food purchases (Nkonya et al., 2003).

Land management has seen little improvement. Inorganic fertilizers are used on less than 2% of plots in the Lake Victoria crescent, while manure or compost is used on only 18% of plots, and mulch on 8% (Nkonya et al., 2002). This appears to be largely due to low profitability, due in part to high transport and other transaction costs of fertilizer and outputs, leading to high input/output price ratios (Woelcke, 2003). Use of these practices is increasing, but remains low (Pender et al., 2001b). As a result, farmers have experienced declining yields for all major crops, with declining soil fertility and increased pest and disease problems cited as major causes (Deininger and Okidi, 2001). Estimates of soil nutrient balances confirm that nutrient depletion is a serious problem in this region (Wortmann and Kaizzi, 1998).

3.1.1.3. Central Ethiopia near Addis Ababa.

In the central highlands of Ethiopia, the dominant farming system is mixed cereal–livestock production, as in much of the rest of the Ethiopian highlands. Dairy production is important and is growing in urban and peri-urban areas around Addis Ababa, but is far less developed than in central Kenya, despite similar agro-ecological conditions and access to a large and growing urban market. As in Kenya, a parastatal organization—the Dairy Development Enterprise (DDE)—acts as a guaranteed buyer of fluid milk at set prices. Due to the high perishability of milk and limited development of transport and marketing facilities, most fresh milk is supplied through informal market channels by producers in or close to the city, or by producers within a few kilometers of a DDE collection center (Ahmed et al., 2003b; Holloway et al., 2000). In a few places, "cooperatives" have been established by a dairy development project, but their coverage is limited. Most farmers who produce a surplus of milk process it themselves into butter or cheese to sell in local markets (Holloway et al., 2000).

High transaction costs have limited the development of dairy marketing, while low administered prices and an overvalued exchange rate have also reduced profitability (Staal and Shapiro, 1996). Farmers' poverty and lack of credit for purchasing dairy animals have also limited the adoption of higher-yielding crossbred cows (Freeman et al., 1999), despite evidence that such animals are profitable (Ahmed et al., 2003a; Holloway and Ehui, 2002). Concerns about susceptibility of crossbred cattle to local diseases are also a constraint on widespread adoption (Holloway and Ehui, 2002). Severe shortages, low quality, and seasonal unavailability of feed similarly remain as major constraints to livestock production in Ethiopia (Ahmed et al., 2003b). The net result has been to limit the development of dairy production, despite evidence that Ethiopia could be efficient in peri-urban production (Staal and Shapiro, 1996). These constraints need to be addressed and technological change should be promoted in order to increase milk production.

Planting of eucalyptus trees is also profitable, especially in areas of good market access, adequate rainfall, and moderate temperatures, as a result of the extreme scarcity of fuelwood and the high value of the poles for construction, ploughing materials, and other purposes (Jagger and Pender, 2003). Holden et al. (2003) predict that allowing farmers to plant eucalyptus on land unsuitable for crop production in a community in North Shewa could increase household incomes by at least 30%, with little impact on soil conservation incentives or erosion of farmland. Okumu et al. (2002) predict that allowing eucalyptus planting on uncultivated slopes in a watershed close to Addis Ababa, along with other technological interventions focusing on improved vertisol management, could increase household cash incomes substantially and reduce soil erosion, even if eucalyptus prices were to fall substantially.

Intensification of cereal crop (especially maize) production is occurring in the central Ethiopian highlands, using high levels of fertilizer and improved seeds, stimulated by the government's agricultural extension and credit program. This program succeeded in increasing cereal production to near record levels in 2000/2001 (together with favorable weather), though the effect on farmers' incomes was less positive because of the fall in cereal prices that ensued. Many farmers' reaction to the price crash was to reduce use of inputs in cereal production in 2002, which was a possible contributing factor to the 2003 famine (Thurow, 2003). This problem resulted from limitations of the marketing and credit system, including the absence of credit for marketing, the fact that credit for inputs must be repaid at harvest, the high costs of transporting cereals from surplus to deficit regions or for export, limited storage facilities, absence of grades and standards or an effective market price information system, the limited use of the cereals as feed for livestock, and other factors leading to high transactions costs and inelastic demand for cereals (Amha and Gabre-Madhin, 2003; Gabre-Madhin and Amha, 2003).

Intensified cereal production, if it can be sustained, is complementary to dairy production and other intensive livestock operations in peri-urban areas, and development of such livestock operations can increase cereal prices and reduce their variability, as well as provide a source of manure that can be used in intensive crop production. Nevertheless, development of this potential remains low, probably in part because of the high variability of cereal supplies and prices, but also because of the need to develop markets, market institutions, and farmers' and traders' technical, financial, and managerial capacity. Concentration of supply chains, transaction costs, and economies of scale in production or marketing may limit the potential of smallholders in particular, to benefit from opportunities to produce intensive livestock and other high-value products (Delgado and Minot, 2003). The evidence cited earlier of smallholder participation in dairy and horticulture production in Kenya suggests that such barriers are surmountable, given the right economic and policy environment.

3.1.2. High-potential areas with poor market access

3.1.2.1. Western Kenya. In areas with high agricultural potential that have less favorable access to a large urban market (or to export markets), as in western Kenya, some of the options available in areas of better market access will be less feasible. Commercial production of high-value perishable commodities such as milk and horticultural crops is less likely the further away from a major market one is located. For example, Staal et al. (2002) estimate that being 10 km further away from Nairobi reduces the probability of Kenyan households using improved dairy cattle (crossbred or exotic breeds) by 0.7%. As a result, dairy production in the western highlands is much less intensive than in the central highlands. Besides the impact of market access, dairy production may also be limited by the high population density in the western highlands, which raises the opportunity cost of devoting land to feed production (Place et al., 2003c). However, both Staal et al. (2002) and Place et al. (2003b) found that market access is a more important factor than population density in determining differences between central and western Kenya in dairy production.

A wide variety of crops are grown in the western highlands, with maize and beans the dominant crops. The share of land allocated to horticultural crops is lower than the central highlands and the share of crop revenue is much lower (Owuor, 1999). Traditional industrial cash crops such as coffee and tea are also grown, though they also account for a substantially smaller share of land and revenue than in the central highlands (Owuor, 1999). Lower access to markets, processing facilities, and credit appear to be the main reasons for differences between central and western Kenya in the adoption of such high-value commodities, since agro-ecological potential, access to technical assistance, and land tenure security are similar in these two regions (Place et al., 2003a).

Private woodlots are common, much more than in the central highlands (Place et al., 2003a), largely because the returns to such investments are not sufficiently high to compete with the high-value perishable commodities in the central highlands, but are still profitable and necessary as sources of scarce fuelwood and construction materials in densely populated areas further from markets. Consistent with this explanation, Place et al. (2003b) found that the allocation of land to woodlots in Kenya is positively associated with higher population density, but not significantly affected by distance to an urban market. As a result, woodlots appear to have a comparative advantage in densely populated areas more distant from a major urban center. Rainfall and altitude also matter; woodlots are more common at higher elevation and where rainfall is higher.

Because of the small farm sizes, rapid population growth, and limited development of high-value agricultural production, many people (especially men) migrate out of the western highlands. Nearly 40% of household income comes from nonfarm income, due in part to such migration (Place et al., 2003a). While this income benefits the household, temporary migration is also associated with problems such as increased prevalence of HIV/AIDS. Thus, despite high population density, there is often a scarcity of labor for agriculture, which can undermine the ability of farmers to use labor-intensive soil conservation methods (Place et al., 2003c). Furthermore, the temporary absence of male household heads may undermine decisions about land management, as there is wide variation in the extent to which women are allowed to make such decisions.

Due to the limited production of cash crops and the relatively low use of hybrid maize, use of fertilizer is relatively low in the western highlands. Only one fifth of the farmers in two survey sites in western Kenya used fertilizer (Place et al., 2003c), and the average amount of fertilizer used is only about one fifth of that used in the central highlands (Owuor, 1999). Cash constraints are a major reason for the low use of purchased inputs (Place et al., 2003c).

Organic soil nutrient replenishment approaches are being promoted by nongovernmental organizations (NGOs), contributing to significant adoption: over 70% of households apply manure and 40% apply compost in the study sites, while 10% to 30% of households have adopted agro-forestry practices where they are being promoted (Place et al., 2003c). Such technologies (e.g., leguminous cover crops in an improved fallow rotation, biomass transfer, manure use) have shown good potential to increase agricultural production in western Kenya and eastern Uganda, but are not attractive to farmers in many circumstances. Where land is scarce, farmers are reluctant to give up crop production on a field for even one season of improved fallow, even if total production over two seasons would be significantly increased (Delve and Ramisch, 2003). High labor requirements also limit the adoption of such practices, especially by labor-scarce households, such as female-headed households. Thus, adoption of such practices remains low except where they have been intensively promoted, despite their potential to increase food production and reduce land degradation. The downward spiral of declining fertility, productivity, and income is thus not likely to be broken by the availability of such technologies, unless maize prices improve relative to input costs or farmers shift to higher-value crops, and demand-driven technical assistance responsive to the opportunities and constraints of farmers in diverse settings becomes broadly available.

3.1.2.2. Highlands of Uganda. In the highlands of eastern Uganda, the dominant livelihoods are production of coffee (high value arabica), cereals (primarily maize), and bananas (Pender et al., 2001b). The comparative advantage of the region in high-value coffee production is due to the high potential of the volcanic soils and the climate. Bananas are produced primarily for subsistence, while maize is both consumed and sold. Cattle raising is important as a secondary occupation, while other annual crops (including vegetables) are of tertiary importance. In the southwest highlands, the dominant activities are banana and cereal production, while other storable annuals, coffee, root crops, and cattle are important secondary or tertiary activities. Perishable horticultural crops and dairy production are somewhat more important in parts of the eastern highlands than in the southwest, probably due to better access to markets (in Kenya). Nonfarm activities are less important than in Kenya, accounting for less than 10% of household income in the eastern highlands and little over 20% of household income in the southwestern highlands (unpublished IFPRI data), probably due to more limited market access in the Uganda highlands.

Adoption of improved land management technologies is even lower than in western Kenya, despite similar high agricultural potential in these regions. Only about one tenth of households use fertilizer in these regions (Pender et al., 2001b). Adoption of improved seeds is much higher than fertilizer adoption, contributing to more rapid soil nutrient depletion as improved seeds increase yields without adequate nutrient replenishment. Soil erosion also contributes substantially to soil nutrient depletion and to diminishing soil depth

(Nkonya et al., 2004). Contributing to erosion problems is cultivation of steep slopes with limited investment in soil and water conservation measures, especially in annual crop systems. The most common measure used is grass strips (used by about one fourth of the farmers surveyed in the southwest highland and one third of those surveyed in the eastern highlands), which are used mainly in perennial systems (Nkonya et al., 2002). Mulching, composting, and manuring are also common, but are also used more for perennials.

The major causes of low adoption of fertilizer in Uganda include low profitability of fertilizer in many places, limited development and coverage of technical assistance programs, cash constraints, and limited availability of credit. The low profitability results from limited impacts of fertilizer on crop production in many environments (Kaizzi et al., 2002; Nkonya et al., 2002; Woelcke, 2003), as well as high transport costs and the limited size and development of the market for fertilizer. Nevertheless, fertilizer is profitable in maize production in higher potential areas of the eastern highlands (Kaizzi, 2002). Such places should be priority for technical assistance and credit focused on promoting fertilizer and complementary soil and water conservation measures. Thus, land degradation and declining agricultural productivity are likely to continue without expansion of participatory technical assistance efforts, improvements in market development, and/or a shift toward higher-value crops. Developing this potential in the highlands may thus represent a "win–win" strategy, increasing production while reducing land degradation.

Livestock production is also increasing in the eastern highlands, including some use of crossbred cows for intensive dairy production (Pender et al., 2001b). Development of livestock production is associated with higher crop production—reflecting complementarities between crop and livestock production—reduced soil nutrient depletion, and higher household incomes (Nkonya et al., 2004). Promotion of livestock development thus offers a potential pathway out of poverty, low agricultural productivity, and land degradation in the highlands of Uganda.

3.1.2.3. Southern, Western, and Central Ethiopia. A large share of the Ethiopian highlands can be considered as high agricultural potential with poor market access. For example, the average walking time to the nearest all-weather road from rural communities in higher rainfall areas of the highlands of Amhara region is nearly three hours, and walking time to the nearest bus service is more than three hours (Pender, 2001).

High-potential agricultural zones in the Ethiopian highlands (having growing periods of more than 180 days) have been classified into the humid high-potential perennials (HPP) zone, mainly in the southern and western highlands (mostly within Oromiya regional state), and the sub-humid high-potential cereals (HPC) zone, mainly in the central and northwestern highlands (within Oromiya and the western part of Amhara regional state) (Technical Committee for Agroforestry in Ethiopia, 1990). The soils are generally of good potential for agriculture, though significant areas of vertisols (heavy clay soils, more common in the HPC zone) face special management problems as they are difficult to plough and prone to waterlogging in the rainy season and cracking in the dry season. Average population density is about 100 persons per square kilometer in these regions, with average farm sizes of less than 2 hectares.

Mixed crop–livestock farming systems are dominant in both zones. In the HPP zone, the dominant crops are coffee, enset (a perennial tuber crop), and cereals, while in the HPC zone, cereals (mainly barley, wheat, maize, teff, and sorghum) dominate. In the HPP zone, enset and coffee are commonly grown near the homestead, while cereals are grown in more distant fields. In the HPC zone, eucalyptus trees are commonly planted in homestead plots. Livestock (local breeds only) are kept as a source of draft power, manure, wealth, and food in these systems.

The primary comparative advantage in the HPP zone has long been in high-quality arabica coffee. However, this advantage has been eroded because of declining world market prices. Many farmers are increasingly growing chat (a tree crop used as a stimulant), due to its relatively high profitability and lower susceptibility to diseases. Nevertheless, coffee will remain the dominant cash crop for the foreseeable future. There is also increasing pressure on coffee production from population growth, leading to increased needs for food production. Given the high costs of transporting food, as well as farmers' lack of access to consumption credit, subsistence food production is important for smallholder coffee producers, implying that land scarcity and increasing food needs are likely to constrain and

may even reduce the area devoted to coffee production. The emphasis of the agricultural extension system on promoting food production may have contributed to the tension between food and coffee production (Westlake, 1998). Significant expansion in the area planted with coffee seems unlikely, especially if world coffee prices remain low.

Smallholders produce more than 80% of Ethiopian coffee in forest, semi-forest, and garden systems. Few coffee farmers use purchased inputs (fertilizer or pesticides). Under shade conditions, coffee production is not very responsive to fertilizer, so the potential for adoption is limited. Given low and uncertain prices, a shift to higher input systems is risky and unlikely. There is more potential to increase production through introduction of higher-yielding improved cultivars resistant to coffee berry disease (CBD), a risk faced by more than half of the coffee planted in Ethiopia. Resistant cultivars have been developed and disseminated that have two to three times the yield potential of traditional varieties. The extent of area replanted with these cultivars and their impacts on coffee yield is not certain, but it has been estimated to have increased national production by perhaps 10% by the late 1990s (Westlake, 1998). Continued development and dissemination of improved cultivars is likely to be the main avenue to increasing coffee yields in the near future.

Development in coffee-producing areas is also affected by management of coffee quality and the efficiency of the marketing system. Despite stringent quality regulation by the Coffee and Tea Authority (CTA), quality problems resulting from farmers picking berries at the wrong time, drying the berries on bare ground, or other problems in handling and storage, are common, and reduce the value of the product (Westlake, 1998). These problems, together with high transaction costs, limited finance, and high risks faced by coffee traders lead to high marketing margins. Although margins declined after liberalization, with reduction of export taxes starting in 1992 (LMC International, 2000), margins remain relatively high, with farm-level prices only about 50–60% of export prices in 2000 and 2001. Marketing costs are higher in Ethiopia than in most other exporting countries, reflecting higher transaction costs and risks in the system. Improvements in infrastructure and transportation, and the development of institutions to reduce traders' risks, such as a forward auction for coffee could help to reduce marketing margins (Schluter, 2003). Improving traders' and farmers' access to marketing credit by the establishment of a warehouse receipts system could also help improve efficiency and reduce margins.

Intensification of food crop production could help to alleviate the tension between food production and coffee production in densely populated areas of the HPP zone. Thus, the agricultural extension program focus on promoting improved seed and fertilizer use in food production can be complementary to the promotion of coffee or other cash crops, while options for improving cash crops, intercropping, and other opportunities to increase the complementarity between food and cash crops should also be identified and promoted.

In the HPC zone, substantial increases in cereal yields have been achieved because of the extension and credit package focusing on improved seeds and fertilizer (Benin, 2003). However, the benefit of this improvement has been limited by limitations in the cereal marketing system, resulting in low cereal prices in 2001 and 2002. Improvements in infrastructure and market institutions, facilitation of cereal exports, and development of new sources of domestic cereal demand, are needed to achieve sustainable development. Development of dairy, other intensive livestock production, and agro-processing industries in areas closer to urban markets will be an important complementary strategy. Given the substantial potential to increase cereal yields further (e.g., maize yields averaged only about 2 tons per hectare in western Amhara in 1999–2000 [IFPRI/ILRI, unpublished data], much less than the potential), this zone could become the breadbasket of East Africa, if development of market institutions and market demand keep pace with technology impacts.

3.1.3. Low-potential areas

3.1.3.1. Northern and Eastern Ethiopia. In areas with lower agricultural potential due, for example, to low and uncertain rainfall and thin soils, as in much of the northern and eastern highlands of Ethiopia, the potential for intensive production of food or cash crops using fertilizer and improved seeds is more limited, except where investments in irrigation, water harvesting, or soil and water conservation enable farmers to overcome soil moisture constraints. In Tigray and

eastern Amhara, for example, substantial efforts have been made by the extension program to promote fertilizer and improved seed use, but these have not been very profitable to farmers in rainfed areas (Benin, 2003; Pender and Gebremedhin, 2004). Therefore, the agricultural extension and credit program has not had much positive impact on farm incomes in these regions.

Investments in small-scale irrigation offer the potential to increase productivity and promote a shift to higher-value crop production. There has been substantial investment in such irrigation schemes in recent years. Irrigation was found to contribute to greater use of fertilizer and labor in crop production, contributing to an average 26% increase in the value of crop production per hectare, controlling for plot quality and other factors (Pender and Gebremedhin, 2004). Problems resulting from limited farmer experience with irrigation, irrigated production of low-value crops such as maize, limited availability of irrigation water when needed, and salinization due to inadequate drainage may have contributed to the limited impact of irrigation (Haile et al., 2002a; Tesfay et al., 2000). Larger productivity impacts of small-scale irrigation were found in the eastern Amhara region (Benin, 2003).

Farmers' private investments in soil and water conservation structures can yield high returns in semi-arid environments by increasing the availability of soil moisture (Herweg, 1993; Shapiro and Sanders, 2001). Econometric analysis of survey data from Tigray indicates that stone terraces increase the value of crop production by 17% on average, implying an average rate of return of 34% (Pender and Gebremedhin, 2004). Using on-farm experimental results, Gebremedhin et al. (1999), estimated even higher rates of return to stone terraces. Clearly, there are substantial economic returns to such investments, in addition to their beneficial impact on controlling soil erosion, which explains their widespread adoption.

Pender and Gebremedhin (2004) also found significant returns to other land management practices, including reduced tillage, reduced burning, and application of manure or compost. The positive impact of reduced tillage is supported by recent experimental evidence from Ethiopia, showing that higher yields are possible with reduced labor use using zero tillage (Aune et al., 2003). This technology could be of great benefit to female-headed households, who face a cultural taboo against ploughing in Ethiopia, and other poor households that lack access to oxen. It could also lessen dependence on oxen, enabling households to raise more profitable dairy animals (especially in areas closer to urban markets), and reducing pressure on grazing lands. In addition, reduced tillage reduces soil erosion and depletion of soil organic matter, contributing to soil fertility and global carbon sequestration (Aune et al., 2003). Reduced burning and application of manure or compost also help improve soil organic matter content (Haile et al., 2002b).

Ownership of cattle also contributes to higher crop productivity in Tigray, due in part to the benefits of manure (Pender and Gebremedhin, 2004). Cattle also contribute substantially to household income, earning an estimated marginal rate of return on investment of 36% (Pender et al., 2002a). Investments in chickens and beehives also earn relatively high returns (over 30%), while returns to sheep and goats are lower (Pender et al., 2002a).

The potential of realizing the benefits of improved livestock production and greater recycling of manure to the soil depends on improved management of communal lands. Because of increasing population pressure, grazing lands are increasingly scarce and degraded in the northern Ethiopian highlands (Pender et al., 2001a). Grazing lands and croplands after harvest are often unregulated open access resources, though many communities have established access restrictions, which are helping to reduce the degradation (Gebremedhin et al., 2002). Econometric evidence from Gebremedhin et al. (2002) suggests that establishment and effectiveness of restrictions is induced by population pressure, but it is not clear whether these responses are optimal. Investments in improving grazing lands (e.g., planting fodder trees and grasses) are also needed, but are rare. In some parts of Tigray, access to grazing lands has been privatized (using social fencing), and this appears to be contributing to better management. Such institutional changes in the free grazing system are becoming increasingly necessary as resource scarcity increases.

Improved management of forests, community woodlots, and degraded lands is also necessary. Farmers in the Ethiopian highlands increasingly burn animal dung and crop residues as fuel sources, due to the limited availability of fuelwood, reducing the recycling of nutrients and organic matter to the soil, and contributing to human health problems. Community woodlots and area "enclosures" (protected areas left to regenerate naturally) have been established in many communities. These have not succeeded in addressing the shortage of fuelwood and poles because households have not been allowed to cut trees or collect fuelwood, even from older woodlots (Gebremedhin et al., 2003). Part of the problem is lack of clarity in the regulations of the regional government: although the government claims that communities are free to use woodlots however they wish, most communities believe that they must have permission from the regional Bureau of Agriculture to cut trees. Empowerment of local communities to decide when and how to use such woodlots could greatly increase the benefits, which could be quite substantial because of the high value of the trees in these woodlots (Gebremedhin et al., 2003). This would, in turn, increase community members' incentive to manage woodlots sustainably.

One way to empower communities and households is to allow allocation of unused degraded land for private tree planting or other economic uses. In the early 1990s, a village in eastern Tigray (Echmare) decided to allocate part of a degraded hillside in very small parcels for private tree planting. All households received an allocation, and were expected to manage the parcels intensively in order to ensure tree survival. The penalty for failing to do so was to lose access to the parcel and to future allocations of such parcels. The results were remarkable: households cleared stones and built stone walls around each parcel and hand-watered the seedlings from a stream one kilometer or more away, resulting in significantly higher survival rates than in most community woodlots (Jagger and Pender, 2003). The regional government was so impressed with the results that it established a pilot program allowing selected communities to allocate degraded lands for private tree planting, and has since established this as a general policy. Results of a recent survey of these and other communities show that tree survival rates are (slightly) higher for private woodlots or village-managed woodlots than for woodlots managed at a higher administrative level, despite much lower labor inputs on the private or village woodlots (Jagger et al., 2003). These results indicate that labor is used more effectively to ensure tree survival when management is at a more local level. The economic benefits of these woodlots have yet to be determined, but could be quite substantial, given the value of poles, fuelwood, and other products. Jagger and Pender (2003) estimated potential internal rates of return from community and private woodlots ranging from about 20% to over 100%, depending upon the access to markets, the growth rate of the trees (determined by agro-climatic factors), and the opportunity cost of land used for the woodlots.

Nonfarm activities are also important in the low potential highlands, accounting for more than one fourth of average household income in the highlands of Tigray in 1998–1999 (unpublished IFPRI/ILRI data). Some nonfarm activities, such as participation in Food for Work (FFW) projects, are more important further from towns (Pender et al., 2002a). This reflects the importance of such projects as sources of employment in food-deficit areas, and their decisions to locate activities away from towns. FFW and cash for work projects accounted for a sizable portion (about 40%) of average nonfarm income in Tigray in 1998–1999, and more than 10% of average total household income (unpublished IFPRI/ILRI data). Further, participants in such projects earned higher incomes than other households (Pender et al., 2002a).

Investments in education are likely to be key for long-term development in low-potential areas, and are a high priority, as evidenced by the rapid increase in school access and participation (Pender et al., 2002b). Education was found to be associated with higher per capita income and increased value of crop production (Pender et al., 2002a). Membership in marketing cooperatives also was found to increase the value of production and incomes substantially (Pender et al., 2002a).

The results of this subsection indicate that profitable investments exist that can increase agricultural production, income, and reduce land degradation in low-potential highland areas. The comparative advantage of such areas is not in intensive food crop production using high levels of fertilizer and improved seeds, as these inputs are risky and not very profitable in this environment, as long as soil moisture is the key limiting factor in production. The government of Ethiopia has recognized this fact in its own agricultural extension program, and is now promoting a broader package of interventions for such areas, including a new emphasis on water harvesting. Other opportunities to increase food crop production without large reliance on purchased inputs include promotion of sustainable land management practices such as reduced tillage, reduced burning, manuring, and composting. Yet the comparative advantage of the low-potential highlands is more in livestock production and tree planting.

4. Conclusion

Despite the fact that two thirds of the 627 million people living in SSA depend on agriculture or agriculture-related activities for their livelihoods, the region has the lowest agricultural productivity in the world. Soil fertility is the principal biophysical constraint for sustaining agricultural production in the region. With rapid population growth, soil nutrient stocks are being mined due to reduction in fallow lengths, cultivation of fragile lands, and limited use of inorganic or organic sources of fertility, lowering agricultural productivity and increasing poverty. Poor policies combined with low public investment have also undermined agricultural growth.

Building on several years of empirical research conducted in East Africa this paper has identified potential strategies for sustainable development in this region. In general, the empirical evidence reviewed confirms that different strategies are needed in different development domains of SSA. Nevertheless, some elements will be common to all successful strategies, including assurance of peace and security, a stable macroeconomic environment, provision of incentives through markets, where markets function, development of market institutions, where they do not, and public and private investment in an appropriate mix of physical, human, natural, and social capital. The differences in strategies across these domains mainly reflect differences in the mix of those investments as influenced by different comparative advantages.

Acknowledgments

This paper was started when the senior author was a staff-member of the International Livestock Research Institute. Writing of this paper was made possible thanks to partial funding from Norway (Royal Ministry of Foreign Affairs), Switzerland (Swiss Development Corporation), the International Livestock Research Institute, and the International Food Policy Research Institute.

The views expressed in this paper represent solely those of the authors and not those of the institutions they work for.

References

Ahmed, M. M., B. Emana, M. A. Jabbar, F. Tangka, and S. Ehui, "Economic and Nutritional Impacts of Market-Oriented Dairy Production in the Ethiopian Highlands," Socio-economics and Policy Research Working Paper No. 51 (International Livestock Research Institute: Nairobi, Kenya, 2003a).

Ahmed, M. M., S. Ehui, and Y. Assefa, "Dairy Development in Ethiopia," Socio-Economics and Policy Research Working Paper 58 (ILRI: International Livestock Research Institute: Nairobi, Kenya, 2003b), p. 47.

Alston, J. M., C. Chan-Kang, M. C. Marra, P. G. Pardey, and T. J. Wyatt, "A Meta-Analysis of Rates of Return to Agricultural Research and Development," Research Report No. 113 (International Food Policy Research Institute: Washington, DC, 2000).

Alston, J. M., P. G. Pardey, and M. J. Taylor; eds., "Changing Contexts for Agricultural Research and Development," in *Agricultural Science Policy: Changing Global Agendas* (Johns Hopkins University Press: Baltimore, 2001).

Amha, W., and E. Gabre-Madhin, "An Analysis of the Structure, Conduct, and Performance of the Ethiopian Grain Market" (International Food Policy Research Institute: Washington, DC, 2003), Mimeo.

Argwings-Kodhek, G., T. Jayne, G. Nyambane, T. Awuor, and T. Yamano, "How Can Micro-Level Household Information Make a Difference for Agricultural Policy Making?" Tegemeo Institute of Agricultural Policy and Development, Kenya Agricultural Research Institute, and Michigan State University (1999), Mimeo.

Aune, J., R. Asrat, D. A. Teklehaimanot, and B. Tulema, "Zero Tillage/Reduced Tillage— The Key to Intensification of the Crop-Livestock System in Ethiopia," Paper presented at the Conference on Policies for Sustainable Land Management in the East African Highlands, UNECA, Addis Ababa, April 24–26, 2002 (revised) (IFPRI: Washington, DC, 2003). Mimeo.

Barrett, C. B., F. Place, A. Aboud, and D. R. Brown, "The Challenge of Stimulating Adoption of Improved Natural Resource Management Practices in African Agriculture," in C. Barrett, F. Place, and A. A. Aboud, eds., *Natural Resources Management in African Agriculture: Understanding and Improving Current Practices* (CABI: New York, 2002).

Benin, S., "Increasing Land Productivity in High versus Low Agricultural Potential Areas: The Case of the Ethiopian Highlands". Paper under review with *Food Policy* (International Food Policy Research Institute: Washington, DC, 2003), Mimeo.

Benin, S., S. Ehui, and J. Pender, *Soil Fertility Management in Sub-Saharan Africa: Smallholder Farms in The Highlands of Northern Ethiopia*, Draft Manuscript (International Livestock Research Institute: Addis Ababa, Ethiopia, 2003a).

Benin, S., J. Pender, and S. Ehui, *Policies for Sustainable Land Management in the East African Highlands: Summary of Papers and Proceedings of a Conference held at the United Nations*

Conference Center, April 24–26, 2002, Socio-economics and Policy Working Paper No 50 (International Livestock Research Institute: Addis Ababa, Ethiopia, 2003b).

Binswanger, H. P., and R. F. Townsend, "The Growth Performance of Agriculture in Sub-Saharan Africa," *American Journal of Agricultural Economics* 82, no. 5 (2000), 1075–1086.

Boserup, E., *The Conditions of Agricultural Growth: The Economics of Agrarian Change under Population Pressure* (Aldine Publishers: New York, 1965).

Carloni, A., "Regional Analysis: Sub-Saharan Africa," in J. Dixon, A. Gulliver, and D. Gibbon, eds., *Global Farming Systems Study: Challenges and Priorities to 2030* (Food and Agriculture Organization of the United Nations: Rome, 2001).

Cleaver, K. M., and G. A. Schreiber, *Reversing the Spiral: The Population, Agriculture and Environment Nexus in Sub-Saharan Africa* (The World Bank: Washington, DC, 1994).

Deininger, K., and J. Okidi, "Rural Households: Incomes, Productivity and Nonfarm Enterprises," *Uganda's Recovery: The Role of Farms, Firms, and Government* (The World Bank: Washington, DC, 2001).

Delgado, C., and N. Minot, "Participation in High-Value Agricultural Markets," *Proposal for a Global and Regional Policy Research Program* (International Food Policy Research Institute: Washington, DC, 2003), Mimeo.

Delve, R., and J. Ramisch, "Impacts of Land Management Options in Western Kenya and Eastern Uganda." Paper presented at the Conference on Policies for Sustainable Land Management in the East African Highlands, UNECA, Addis Ababa, 24–26 April 2002, Washington, DC, Mimeo.

Dixon, J., A. Gulliver, and D. Gibbon, *Global Farming Systems Study: Challenges and Priorities to 2030, Synthesis and Global Overview* (Food and Agriculture Organization of the United Nations, Rural Development Division. Sustainable Development Department, Rome, 2001).

Dorward, A., J. Kydd, J. Morrison, and I. Urey, "A Policy Agenda for Pro-Poor Agricultural Growth," *World Development* 32, no. 1 (2004), 73–89.

FAO (Food and Agriculture Organization of the United Nations), *The State of Food and Agriculture. Part II.* Regional Review. (FAO: Rome, 2001).

Freeman, H. A., S. K. Ehui, and M. A. Jabbar, "Credit Constraints and Smallholder Dairy Production in the East African Highlands: Application of a Switching Regression Model," *Agricultural Economics* (1999).

Gabre-Madhin, E., and W. Amha, "Getting Markets Right in Ethiopia: Institutional Changes to Grain Markets" (International Food Policy Research Institute: Washington, DC, 2003), Mimeo.

Gebremedhin, B., J. Pender, and G. Tesfay, "Collective Action for Grazing Land Management in Mixed Crop-Livestock Systems in the Highlands of Northern Ethiopia," *Agricultural Systems* 82, no. 3 (2002), 273–290.

Gebremedhin, B., J. Pender, and G. Tesfay, "Community Natural Resource Management: The Case of Woodlots in Northern Ethiopia," *Environment and Development Economics* 8 (2003), 129–148.

Gebremedhin, B., S. M. Swinton, and Y. Tilahun, "Effects of Stone Terraces on Crop Yields and Farm Profitability: Results of

On-Farm Research in Tigray, Northern Ethiopia," *Journal of Soil and Water Conservation* 54, no. 3 (1999), 568–573.

Haile, M., T. Adhanom, K. Witten, M. Yohannes, P. Byass, and S. Linsay, "Water Harvesting in Micro Dams: Its Impact on the Socio-Economic Condition of the Community and the Salinity of the Irrigated Fields in Tigray," Paper presented at the workshop on Policies for Sustainable Land Management in the Highlands of Tigray, Northern Ethiopia 28–29 March, Mekelle, Ethiopia. Mekelle University (2002a), Mimeo.

Haile, M., B. Gebremedhin, and A. Belay, "The Status of Soil Fertility in Tigray," Paper presented at the workshop on Policies for Sustainable Land Management in the Highlands of Tigray, Northern Ethiopia, 28–29 March, Mekelle, Ethiopia. Mekelle University (2002b), Mimeo.

Herweg, K., "Problems of Acceptance and Adoption of Soil Conservation in Ethiopia," *Topics in Applied Resource Management* 3 (1993), 391–411.

Holden, S., S. Benin, B. Shiferaw, and J. Pender, "Tree Planting for Poverty Reduction in Less-Favoured Areas of the Ethiopian Highlands," *Small-Scale Forest Economics, Management and Policy* 2, no. 1 (2003), 63–80.

Holden, S. T., and H. P. Binswanger, "Small-Farmer Decision Making, Market Imperfections, and Natural Resource Management in Developing Countries," in E. Lutz, ed., *Agriculture and the Environment: Perspectives on Sustainable Development.* A World Bank Symposium (The World Bank: Washington, DC, 1998).

Holloway, G., and S. Ehui, "Expanding Market Participation among Smallholder Livestock Producers," Socio-economics and Policy Research Working Paper No. 48 (International Livestock Research Institute: Nairobi, Kenya, 2002).

Holloway, G., C. Nicholson, C. Delgado, S. Staal, and S. Ehui, "How to Make a Milk Market: A Case Study from the Ethiopian Highlands," Socio-Economics and Policy Research Working Paper No. 28 (International Livestock Research Institute: Nairobi, Kenya, 2000).

International Livestock Research Institute (ILRI), *ILRI 1997: Livestock, People, and the Environment* (ILRI: Nairobi, Kenya, 1998).

Jaffee, S., "The Many Faces of Success: The Development of Kenyan Horticultural Exports," in S. Jaffee, and J. Morton, eds., *Marketing Africa's High-Value Foods: Comparative Experiences of an Emergent Private Sector* (The World Bank: Washington, DC, 1995).

Jagger, P., and J. Pender, "The Role of Trees for Sustainable Management of Less-Favored Lands: The Case of Eucalyptus in Ethiopia," *Forest Policy and Economics* 3, no. 1 (2003), 83–95.

Jagger, P., J. Pender, and B. Gebremedhin, *Woodlot Devolution in Northern Ethiopia: Opportunities for Empowerment, Smallholder Income Diversification, and Sustainable Land Management*, Environment and Production Technology Division Discussion Paper No. 107 (International Food Policy Research Institute: Washington, DC, 2003).

Kaizzi, C. K., "The Potential and Benefit of Green Manures Inorganic Fertilizers in Cereal Production on Contrasting Soils in Eastern Uganda." Ph.D. Dissertation (University of Bonn, Center for Development Research [ZEF]: Bonn, 2002).

Larson, B., and G. Frisvold, "Fertilizers to Support Agricultural Development," *Food Policy* 21, no. 2 (1996), 509–525.

Lipton, M., "Urban Bias: Of Consequences, Classes and Causality," *Journal of Development Studies* 29 (1993), 229–258.

LMC International, *ICO/CFC Study of Marketing and Trading Policies and Systems in Selected Coffee Producing Countries: Ethiopia Country Profile* (LMC International Ltd: Oxford, UK, 2000).

Murithi, F., "Economic Evaluation of the Role of Livestock in Mixed Smallholder Farms of the Central Highlands of Kenya," Ph.D. Thesis, Department of Agriculture, University of Reading (1998).

Ngigi, M., "Smallholder Dairy in Kenya," in Haggblade, ed., *Building on Successes in African Agriculture*, 2020 Focus 12, Brief 6 of 10 (International Food Policy Research Institute: Washington, DC, 2004).

Njuki, J., and F. Verdeaux, "Changes in Land Use and Land Management in the Eastern Highlands of Kenya: Before Land Demarcation to the Present" (International Centre for Research in Agroforestry: Nairobi, 2001), Mimeo.

Nkonya, E., J. Pender, P. Jagger, D. Sserunkuuma, C. K. Kaizzi, and H. Ssali, "Strategies for Sustainable Land Management and Poverty Reduction in Uganda," Research Report No. 133 (International Food Policy Research Institute: Washington, DC, 2004).

Nkonya, E., J. Pender, D. Sserunkuuma, and P. Jagger, "Development Pathways and Land Management in Uganda," Paper presented at the Conference on Policies for Sustainable Land Management in the East African Highlands, UNECA, Addis Ababa, April 24–26, (IFPRI: Washington, DC, 2003), Mimeo.

Okumu, B. N., M. A. Jabbar, D. Colman, and N. Russell, "A Bio-Economic Model of Integrated Crop-Livestock Farming Systems: The Case of the Ginchi Watershed in Ethiopia," in C. B. Barrett, F. Place, and A. A. Aboud, eds., *Natural Resources Management in African Agriculture: Understanding and Improving Current Practices* (CAB International: Wallingford, UK, 2002).

Owuor, J., "Determinants of Agricultural Productivity in Kenya." Tegemeo Institute of Agricultural Policy and Development, Kenya Agricultural Research Institute and Michigan State University, Mimeo.

Palm, C. A., R. J. K. Myers, and S. M. Nandwa, "Combined Use of Organic and Inorganic Nutrient Sources for Soil Fertility Maintenance and Replenishment," in R. J. Buresh, P. A. Sanchez, and F. Calhoun, eds., *Replenishing Soil Fertility in Africa* (Soil Science Society of America: Madison, WI, 1997).

Pender, J., "Development Strategies for the East African Highlands," in J. Pender, and P. Hazell, eds., *Promoting Sustainable Development in Less-Favoured Areas.* 2020 Focus 4 (International Food Policy Research Institute: Washington, DC, 2000). Brief 7, p. 2.

Pender, J., "Sustainable Land Management in the Ethiopian Highlands: Problems, Policies, and Impacts in the Amhara Region," Presentation at the Amhara Regional Workshop of the project "Policies for Sustainable Land Management in the Ethiopian Highlands," 30–31 May, Bahir Dar (IFPRI: Washington, DC, 2001), Powerpoint presentation.

Pender, J., and B. Gebremedhin, "Impacts of Policies and Technologies in Dryland Agriculture: Evidence from Northern Ethiopia," in S. C. Rao, ed., *Challenges and Strategies for Dryland Agriculture*, American Society of Agronomy and Crop Science Society of America, CSSA Special Publication 32 (Madison, WI, 2004).

Pender, J., B. Gebremedhin, S. Benin, and S. Ehui, "Strategies for Sustainable Development in the Ethiopian Highlands," *American Journal of Agricultural Economics* 83, no. 5 (2001a), 1231–1240.

Pender, J., B. Gebremedhin, and M. Haile, "Livelihood Strategies and Land Management Practices in the Highlands of Tigray," Paper presented at the Conference on Policies for Sustainable Land Management in the East African Highlands, UNECA, Addis Ababa, 24–26 April (2002a). Mimeo.

Pender, J., P. Jagger, B. Gebremedhin, and M. Haile, "Agricultural Change and Land Management in Tigray: Causes and Implications," Paper presented at the workshop on "Policies for Sustainable Land Management in the Highlands of Tigray, Northern Ethiopia," March 28–29 (IFPRI: Mekelle, Ethiopia, 2002b). Mimeo.

Pender, J., P. Jagger, E. Nkonya, and D. Sserunkuuma, "Development Pathways and Land Management in Uganda: Causes and Implications," Environment and Production Technology Division Discussion Paper No. 85 (IFPRI: Washington, DC, 2001b).

Pender, J., F. Place, and S. Ehui, "Strategies for Sustainable Agricultural Development in the East African Highlands," Environment and Production Technology Division Discussion Paper No. 41 (International Food Policy Research Institute: Washington, DC, 1999).

Place, F., "Technologies for the East African Highlands," in J. Pender, and P. Hazell, eds., *Promoting Sustainable Development in Less-Favoured Areas*, 2020 Focus 4 (IFPRI: International Food Policy Research Institute: Washington, DC, USA, 2000). Brief 8, p. 2.

Place, F., C. B. Barrett, H. A. Freeman, J. J. Ramisch, and B. Vanlauwe, "Prospects for Integrated Soil Fertility Management using Organic and Inorganic Inputs: Evidence from Smallholder African Agricultural Systems," *Food Policy* 28 (2003a), 365–378.

Place, F., P. Kristjanson, S. Staal, R. Kruska, T. deWolff, R. Zomer, and E. C. Njuguna, "Development Pathways in Medium-High Potential Kenya: A Meso Level Analysis of Agricultural Patterns and Determinants," Paper presented at the Conference on Policies for Sustainable Land Management in the East African Highlands, UNECA, Addis Ababa, 24–26 April, 2002 (Revised) (IFPRI: Washington, DC, 2003b), Mimeo.

Place, F., J. Njuki, F. Murithi, and F. Mugo, "Agricultural Land Management by Households in the Highlands of Kenya," Paper presented at the Conference on Policies for Sustainable Land Management in the East African Highlands, UNECA, Addis Ababa, 24–26 April 2002 (Revised) (World Agroforestry Center: Nairobi, 2003c), Mimeo.

Reijntjes, C., B. Havercoft, and A. Waters-Bayer, *Farming the Future: An Introduction to Low External Input and Sustainable Agriculture* (Macmillan: London, 1992).

Rosegrant, M., M. Paisner, S. Meijer, and J. Witcover, *Global Food Projections to 2020: Emerging Trends and Alternative Futures* (International Food Policy Research Institute: Washington, DC, 2001), p. 207.

Schluter, J., "A Study of the Coffee Market and Proposal for a Forward Auction and Coffee Exchange" (IFPRI: Washington, DC, 2003), Mimeo.

Shapiro, B. I., and J. H. Sanders, "Natural Resource Technologies for Semi-Arid Regions of Sub-Saharan Africa," in C. B. Barrett, F. Place, and A. A. Aboud, eds., *Natural Resources Management in African Agriculture: Understanding and Improving Current Practices* (CAB International: Wallingford, UK, 2001).

Smaling, E. M. A., S. M. Nandwa, and B. H. Janseen, "Soil Fertility in Africa is at Stake," in R. J. Buresh, P. A. Sanchez, and F. Calhoun, eds., *Replenishing Soil Fertility in Africa* (Soil Science Society of America: Madison, WI, 1997).

Staal, S. J., J. Baltenweck, M. M. Waithaka, T. de Wolff, and L. Njoroge, "Location and Uptake: Integrated Household and GIS Analysis of Technology Adoption and Land Use, with Application to Smallholder Dairy Farms in Kenya," *Agricultural Economics* 27 (2003), 295–315.

Staal, S. J., and B. I. Shapiro, "The Economic Impact of Public Policy on Smallholder Periurban Dairy Producers in and around Addis Ababa," Ethiopian Society of Animal Production Publication No. 2, Addis Ababa (1996).

Stoorvogel, J. J., and E. M. A. Smaling, *Assessment of Soil Nutrient Depletion in Sub-Saharan Africa, 1983–2000*. Report No 28 (DLO Winand Staring Centre for Integrated Land, Soil and Resources: Wageningen, The Netherlands, 1990).

Technical Committee for Agroforestry in Ethiopia, "Agroforestry: Potentials and Research Needs for the Ethiopian Highlands. No. 21" (International Centre for Research on Agroforestry: Nairobi, Kenya, 1990).

Tegene, A., and K. Wiebe, "Resource Quality and Agricultural Productivity in Sub-Saharan Africa with Implications for Ethiopia," Paper presented at the International Symposium on Contemporary Development Issues in Ethiopia, Ghion Hotel, 11–12 July, Addis Ababa, Ethiopia (2003).

Tesfay, G., M. Haile, B. Gebremedhin, J. Pender, and E. Yazew, "Small-Scale Irrigation in Tigray: Management and Institutional Considerations," in M. A. Jabbar, J. Pender, and S. Ehui, eds., Policies for Sustainable Land Management in the Highlands of Ethiopia: Summary of Papers and Proceedings of a Seminar Held at ILRI, Addis Ababa, Ethiopia, 22–23 May. Socioeconomics and Policy Research Working Paper 30 (ILRI: Nairobi, 2000).

Thurow, R., "Behind the Famine in Ethiopia: Glut and Aid Policies Gone Bad," *Wall Street Journal* (2003).

Voortman, R. L., B. G. J. S. Sonneveld, and M. A. Keyzer, "African Land Ecology: Opportunities and Constraints for Agricultural Development," CID Working Paper No. 37. Center for International Development, Harvard University: Cambridge, MA (2000).

Westlake, M., "Strategy for Development of the Coffee Sector," Project Report PA-6. Decentralization Support Activity Project, Ministry of Finance and Ministry of Economic Development and Cooperation, Addis Ababa, Ethiopia (1998).

William, T. O., and Ndjeunga, *Constraints and Opportunities for Improved Agricultural Productivity in the Sahelian Farming Systems*. Paper presented at the CORAF/IAC consultation workshop on "Science and Technology Strategies for Improved Agricultural Productivity and Food Security in West and Central Africa," 10–12 February 2003, Dakar, Senegal (2003).

Woelcke, J., "Bio-Economics of Sustainable Land Management in Uganda," in F. Heidhues, and J. von Braun, eds., *Development Economics and Policy* (Peter Lang: Frankfurt, 2003).

World Bank, *Adjustment in Africa: Reform, Results, and the Road Ahead*, A World Bank Policy Research Report (The World Bank: Washington, DC, 1994).

World Bank, *World Development Indicators CDROM* (World Bank: Washington, DC, 2000a).

World Bank, *Can Africa Claim the 21st Century* (World Bank: Washington, DC, 2000b).

World Commission on Environment and Development, *Our Common Future* (Oxford University Press: Oxford, 1987).

Wortmann, C. S., and C. K. Kaizzi, "Nutrient Balances and Expected Effects of Alternative Practices in Farming Systems of Uganda," *Agriculture, Ecosystems and Environment* 71, nos. 1–3 (1998), 115–130.

Economic development and environmental management in the uplands of Southeast Asia: challenges for policy and institutional development

Agnes C. Rola*, Ian Coxhead**

Abstract

Using a Philippine case study site, the forces driving the recent evolution of economic behavior and institutional arrangements in upland and forest margin areas of Southeast Asia are considered. In early modern development, subsistence agriculture using long-phase forest–fallow rotations and regulated by customary law was replaced by more intensive, commercially oriented farming systems, a process heavily influenced by internal migration to the agricultural frontier. Traditional land and resource use institutions were quickly displaced during this shift—both *de facto*, through the actions of colonists, and *de jure*, through the state's assertion of ownership over forests and uplands and the introduction of private title to agricultural lands. Yet the effective implementation of natural resource use constraints lagged substantially behind the pace of agricultural development and forest exploitation, resulting in a period in which high demand for such resources coincided with virtually open access. These processes were noticeably subject to the influence of policies and reforms affecting markets, prices, and institutions.

JEL classification: H73, O13, Q56, Q58

Keywords: institutional evolution; economic development; uplands; resource management; devolution

1. Introduction

...Once upon a time there were traditional patterns of order and balance between a people's needs and the sustainability of their resources. Authority and accountability were close to the source of need and nature. Then came a period of disorder and destruction as resources were redefined to meet centralized, commercial goals of distant accountability and whimsical market forces.

> —William R. Burch, Foreword to *Keepers of the Forest* (Poffenberger, 1990)

In the past 50 years the uplands of Southeast Asia have been subject to timber extraction, intensified subsistence farming, plantation establishment, and other commercial activities that increasingly include highly capital-intensive horticulture and livestock-rearing operations. This expansion and commercialization has been driven by demand factors, especially the search for land and forest as complements to labor; supply factors, including unregulated access to many resources, and reduced transactions costs due to the expansion of roads and other infrastructure. In the absence of appropriate policy and institutional support for environmental management, these activities have denuded forest cover, polluted rivers, eroded soils, and diminished biodiversity. In many areas, the deterioration of the natural environment has reached the point at which the viability of future production activities on the same resource base is in question.

Case studies of resource depletion associated either with rapid economic growth or with a poverty-driven "optimal path to extinction" (Perrings, 1987) abound in the literature. Many of these identify "institutional failures" for creating open access to natural resources as an enabling condition. A longer-time perspective, however, reveals that open access was not always the

---

* *University of the Philippines-Los Banos, Laguna, Philippines.*
** *University of Wisconsin-Madison, Madison, WI, USA.*

norm in the uplands; nor, if the experience of comparable areas in wealthy countries is a guide, does this institutional failure persist as economies develop. Clearly, the ways in which economic growth interacts with institutions and influences their evolution are potentially important determinants of the uses of natural resources and thus (through biofeedback mechanisms) of the health of upland economies.

Historically, economic development in the uplands was accompanied by distinct phases of institutional development (Table 1). Prior to colonization or massive internal migration, when upland populations were sparse and production was primarily for subsistence, customary laws governed natural resource use. Tribes and communities managed the resources over which they had control. These institutions were effective as a means of governing the commons, not least because demand for the resources was low, technologies for their exploitation limited, and transport infrastructure was poor. Long-rotation bush farming fallow systems, typical of this era of development, are widely regarded as "sustainable."

Economic development and population growth in coastal cities and lowland rural areas quickly changes the upland economy. Commercialization (driven by the expansion of domestic and global markets), migration and natural population increase, and the introduction of new technologies all create new pressures on the resource base. Customary law cannot accommodate

such conflicts. In this second phase of institutional development, traditional resource use institutions are swept aside; the state assumes the lead role in controlling resource use and access, and new resource management institutions are imposed from the outside. However, even though local governments and local offices of national resource management agencies may be established, these have no autonomy and little effective authority. State power is thus low at the frontier; the resource base becomes, in effect, open access. Moreover, not only is there no political will for environmental or resource use measures that might reduce the current income-generating opportunities of people living in severe poverty amidst a perceived abundance of natural resources, the state typically *promotes* resource depletion as a means of generating household income and fiscal revenues. What follows is rapid deforestation, shortening of fallow periods, and general degradation of soil and water resources.

This "period of disorder and destruction" was characteristic of many upland regions of Southeast Asia from the 1960s through to the 1990s, during which time the region had the highest annual deforestation rate in the world (FAO, 2001), even as its leading economies grew at historically unprecedented rates. In highly repressed economies such as Myanmar and Indochina the state itself became the primary agent and entrepreneur of resource depletion, having closed

Table 1
Economic development and institutional evolution for environmental management: A framework of analysis

	Prehistory/subsistence economy	Early development	Late development
Economy and resource use	Low population growth Resource abundance Subsistence Slash and burn, long fallow period considered "sustainable"	High population growth, migration Increased competition of resource use Intensive agriculture Commercializing economy Shortening of fallow period	Declining population density Highly commercial economy Declining resource use intensity (i.e., reforestation programs)
Institutional evolution	Customary law; community resource management	State-designed institutions with no practical checks Property rights not well defined, thus resulting to "land grabs	Ideally, Central and local institutional innovations increased Community stewardship of environment Property rights well defined Alternately, Local elite gains power Incentives not compatible with benefits

off most other means of generating jobs, government revenue, and foreign exchange.

In Southeast Asia's more open economies, economic maturity and the end of the demographic transition has ushered in a third phase, one in which lower labor force growth and rapid job creation outside of agriculture greatly reduces population-driven demand for upland land. This has potentially large implications for the governance of resources. This phase may result either in continued rapid resource depletion or in a shift to more conservative strategies. Which of these is more likely to occur will depend on the speed with which institutions catch up with the pace of economic growth. The countries of the region currently present a fascinating array of trends, with no clear, single pattern having emerged.

In this third phase, there is growing community demand for environmental quality and resource conservation. This trend is complemented by a more general decentralization of power and authority, as currently is taking place through formal means in all the large economies of the region. In the best situations, decentralization plus local demands for more ecologically friendly development are complemented by national laws and policies; in the best outcomes, national agencies, local governments, and community groups collaborate to design (and more importantly, to implement) resource management and policies. In the worst cases, however, with reduced power at the national government level and a "business as usual" attitude on the part of local elites, Burch's "disorder and destruction" continues, or even worsens.

In this paper the focus is on the transition to the third phase. The impact of economic development on the upland environment is examined, given the dynamics of decentralizing functions for environmental management. The paper, in contrast with aggregative, cross-country studies, is founded in extended study of a specific region in the Philippines.[1]

The paper is divided into six sections. Section 2 discusses economy–environment linkages as gleaned

from Southeast Asian experience. In section 3 the policy and institutional context of forest and upland land management is investigated. In the remainder of the paper the focus is on decentralization as a new and important factor in upland resource use. In section 4 the outcome of this decentralization policy on institutional structures and environmental governance in selected Asian countries is summarized. Subsequently, in section 5 a Philippine case study of the realities and challenges of resource management in a decentralized setting is presented and discussed. The final section presents a brief conclusion.

2. Economy–environment linkages

Production almost invariably generates environmental damage in the form of pollution and/or natural resource depletion, and it follows that such damage increases as an economy expands, other things equal. It is well known, however, that the environment–economy relationship can be nonlinear—and indeed, nonmonotonic, a concept widely referred to as the Environmental Kuznets Curve (Grossman and Krueger, 1993). Changes in the economic structure occurring in the course of economic growth alter both the valuation of and demand for environmental assets, and if sectors differ in their propensity to pollute or to use depletable resources, it follows that emissions and/or depletion rates will also change, a phenomenon that has been termed the *composition effect*.[2] The net environmental impacts of structural change can either be harmful or benign; more importantly, perhaps, the aggregate composition effect has many components, each with its own specific set of underlying economic and institutional determinants. For land and forest issues, this analytical approach enables research to move beyond trivially true assertions that deforestation and upland land degradation are the consequences

---

[1] Although the paper rests on a broad claim regarding the generality of the conclusions, there is no attempt to provide a complete accounting for the linkages between economy and environment in the uplands. The specific circumstances of the study site lead to an emphasis on some aspects of development and environment rather than on others.

[2] The composition effect is one of three normally identified in analyses of growth–environment linkages. The other two are the *scale effect* (the additional demand on environment and natural resources due to economic expansion), and the *technique effect*, capturing secular changes in technology and nonhomothetic preferences, due to changes in the capital stock and changes in incomes, respectively. Unlike the composition effect, the signs of the scale and technique effects are hypothesized to be unambiguous: scale effects increase environmental damages, and technique effects lessen them. See Antweiler et al. (2001).

of population pressure and "market forces." One thing that quickly becomes clear is that while *total* population growth in a country may justifiably be regarded as exogenous, its spatial distribution, and the incentives that upland populations have when making resource use decisions, are heavily conditioned by government policies, both those directly targeting such populations and activities, and also those operating at the broadest level of the economy.[3]

The economies of Asia's uplands differ both in structure and level of development from lowland zones. They are less densely peopled and more dependent on agriculture and other resource-based industries, their populations are poorer, less healthy, and less well educated, and market access is constrained by higher transport and transactions costs. The population living on "fragile" lands in Asia and the Pacific is currently estimated at 469 million, or 25% of total population (World Bank, 2003).[4]

As recently as a generation ago, upland populations were isolated from lowland and nonfarm economies by infrastructural constraints, travel costs, and even ethnic and political divisions. However, upland population trends, markets, policies, and institutions are now strongly influenced by the development of the overall economy. Roads and telecommunications integrate upland markets with the national economy. As markets expand they create new economic opportunities, which upland and migrant populations are generally quick to seize. In so doing, they also alter the value of immovable resources such as forests and land. In a subsistence economy, such resources (and even labor) have values derived only from the requirements of the local economy, but market integration requires that resource valuations reflect returns obtainable in new uses.

Development policies have direct effects on Southeast Asian uplands largely through infrastructure provision. The impact of road construction is huge, since it reduces transport costs and accelerates flows of migrants and information. As elsewhere in the developing world, road construction has a strong association with

deforestation and the spread and intensification of agriculture (Cropper et al., 1997; Andersen et al., 2002).

The expansion of roads and markets, however, also conveys the *indirect* effects of policy distortions to upland resource use decisions, so economic and policy trends in industry and lowland agriculture become central to upland development (Coxhead and Jayasuriya, 2003a), and national trends in food demand, agricultural technology, and food policy can all have significant environmental consequences. Most obviously, agricultural support policies stimulate the expansion of cultivated area at the expense of the forest. The mechanisms vary from country to country and over time, with contributions from state-sponsored land clearing for settlement programs, commercial forestry, and subsequent land conversion by corporate agribusiness enterprises, and deforestation and land clearing (as well as the intensification of bush fallow rotation systems) by "subsistence" farmers (Angelsen, 1995). All land colonization, however, is driven by a combination of opportunity and necessity, and encouraged by the absence of well-defined and effectively enforced property rights over forest-covered land. The property rights problem itself is partly an artifact of government policies that identify forest-covered land (or land so designated, including cleared land above a certain slope or altitude) as a public resource, neither alienable nor disposable, without providing adequately for its protection from encroachment.

3. The institutional context

Through markets and migration, policies directed at specific "lowland" sectors can also affect upland resource valuations, patterns of land use and production, and thus environmental outcomes (Coxhead and Jayasuriya, 2003a), but the effects are conditioned by the policy and institutional context of upland environmental management. This aspect is examined in this section, focusing on those institutions most critical to forests and agricultural land resources.

3.1. Property rights over land and forest

The centralization of control over natural resource assets has long been a feature of governance in developing countries. In the Philippines, Spanish-era land

---

[3] See Binswanger (1991) for an early analysis of policy influences on deforestation in the Brazilian Amazon.

[4] "Fragility" is defined by criteria relating to aridity, slope, forest cover, and soil type.

law asserted the state's ownership of all land unless a decree was issued to the contrary. This doctrine persisted through the American administration and independence; though the area of declared public land had shrunk to only 62% of the total by the mid-1980s, it still covered 90% of upland (Lynch, 1987). Moreover, a 1975 presidential decree explicitly prevented occupants of uplands from acquiring private property rights at the same time as it declared that existing occupants of public lands were immune from prosecution (Lynch, 1987, p. 284). This act, given the lack of documentation and difficulty of enforcement, effectively legislated open access to forestlands by individuals.

In Indonesia, Dutch colonial practice persisted in the Basic Forestry Law (1967), which designated 74% of the total area (90% of the Outer Islands) as "state forestland" under the control of the Ministry of Forestry, nullifying traditional law or *adat*. Other countries passed similar laws around this time.

Though the primary incentive for control over land and natural resources was its exploitation as an input to growth, motivation for continued central ownership and control was as much political as economic or managerial. Bromley (1991, p. 127) has observed that:

> The new independent nation-states that arose following the Second World War have shown little interest in revitalizing local-level systems of authority... (they) do not relish the thought of local political forces that might challenge the legitimacy and authority of the national government. This means that natural resources have become the "property" of the national governments in acts of outright expropriation when viewed from the perspective of the residents of millions of villages. This expropriation is all the more damaging when national governments lack the rudiments of a natural resource management capability.

National planners regarded natural forests as resources to be exploited for national development. Exploitation of forests (and in trade, of comparative advantage in timber and forest products) was a means to finance modern agricultural and economic growth (e.g., ADB, 1969).[5]

---

[5] As Andersen et al. (2002) argue, deforestation's profitability can be seen in the long term, if the land that is converted is devoted to sustainable agriculture to support food security in the urban centers.

3.2. Forest management

The question of who governed the forests during this period is more complex than the applicable laws indicate. While most forest resources were claimed by the state and governed at the national level, in reality these resources were controlled by a number of actors, with or without the blessing of the national government. Alliances and conflicts among different national agencies, the military, local elites, and domestic and foreign timber corporations often determined forest access. Nevertheless, in some cases local communities successfully maintained at least some degree of influence, usually due to difficulty of access. Examples include the highlands of Vietnam, Laos, and Thailand (Poffenberger, 2000). In the Philippines and Indonesia, on the other hand, the central government and private sector timber corporations have had more dominant power in the use of forest resources (Kummer, 1992), contributing to high rates of forest clearing.

Forest and land management has also been affected indirectly through other policies such as the various internal migration (or "transmigration") initiatives of the 1960s to the 1980s, in which government agencies cleared and developed virgin land at the frontier to house and sustain sponsored migrants. Under these programs large areas, most notably in Malaysia and Indonesia, were converted to plantations and upland fields, the latter sometimes supplied with irrigation at considerable expense, and with mixed results. The degree of control over land use exerted by government agencies in transmigration areas varied, however. Federal land development agencies in Malaysia, where the main focus was on the development of rubber and other plantation crops, sustained a relatively high degree of central control, whereas in Indonesia national agencies had little power (Gérard and Ruf, 2001).

Forest policies began to change in the late 1980s as commercially exploitable stocks dwindled, the damaging side effects of deforestation became more readily apparent, and domestic and foreign conservation movements gained voice. The policy approach, however, continued to rely largely on central control. Thailand and the Philippines both imposed bans on commercial logging and on the export of raw timber around 1990 (Indonesia had imposed a similar ban in 1981 to promote its plywood industry rather than to conserve timber resources). However, these bans were difficult to implement in the absence of appropriate incentives for timber harvesters and exporters, and were widely circumvented. Efforts to provide incentives for sustainable forest management met with only limited success.[6]

3.3. The shift to local ownership and management

The evolution of forest policy away from centralized exploitation (and more recently, from centralized attempts at conservation) toward giving communities and their representatives a more central role is more recent. The recognition of ownership by upland communities is even more so, in spite of earlier legal steps in the same direction, and remains incomplete.

Decentralized control over resources, including community-based forest management (CBFM) programs, was widely adopted in most Southeast Asian countries only in the 1990s—usually coinciding with broader programs of political decentralization. Yet the shift has been incomplete, and the devolution of aspects of forest ownership and/or management has not provided a silver bullet to manage forest degradation and depletion. Policy conflicts, fueled both by economic pressures and mismatches between central and local powers, have continued. In Thailand, for example, religious movements have played an important role in the formation and expression of community opinions on environmental management, often bringing them into conflict with national government programs. In post-Suharto Indonesia, local administrations have frequently augmented *de jure* steps toward decentralized control taken in 1999 with *de facto* assertions, gambling that central authorities will not provide effective countermeasures.

The period of decentralization has seen governments commit to giving more power to communities to manage environmental resources, but the legal basis for

such actions is only weakly established. For instance, community involvement is encouraged in Cambodia, but the absence of documented forest rights and responsibilities leaves the community with no authority to protect local forests (Poffenberger, 2000). Similarly, Thailand has implemented a National Forest Policy but does not specify environmental rights and responsibilities for communities, and the legal basis for community resource management is still lacking. In Vietnam, the 1993 Land Law conferred use rights over agriculture and forestlands to communities and individuals, but ownership still rests with "the people", i.e., the state (Tachibana et al., 2001).

The Philippines is arguably the regional leader in institutional strengthening for environmental management, especially since the passage of the 1991 Local Government Code. Implementation of the Integrated Social Forestry (ISF) Program (part of the Comprehensive Agrarian Reform Program [CARP] passed in 1987 and covering all agricultural lands, including public alienable and disposable lands) has also facilitated giving of tenure to forest occupants. Upland dwellers can secure a Certificate of Stewardship Contract (CSC) giving them exclusive use and occupancy rights to public forestlands for a period of 25 years. Individuals, families, and local communities enter into a contract with the government for this purpose, whereby it is the duty of the CSC holders to engage in the application of soil conservation, suppression of forest fires, and conservation of forest growth in their areas of responsibility (Magno, 2003). Several more seminal laws were passed in the 1990s to manage forest resources. Among these is the Indigenous People's Rights Act (IPRA), passed in 1997, which purports to confer title to ancestral domain and land claims on indigenous communities. However, ambiguities in the provisions of this law persist, and its implementation has not been smooth.

In summary, policies that support community-based stewardship of natural resources are slowly evolving in Southeast Asia. Yet the structures that are emerging are quite different from earlier customary governance systems, not least in that there is collaboration between the state (national or local governments) and civil society. Incomplete decentralization of other areas of administrative control may have complicated the forest and land management process. In many cases, perhaps the majority, local control does not imply governance by

---

[6] In the Philippines, a program called the Industrial Forestry Management Agreement (IFMA) was implemented in 1991. This was a 25-year renewable contract between the Department of Environment and Natural Resources (DENR) and private business or legal commercial forest users wherein the latter will manage part of the denuded forest by planting and having the right to harvest trees (Vitug, 2000). This was suspended in 1995 for political reasons. A similar program in Indonesia (Timber Plantation Development) was also promoted, but its sustainability objective is falling short of expectations (Resosudarmo, 2002).

the community; rather, there is an ongoing evolution of local power that mimics the centralized system it has replaced. Local elites, foreign interests, and other actors now also have access to resources—sometimes with less oversight than before. Whether decentralization is a better arrangement, therefore, is still to be determined.

4. Decentralization and environmental management

Decentralization undermines the traditional environmental management model based on regulatory constraints designed and implemented by central government agencies. After years of failed attempts at centralized control over the management of natural resources, opinion has now turned decisively in favor of local approaches. Can local governments do a better job of resource management than central governments? Some advantages are clear: local administrations can be expected to have specialized knowledge of environmental and economic conditions, and should therefore have the ability to fine-tune policy. But there are disadvantages as well.

Externalities. Jurisdictional boundaries do not typically coincide with relevant natural resource boundaries (such as watersheds), leading to problems of horizontally overlapping control areas and unresolved externalities. When the correspondence between the boundaries of political jurisdictions and the optimal units for natural resource management is inexact, environmental policy managers have an incentive to over-exploit the resource. In Southeast Asia this is most clearly visible in the management of watersheds and river basins, and is compounded when economic and population growth increase local demands on water and land resources. In Vietnam, for instance, conflicts have emerged between upstream coffee irrigators and downstream rice producers, and in China, development activities in upstream locations lead to social conflict with neighboring villages over pollution, erosion, and siltation in downstream locations (Dupar and Badenoch, 2002).

Accountability of local administrations is a critical constraint to socially beneficial local decision making. At the macro level, accountability requires institutional checks and balances on the actions of local

governments, private businesses, and even NGOs, and a strong external audit system is critical to ensuring accountability (Manasan et al., 1999). Local governments could be more fiscally responsible and accountable if they were given more taxing powers because they are closer to the constituents that they tax, but in countries like the Philippines a large portion of a typical local government's income is still controlled by central government. When local funds are limited, fiscal incentives exist for local governments to promote polluting industries and accelerate resource extraction (Rola et al., 2003). This has also been evident in Indonesia since the fall of the New Order regime (World Bank, 2000a).

At the micro level, accountability is determined by the availability of constitutional and practical instruments by which communities acquire a "voice" in the formation and implementation of local policy. While decentralization "does not guarantee that local communities will reap more benefits and be more interested in sustainable environmental management, it does increase the chances that this will happen" (Manasan, 2002). Participatory approaches to natural resource management, involving the community directly as well as the local government, improve the likelihood that local governments will be held accountable for resource management decisions.

Assignation of functions. A third and related problem arises from incomplete and uneven decentralization of functions. This often means that the mandate assigned to a local agency may not be matched by the authority vested in it and that policies applied by one agency may counter the effects of policies applied by others.[7] One of the most common problems is inadequate coordination between line agencies and elected local authorities that have assumed the management of their environmental resources. In Philippines, Thailand, and Indonesia, decentralization laws have failed to provide a clear division of responsibilities between local government and line agencies, for example, by increasing budgetary transfers without allocating new expenditure responsibilities (World Bank, 2000a, 2000b). In countries like Vietnam and Laos, regulatory policies still preclude locally preferred options. Line agencies at the local level are still committed to national programs even if these are not consistent with local development

---

[7] See Manasan (2002).

goals. Hence, one challenge of decentralization is to "embed efforts in a framework that promotes overall national goals of economic and administrative integration, environmental quality and revenue generation while allowing sufficient flexibility in local implementation to meet unique local conditions" (Dupar and Badenoch, 2002).

Capacity constraints. Finally, local governments face capacity constraints in the conduct of analysis, policy, and fiscal powers needed to implement some measures (Coxhead, 2002). Another perceived stumbling block of a decentralized system of environmental governance and management is the lack of capacity to do so at the local level. In most countries, environmental databases are weak, which undermines the ability of national and local policy makers to plan and prioritize projects (World Bank, 1995). Local officials, in addition, lack the technical capacity for policy design and implementation, and for monitoring and evaluation of environmental projects.

5. Decentralized resource management in a Philippine municipality

5.1. Study site

The study municipality, Lantapan in the province of Bukidnon, is located in the Southern Philippine island of Mindanao. It has a landscape that climbs from river flats (300–600 m above sea level) through a rolling middle section (600–1,100 m) to high-altitude, steeply sloped mountainsides (1,100–2,900 m). It hosts the headwaters of a major river (the Manupali) that runs into a dam that diverts flow into a network of canals comprising a 4,000-hectare irrigation system constructed by the Philippines' National Irrigation Authority in 1987. The entire system ultimately drains into the Pulangi River, one of the major waterways of Mindanao Island, about 50 kilometers upstream from the Pulangi IV hydroelectric power generation facility, one of the six largest in the country.

During the 1980s the population in the study area grew at an annual average of 4%, much higher than the Philippine average of 2.4%, mainly due to in-migration. During the 1990s the rate declined to about the same level as the national average of 2.3%.

Agriculture continues to dominate the economy of the municipality and of the province, and more than half of the land area is planted with annual crops. The lower footslopes produce corn and sugarcane, while corn is dominant in the upper footslopes that make up the largest area of the watershed. Coffee is an important secondary crop at middle altitudes, while at higher elevations corn is planted alongside both coffee and temperate-climate vegetables. Starting in 1998, at least ten commercial hog and poultry firms went into business; in 1999, two banana companies were established in both the upper and lower parts of the town. In the earlier times both logging and forest fires facilitated agricultural expansion. In recent decades, however, the profitability of commercial vegetable cultivation has been the primary impetus for forest encroachment, with decisive contributions from road development and the lack of well-defined and enforced property rights in land (Cairns, 1995). The expansion of vegetables and plantation crops in lieu of cereal crops in the area is also a result of favorable price and trade policies.

5.2. Environmental consequences of development

Commercial loggers opened Lantapan's forestland after being granted a Timber License Agreement (TLA) by the national government. Agriculture followed the loggers, with migrants from other parts of the Philippines contributing to the expansion in the cultivated area. Rapid deforestation took place from the 1950s to the 1990s. The uplands were seen as a source of "green gold" by lowlanders (Malanes, 2002), and the spread of intensive upland agriculture was largely driven by market opportunities (Coxhead et al., 2001). Fallow periods were short, and the absence of soil conservation measures resulted in significant land degradation. During 1974–1994, primary forest cover was reduced from half to less than one third of the municipal land area, being replaced mainly by corn and corn-based farming systems (Li, 1994). Resource management decisions were the responsibility of national agencies, although in practice land use rights were allocated through local and informal mechanisms. In-migrants (some relocated from Northern Luzon highland areas in government programs) acquired land from indigenous people in exchange for small sums or in barter trade (Paunlagui and Suminguit, 2001).

The data show that the rate of deforestation has decreased in the past decade and that, at about 0.6% annually during 1994–2001, it is less than half of the national estimate of 1.4% per annum during same period. Out-migration has also been observed, especially although not exclusively during the drought years 1997–1998 (Rola et al., 1999). On the other hand, agricultural intensification continues, driven by opportunities in domestic and international markets. Land use data also confirm the spread of intensive cultivation of annual crops to the upper watershed. Water quality monitoring since 1994 in several watersheds shows that measures of total suspended solids (TSS) are considerably higher in those areas where agricultural cultivation is more widespread, in spite of lower average slope, and that seasonal TSS peaks appear to coincide with months of intensive land preparation activity (Deutsch et al., 2001).

Other consequences of rapid and increasing soil erosion rates as a result of agricultural intensification can be seen in the deterioration of the two water impoundment structures,[8] the MANRIS diversion dam and the Pulangi IV hydro power installation on the Pulangi River, located a few kilometers below the junction of the Manupali. Although forest management policies and strategies to reduce deforestation were adopted in the 1990s, and, despite a ban on commercial logging, policies to promote sustainable upland management have yet to translate into better environmental health.

5.3. Policies and institutions in transition

Environmental management mandates and implementation of environmental programs/plans are distributed across all levels of governance in the Philippines and the still-evolving multilevel/multisectoral institutions, while funding and other administrative support are facilitated through Congress and the various local government levels. Technical support in the preparation of provincial and municipal plans, comes mostly from the Department of the Environment and Natural Resources (DENR), with some assistance from academics and other local government units (LGUs). The implementation of the plans and monitoring and evaluation are placed under the now evolving multilevel, multisectoral institutions.

5.3.1. The national level

Philippine law states that it is the duty of the national government to maintain ecological balance. The DENR is tasked to lead in this function, although its mandate naturally overlaps with that of the Department of Agriculture, the National Irrigation Administration, and several other bodies. Following the Earth Summit in 1992, the national government created the Philippine Council for Sustainable Development (PCSD).[9] Local councils were also organized in the various regions of the country. The PCSD is mandated to oversee and monitor the implementation of the Philippine Agenda 21 (PA 21), the Philippines' blueprint for sustainable development, by providing the coordinating and monitoring mechanisms for its implementation.

The 1991 Local Government Code (LGC) initiated a period of devolution of national mandates, including some of the DENR. Currently, therefore, the system is in transition while national agencies are slowly devolving responsibilities to local governments, but the DENR remains the least decentralized of all Philippine government agencies, with only 4% of its staff and 9% of its budget located outside Manila, in contrast with averages of 51% and 12%, respectively, for all agencies (Manasan, 2002).

5.3.2. Subnational level

Much of the devolution in the LGC bypasses the provinces, moving power directly to the municipalities. In practice, however, provincial governors and their administrations retain considerable influence over local decision making through the exercise of their supervisory and coordination functions.

Bukidnon has two provincial offices for environment and natural resources. The national DENR exercises its line functions through the Provincial Environment and Natural Resources Office (PENRO), i.e., for forest protection and titling, while the local government maintains the Bukidnon Environment and Natural Resources Office (BENRO) and takes care of devolved functions, e.g., small-scale mining.[10]

---

[8] See also Pingali (1997), for similar evidence from elsewhere in Asia.

[9] This is headed by the Director General of the National Economic and Development Authority (NEDA) as chairperson, the Secretary of the DENR as the Vice-Chairperson, and with the membership coming from both government and NGOs.

[10] There is also the city ENRO and the community-level ENRO (or CENRO). The latter's function is mainly focused on protected areas where these areas span two or more LGUs.

On the other hand, municipal government is mandated to undertake water and soil resource utilization and conservation projects, implement community-based forestry projects, manage and control commercial forests with an area not exceeding 50 square kilometers, and establish tree parks, green belts, and similar forest development projects. All these functions should be *pursuant to national policies and subject to supervision and review of the DENR* (DILG, 1991). Thus, the national office still controls the environmental programs of local governments. Further, low-income upland municipalities such as Lantapan do not usually have an environment and natural resources office, while all municipalities have a municipal agricultural office (MAO). As intensive agriculture has a direct impact on the state of environmental resources, it is worth considering that these functions be merged.

Furthermore, the village or *barangay* government has no legal mandate to conduct environmental programs, even though village governments are ideally positioned given their proximity to communities and households. In Lantapan, village-level environmental programs such as soil conservation, tree planting, monitoring the buffer zone, etc. are initiated by external organizations, especially nongovernmental organizations (NGOs), whose sustainability must be questioned.

5.3.3. Multilevel, multistakeholder institutions

Some institutions with an environmental mandate cut across the different layers of governance, and across different sectors. The rationale for this institutional innovation is to fully capture the benefits and costs in the process of managing the resource, and to be able to get as much representation from the different stakeholders.

5.3.3.1. The Protected Area Management Board (PAMB).

In the Philippines, protected areas are governed by a management board whose membership consists of representatives from the various sectors and the different governance levels. Bukidnon hosts the Mount Kitanglad Range Natural Park, a protected area, managed by the PAMB. The chair of the Board is the regional director of the DENR.[11] The PAMB meets

regularly to provide oversight and guidance to field implementers.

The Council of Elders (COE) reinforces the natural park management by applying customary law in dealing with violators. The Kitanglad Guard Volunteers (KGV), a people's organization patrols and monitors illegal activities, with offenders undergoing a *sala* (a cleansing ritual), so as not to commit similar acts in the future. The involvement of the COE is also as stipulated in the Indigenous People's Rights Act (IPRA). In the IPRA, indigenous peoples (IPs) are by law now given the authority to practice their customary rules in the management of natural resources. It is, however, observed that local tribal leadership through this COE is "emerging and exerting influence to the management of the protected area beyond the terms of the legal prescriptions." Hence, "there is a need to clarify the management implications between legally instituted and tradition bound structures" (Mount Kitanglad Range Natural Park Management Plan, no date).[12]

5.3.3.2. Watershed Protection and Development Councils at the local level.

Watershed Protection and Development Councils exist for both the province and the study municipality. At the provincial level, this is a multisectoral body composed of national and local agencies, academics, local government units, and representatives from NGOs. This provincial body was created "in order to fully protect and preserve the remaining forests in the Bukidnon Watersheds and rehabilitate open areas within their headwaters."[13] Representatives of resource agencies in the province make up the membership of the Technical and Advisory Committee (TAC). This body has the potential to allow local actors and stakeholders the opportunity to realign their efforts toward wider watershed-level programs.

At the municipal level, the Watershed Management Council was formed in part as a response to advocacy efforts of the SANREM CRSP research project.

---

[11] This is a generic rule as protected areas could be spread across more than one province. As the Mount Kitanglad range is contained

in Bukidnon, the provincial government is actively involved in its management.

[12] Department of Environment and Natural Resources (DENR) and Kitanglad Integrated NGO (KIN) 2000. Mount Kitanglad Range Natural Park Management Plan. Protected Area Superintendent Unit (PASu), Malaybalay, Bukidnon, Philippines.

[13] This is contained in President Fidel V. Ramos' Presidential Memorandum Order (PMO), 270 dated March 1995 (Sumbalan, 2001).

This is a multisectoral group comprising representatives from agribusiness, NGOs, peoples' organizations, the municipal legislative council, and provincial agencies. The Municipal Watershed Management Plan, just recently approved, will be used as a guide in watershed management in the various micro-watersheds in the municipality. It is an input into the provincial watershed management plan.[14]

The findings reveal, therefore, that policy and institutional initiatives are slowly evolving but are not yet sufficient to arrest the resource degradation brought about by agricultural intensification.

6. Toward a win–win scenario in the late development period

In the terms of the framework of analysis, Lantapan and many other Asian upland villages are in transition to the late period of institutional development. In this setting, two alternative pathways provide two extreme environmental outcomes. In the best possible world, institutions evolve to catch up with the pressures brought about by economic development. Alternatively, institution development lags so far behind economic development pressures that growth is compromised by pervasive and possibly severe environmental damages. The good news in Lantapan is that deforestation may have largely ceased, partly due to the inaccessibility of the remaining commercial forest stands and partly because of forest protection policies and institutions. Yet agricultural intensification, driven largely by the market, continues. Poorly defined property rights in land and the nonpoint nature of water pollution originating in soil erosion and chemical runoff mean that land degradation and water pollution continue to worsen in spite of institutional strengthening and devolution of responsibilities. Development of institutional and legal frameworks for land and watershed protection lags far behind development pressures. How can upland villages like Lantapan attain the best outcomes? The following are some recommendations:

1. *Better accountability requires genuine decentralization.* Genuine decentralization implies autonomy

with respect to fiscal, administrative, and political powers. From the case study, increasing viability of village and community institutions, and increasing willingness of the provincial government to invest in environmental programs can be observed (Sumbalan, 2001). At the lowest levels of government (municipal or village levels), appreciation of the benefits of good environmental management is still lacking, and capacity building is needed if environmental indicators are to be properly monitored. National regulatory policies could be implemented at the very local level, once capacity is present. Thus, while national government plays an important role in orchestrating broader environmental policies, it should help build capacity at the provincial and municipal levels.

Governance must be transparent and responsible, and thus accountable. Broad-based participation by stakeholders can also increase accountability—a "voice" is important, and trust in the leaders and other community members can encourage active participation. The essence of social capital is important; increased social capital with its greater degree of horizontal connection is seen to improve governments and can lead to increased community cooperative action and solve "common property" problems (Narayan and Pritchett, 1997). In the case of Lantapan, it was shown that communities with strong social capital were more accountable to their members (Paunlagui et al., 2003).

The more important effect of genuine decentralization is the ability of local governments to deal with their neighbors in cases where resource management issues spill across jurisdictional boundaries and agency mandates. A strong local government can command collaboration among other LGUs and line agencies for environmental management.

2. *Addressing the externality problem requires watershed-based institutions and policies.* In spite of the intrinsic desirability of decentralization, careful attention must be paid to the appropriate assignment of functions to each level of governance. Given the externalities involved, there is a compelling case for the management of forest, land, and water resources at the watershed or river basin level (Dixon and Easter, 1986; Francisco, 2002). This means defining the hydrologic unit and providing for an administrative structure for this level of

---

[14] In 2003, the province of Bukidnon was the grand prizewinner in the Philippines' *Clean and Green Contest* in recognition of, among others, its innovative watershed management plan. This contest is part of the environmental management incentive program of the Arroyo administration.

management.[15] Both the province of Bukidnon and the municipality of Lantapan have started structuring the watershed management bodies and preparing their watershed management plans accordingly. However, strong cooperation among villages and among local government units is needed, especially in terms of regulatory and fiscal functions. Most importantly, the watershed management plan should be properly funded. A seemingly weak or total lack of a watershed-level organizational structure is seen to be a factor in watershed degradation in the Philippine case and throughout the region.[16]

3. *Market-based mechanisms can support sustainable upland management, but only if the appropriate institutions are in place.* An important lesson from Lantapan is that market expansion is seldom the only cause of unsustainable development. Property rights failures, externalities, and incomplete markets are usually present, and their effect is to distort market-driven resource allocation and conservation decisions (Barbier, 1990; Pagiola et al., 2002). In this setting, first-best solutions require addressing the failures at source, and then designing policies for agricultural development. Uncontrolled market expansion without property rights or other corrective and regulatory institutions results in a "third phase" of development in which resource depletion and environmental destruction is rapid, pervasive, and potentially disastrous.[17]

Market-based incentives can also be used by local governments to encourage sustainable land use decisions. Subsidies can be offered to farmers who may be practicing soil conservation measures or planting perennials. The value of these can be equal to the "true" cost of soil erosion, i.e., including downstream effects. In reality, there is a tendency to overstate cost if subsidies are given (Goodstein, 1999); but the incentive-compatible approach can encourage truthful behavior (Kwerel, 1977 as cited by Goodstein, 1999). These are areas for future research.

In the study municipality, one recently adopted municipal ordinance stipulates that if farmers practice soil conservation measures, they will be favored as participants in government programs. Tree farming and agroforestry, while promoted in Lantapan, may not be on a scale sufficient to arrest erosion problems. Nevertheless, recent studies underscore the importance of price policy intervention for tree-growing in promoting tree-based farming systems for small holders (Predo, 2002).

7. Conclusions

The findings of this research show that for the past decade, deforestation rates in the upland areas of Southeast Asia have been reduced, mainly due to state and communities working together. The sustainable management of upland villages like Lantapan can best be achieved if policies and institutions catch up with the pressures of economic development. Genuine decentralization involves transparency in the use of resources, responsible governance, and accountability.

Broad-based participation of stakeholders can flourish if social capital is strong. More innovative management structures for upland resources should include the lowlands and the broader economic environment; thus a watershed approach is considered ideal. Incentives for practicing sustainable agricultural land use practices such as soil conservation, and tree farming should be compatible with the benefits to be derived by society as a whole. In the long term, price, trade, and wage policies can be used as instruments in promoting sustainable upland resource management.

What lessons have been learned? In this later phase of economic development, people and social structures are the most important features for responsible environmental management. Markets, communities, and the state will need to work together and can do so only in an atmosphere of mutual trust. "Seeking sustainability" is difficult, as field experiences reveal, yet future research strategies should take on participation by actors at all levels, should aim to build capacity, understand institutions, and should aid in the environmental policy decision-making process.

---

[15] This is a problem addressed in wealthy countries by the creation of single-purpose jurisdictions such as school, fire, and policy districts, which may be subsets of political jurisdictions or may overlap several such units.

[16] Kerr (2002) provides examples of good watershed-based management, such as the popularly cited case of Sukhomajri, India, that other upland villages can learn from and emulate.

[17] As illustrated by such catastrophes as the 1992 Ormoc mudslides in the central Philippines, where denuding of upland lands created the conditions for a monsoon-rain fueled landslide that wiped out an entire town.

References

Andersen, L. E., C. W. J. Granger, E. J. Reis, D. Weinhold, and S. Wunder, *The Dynamics of Deforestation and Economic Growth in the Brazilian Amazon* (Cambridge University Press: Cambridge, UK, 2002).

Angelsen, A., "Shifting Cultivation and 'Deforestation': A Study from Indonesia," *World Development* 23, no. 10 (1995), 1713–1729.

Antweiler, W., B. R. Copeland, and M. Scott Taylor, "Is Free Trade Good for the Environment?" *American Economic Review* 91, no. 4 (2001), 877–908.

Asian Development Bank, *Asian Agricultural Survey* (University of Tokyo Press and the University of Washington Press: Manila, 1969).

Barbier, E. B., "The Farm-Level Economics of Soil Conservation: The Uplands of Java," *Land Economics* 66, no. 2 (1990), 199–211.

Binswanger, H. P., "Brazilian Policies That Encourage Deforestation in the Amazon," *World Development* 19, no. 7 (July 1991), 821–829.

Bromley, D. W., *Environment and Economy: Property Rights and Public Policy* (Basil Blackwell: Oxford, UK, and Cambridge, MA, 1991).

Cairns, M., "Ancestral Domain and National Park Protection: Mutually Supportive Paradigms? A Case Study of the Mt. Kitanglad Range National Park, Bukidnon, Philippines," Paper presented at a workshop on Buffer Zone Management and Agroforestry, Central Mindanao University, Musuan, Philippines (August 1995), mimeo.

Coxhead, I., "It Takes a Village to Raise a Pigovian Tax. . .or Does It Take More? Prospects for Devolved Watershed Management in Developing Countries," Paper presented at a conference on Sustaining Food Security and Managing Natural Resources in Southeast Asia: Challenges for the 21st Century, Chiangmai, Thailand (January 2002).

Coxhead, I., and S. K. Jayasuriya, *The Open Economy and the Environment: Development, Trade and Resources in Asia* (Edward Elgar: Cheltenham, UK and Northampton, MA, 2003a).

Coxhead, I., A. C. Rola, and K. Kim, "How Do National Markets and Price Policies Affect Land Use at the Forest Margin? Evidence from the Philippines," *Land Economics* 77, no. 2 (2001), 250–267.

Cropper, M. C., C. Griffiths, and M. Mani, "Roads, Population Pressures and Deforestation in Thailand, 1976–1989," Policy Research Working Papers (World Bank: Washington, DC, 1997).

Deutsch, W. D., A. L. Busby, J. L. Orprecio, J. P. Bago-Labis, and E. Y. Cequiña, "Community-Based Water Quality Monitoring: from Data Collection to Sustainable Management of Water Resources," in I. Coxhead and G. Buenavista, eds., *Seeking Sustainability: Challenges of Agricultural Development and Environmental Management in a Philippine Watershed* (PCARRD: Los Baños, Philippines, 2001), pp. 138–160.

DILG, *Local Government Code of the Philippines* (Republic of The Philippines: Manila, 1991).

Dixon, J. A., and K. W. Easter, "Integrated Watershed Management: An Approach to Resource Management," in K. W. Easter, J. A. Dixon and M. M. Hufschmidt, eds., *Watershed Resources Management* (Westview Press: Boulder, 1986).

Dupar, M., and N. Badenoch, *Environment, Livelihoods, and Local Institutions: Decentralization in Mainland Southeast Asia* (World Resources Institute: Washington, DC, 2002).

FAO (Food and Agriculture Organization of the United Nations), *Global Forest Resources Assessment 2000* (FAO: Rome, 2001). Accessed at: http://www.fao.org/forestry/fo/fra/main/pdf/main_report.zip, November 2001.

Francisco, H. A., "Why Watershed-Based Water Management Makes Sense," PIDS Policy Notes 2002-09. Philippine Institute for Development Studies, Makati, Manila, Philippines (2002).

Gérard, F., and F. Ruf, eds., *Agriculture in Crisis: People, Commodities and Natural Resources in Indonesia, 1996–2000* (CIRAD: Montpellier, France and Curzon: Richmond, UK, 2001).

Goodstein, E. S., *Economics and the Environment* (Prentice Hall: Upper Saddle River, NJ, 1999).

Grossman, G., and A. B. Krueger, "The Environmental Impacts of a North American Free Trade Agreement," in P. Garber, ed., *The U.S.-Mexico Free Trade Agreement* (MIT Press: Cambridge, MA, 1993), pp. 13–56.

Kerr, J., "Sharing the Benefits of Watershed Management in Sukhomajri, India," in S. Pagiola, J. Bishop, and N. Landell-Mills, eds., *Selling Forest Environmental Services: Market Based Mechanisms for Conservation and Development* (Earthscan Publications: Sterling, VA, 2002).

Kummer, D., *Deforestation in the Philippines* (Ateneo de Manila Press: Manila, 1992).

Kwerel, E., "To Tell the Truth: Imperfect Information and Optimal Pollution Control," *Review of Economic Studies* 44, no. 3 (1977), 595–601.

Li B., "The Impact Assessment of Land Use Changes in the Watershed Area Using Remote Sensing and GIS: A Case Study of the Manupali Watershed, the Philippines," Unpublished Masters' Thesis (Asian Institute of Technology: Bangkok, 1994).

Lynch, O. J., "Philippine Law and Upland Tenure," in S. Fujisaka, P. Sajise and R. del Castillo, eds., *Man, Agriculture and the Tropical Forest: Change and Development in the Upland Philippines* (Winrock International: Bangkok, 1987), pp. 269–292.

Magno, F. A., "Forest Devolution and Social Capital State-Civil Society Relations in the Philippines," in A. P. Contreras, ed., *Creating Space for Local Forest Management in the Philippines* (CIFOR and La Salle Institute of Governance: Manila, 2003).

Malanes, M., *Power from the Mountains: Indigenous Knowledge Systems and Practices in Ancestral Domain Management—The Experience of the Kankanaey-Bago People in Bakun, Benguet Province, Philippines* (ILO, 2002).

Manasan, R., "Devolution of Environmental and Natural Resources Management in the Philippines: Analytical and Policy Issues," *Philippine Journal of Development*, Number 53, XXIX, no. 1 (2002), 33–54.

Manasan, R. G., E. T. Gonzalez, and R. B. Gaffud, "Indicators of Good Governance: Developing an Index of Governance Quality at the LGU Level," *Journal of Philippine Development*, Number 48, XXVI, no. 2 (1999), 149–212.

Narayan, D., and L. Pritchett, "Cents and Sociability: Household Income and Social Capital in Rural Tanzania," Policy Working Paper 1796 (The World Bank: Washington, DC, 1997).

Pagiola, S., N. Landell-Mills, and J. Bishop, "Making Market Based Mechanisms Work for Forests and People," in S. Pagiola, J. Bishop and N. Landell-Mills, eds., *Selling Forest Environmental Services: Market Based Mechanisms for Conservation and Development* (Earthscan Publications: Sterling, VA, 2002).

Paunlagui, M. M., M. R. Nguyen, and A. C. Rola, "Social Capital, Ecogovernance and Natural Resource Management: A Case Study in Bukidnon, Philippines," ISPPS Working Paper No. 03-04, University of the Philippines, Los Banos College, Laguna, Philippines (2003).

Paunlagui, M. M., and V. Suminguit, "Demographic Development in Lantapan," in I. Coxhead and G. Buenavista, eds., *Seeking Sustainability: Challenges of Agricultural Development and Environmental Management in a Philippine Watershed* (PCARRD: Los Baños, Philippines, 2001), pp. 138–160.

Perrings, C., "An Optimal Path to Extinction? Poverty and Resource Degradation in an Open Agrarian Economy," *Journal of Development Economics* 30 (1987), 1–24.

Pingali, P. L., "Agriculture-Environment Interactions in the Southeast Asian Humid Tropics," in S. Vosti and T. Reardon, eds., *Sustainability, Growth and Poverty Alleviation: A Policy and Agroecological Perspective* (Johns Hopkins University Press for the International Food Policy Research Institute: Baltimore, 1997).

Poffenberger, M., "The Evolution of Forest Management Systems in Southeast Asia," in M. Poffenberger, ed., *Keepers of the Forest. Land Management Alternatives in Southeast Asia* (Ateneo de Manila University Press: Quezon City, 1990).

Poffenberger, M., ed., *Communities and Forest Management in Southeast Asia: A Regional Profile of the Working Group on Community Involvement in Forest Management* (IUCN: Geneva, 2000).

Predo, C., "Bioeconomic Modeling of Alternative Land Uses for Grassland Area and Farmers' Tree-growing Decisions in Misamis Oriental, Philippines," Unpublished Ph.D. dissertation, University of the Philippines Los Banos College, Laguna, Philippines (2002).

Resosudarmo, I. A. P., "Timber Management and Related Policies," in P. Colfer, J. Carol, and I. A. P Resosudarmo, eds., *Which Way Forward? People, Forests, and Policy Making in Indonesia* (Resources for the Future, CIFOR, and Institute of Southeast Asian Studies: Singapore, 2002).

Rola, A. C., D. D. Elazegui, C. A. Foronda, and A. R. Chupungco. "The Hidden Costs of Bananas: Imperatives for Regulatory Action by Local Governments," CPAf Policy Brief. No. 03-01. College of Public Affairs, UPLB College, Laguna, Philippines (2003).

Rola, A. C., C. O. Tabien, and I. B. Bagares, "Coping with El Nino, 1998: An Investigation in the Upland Community of Lantapan, Bukidnon," ISPPS Working Paper 99-03. UPLB College, Laguna (1999).

Sumbalan, A. T., "The Bukidnon Experience on Natural Resource Management Decentralization," Paper presented at the SANREM conference, ACCEED Makati (May 2001).

Tachibana, T., T. M. Nguyen, and K. Otsuka, "Management of State Land and Privatization in Vietnam," in K. Otsuka and F. Place, eds., *Land Tenure and Natural Resource Management: A Comparative Study of Agrarian Communities in Asia and Africa* (Johns Hopkins University Press for IFPRI: Washington, DC, 2001), pp. 234–272.

Vitug, M. D., "Forest Policy and National Politics," in Peter Utting, ed., *Forest Policy and Politics in the Philippines: The Dynamics of Participatory Conservation* (Ateneo de Manila University Press: Quezon City, 2000).

World Bank, *Mainstreaming the Environment.* A summary, (Washington, DC, 1995).

World Bank, *Indonesia: Public Spending in a Time of Change* (2000a). Accessed at http://lnweb18.worldbank.org/eap/eap.nsf, August 2003.

World Bank, *Thailand: Public Finance in Transition* (2000b). Accessed at http://www.worldbank.or.th/economic/index.html, August 2003.

World Bank, *World Development Report 2003* (Oxford University Press for the World Bank: Washington, DC, 2003).

IAAE Synopsis: Reshaping Agriculture's contribution to society

1. Introduction

With the synoptic session we brought to a close the 25th conference of the International Agricultural Economics Association. This was only the third time that the IAAE had met in Africa, and the conference rightfully highlighted the food and agricultural situation in Africa. At the start of the meeting we were challenged by the South African Minister of Agriculture to make our work relevant to the problems of poor countries and poor people. We were also warned that we were in danger of becoming an endangered species. Yet, I found listening to the papers, the panels, symposia, discussion groups and the poster sessions, that we continue to be a vibrant profession, that is continually renewing itself, and continues to be highly relevant.

In this conference we took stock of where we were on age-old issues, re-examined our positions on perennial debates, and charted course for venturing into new and urgent issues facing the global community. As we prepared to leave Durban, I was convinced that we continue to make major contributions to the issues of today and the issues of the future: such as globalization, trade integration, urbanization, and environmental conservation. I should also note that at this conference we saw the birth of the African Association of Agricultural Economists. The AAAE can become a crucial forum for highlighting the unique problems of African agriculture and a useful network for increasing numbers of agricultural economists in the region.

Some of the major messages which I heard in this conference naturally related to the four sub-themes that the conference was structured around: 1. Strategies for reducing poverty; 2. Efficiency in food and farming systems; 3. Environmental stewardship; and 4. Food safety and security.

2. Strategies for reducing poverty

We were reminded several times during the conference that despite significant gains, the number of absolute poor and food insecure people continues to be inexcusably high, particularly in Africa and South Asia. We were also reminded that the problem of poverty and food insecurity persists even in the rich countries, such as the USA. Economic growth is a necessary condition for poverty alleviation, but accompanying policies that address the underlying structural causes of poverty are crucial if we hope to make significant progress in reducing the numbers of poor and undernourished.

Our conceptual frameworks for understanding the dimensions of poverty have improved significantly, we have the tools to track the extent of poverty both temporally and spatially, and we have a menu of policy interventions that could help the poor climb out of poverty. These interventions could be in the form of safety nets or cargo nets (the former for transient poverty and the latter for removing the structural and institutional obstacles that contribute to chronic poverty, such as: lack of credit, access to markets, etc).

Even when we know the extent of the problem and how to deal with it, we have not been very successful in beating the problem of persistent poverty and food insecurity. Is this because the problem is not that we do not know what to do, but rather in being unable to do it consistently? What is the political economy behind the slow implementation of pro-poor policies in the developing world? We as a profession ought to pay a lot more attention to the political economy, and institutional factors that govern the promulgation, implementation and effectiveness of anti-poverty programs. Important ways in which economists can contribute to the growing right to food debate emerged at the conference. We explored the political, institutional and legal dimensions of incorporating a "rights based" approach in the national and international fight against hunger.

3. Efficiency in food and farming systems

Several papers in the conference re-visited the issue of farm size and productivity, and questioned the ability of small farmers to use agriculture as a means of climbing out of poverty. Productivity growth on small farms, especially in intensive farming systems has been well established, and so has their commercial viability. The green revolution experience has shown quite conclusively that small farms can indeed act as an engine of overall economic growth in developing countries. Where small farms have not been successful, it has been because structural factors, both agro-climatic and socioeconomic have constrained their growth.

The debate on the viability of small farmers has surfaced again in the context of agricultural commercialization trends that are induced by globalization and rapid urbanization across the developing world. Global and national food systems are changing fast; vertical integration in procurement, processing, distribution and retailing is fast becoming a reality across the developing world. We agreed that the transformation of agricultural systems will lead to dramatic changes in the small farm sector. Several examples were presented of small farmers successfully integrating into the global economy, however, there are an equal number of cases of them losing out in the process. Agricultural commercialization is certainly not a frictionless process and the short term adjustment costs can be enormous. We agreed that we need to do a lot more work to understand who wins and who loses, and how we can help the losers in the transition process. "Retooling" smallholders with appropriate technology and knowledge that makes them able to face the requirements posed by commercial markets will be a formidable challenge for research and extension systems in developing countries.

Our research supports the view that increased agricultural trade liberalization and enhanced access to OECD markets would be beneficial to developing countries. But our work also cautions that the gains from trade have limits too. It is important for us not to forget that complimentary investments in infrastructure and human capital are crucial in order to "make markets work for the poor". We need more research on volatility effects and distributional impacts on small farm households, particularly those in marginal production environments. We need to invest a lot more in capacity building so that developing countries can more effectively participate in the negotiation process.

The conference examined the role of emerging technologies in helping improve the efficiency of small

farm production systems. There were several papers on biotechnology addressing the prospects, the constraints and the consequences of adoption. The focus of biotechnology research is shifting from whether or not to promote/adopt these novel inputs to how best to target biotechnology to the needs of the poor. While biotechnology has captured the interest and imagination of our profession, information technology and its consequences, which I judge to be equally important, received scant attention in the conference and by has been neglected by agricultural economists in general.

4. Environmental stewardship

The state of the environment was discussed quite extensively, with a particular emphasis on the identification of potential win-win solutions for managing local and global resources. Examples, such as conservation tillage, better water management, etc., were cited as opportunities for enhancing small farm productivity while at the same time contributing to sustaining the local and global resource base. The role of policy reforms, particularly the removal of input subsidies and protectionist policies, were examined in the context of improving incentives and/or removing disincentives for the adoption of productivity enhancing and resource conserving technologies.

The conference highlighted the need for aligning local incentives with the goals and desires of global treaties. Specifically, the question that was asked was: what incentive do local communities and producers have to change their behavior in order to comply with global treaty requirements? The Kyoto Protocol on climate change and the Clean Development Mechanism (CDM) that emerged from it was widely discussed in this context. We were left with several questions on the feasibility of such payment mechanisms to mitigate climate change while at the same time addressing development and poverty alleviation goals.

Several empirical studies tested the relationship between poverty and environmental degradation. These studies questioned the common belief that the poor degrade their environment. Where incentives exist poor households have made the necessary investments and modified their behavior in order to conserve the environment. These studies call for concerted efforts at policy and institutional reform that generate incentives for sustainable management of the resource base.

5. Food safety and security

Effective mechanisms for improving food safety standards is an emerging agenda for the developing world, not just in terms of complying with SPS standards for commodity exports, but also for the domestic food supply chain, especially given the burgeoning task of feeding the cities. Urbanization is likely to increase the "effective demand" for food safety. In developing countries, the informal sector is often a significant producer, processor, distributor and preparer of food and food products (e.g. street foods).

Food safety is a very important dimension of food security, and ensuring the provision of safe food is as important for the developing countries as it is for the high income countries. However, we do need to address the trade-off question. When imposing standards that are difficult and costly to achieve, policymakers need to be wary of the implications for low-income food producers, sellers and consumers. Regulation has to be accompanied by capacity building, nutrition education and other means of support. This is a huge task.

The conference highlighted the fact that in today's world there are more people that are overfed than underfed. It was said that "over-consumption of food is not safe consumption of food". The problem of over-consumption is becoming an increasingly important phenomenon in developing countries, particularly among the middle classes in urban areas. Already a number of countries experience what is termed "epidemiological transition", i.e. a gradual shift from a prevalence of infectious diseases to the prevalence of chronic ones associated with changing diets and a sedentary lifestyle. The paradox lies in the fact that high incidence of nutrition related diseases can occur alongside high prevalence of hunger and malnutrition in the same country. Dealing with the co-existence of obesity and under nutrition at a national, regional, and household level is a increasingly important challenge that we will face as we look ahead.

6. Concluding remarks

We concluded the conference with a sense of satisfaction, that we as a profession had actively contributed to the debates of the day, and with a sense of expectation for the enormous agenda ahead of us, both for the problems of the developing world as well as those of the developed world.

Our toolkit is well developed and the issues we are applying our tools to are current and highly relevant, yet our influence on policy debates could be improved through better communication with those responsible for policy at the global, national and local levels. The translation of results derived from rigorous analysis to effective policy is a challenge that we need to face up to if we want to enhance our relevance.

As agriculture is back on the agenda of the policy community, we need to ensure that we as agricultural economists have a place at the table. I see this decade as a period of renewal and renaissance of agricultural economics and not one of irrelevance and of being an endangered species.

Prabhu Pingali

I want to address here one key question: what, in this conference, did we learn that is relevant to addressing the major issues faced by our societies in the field of agriculture, very broadly defined? Admittedly, this formulation is very broad and my time is limited. Thus, I deliberately accept the risk of being perceived as superficial as I present here a set of subjective and personal assessments. But such is the price to be paid if one wants to be synthetic, as one must try to be in a synoptic session.

Totally accepting the challenge, raised by Minister Thoko Didiza at the opening of our conference, that "professionals in the field of agricultural economics should guide governments," I will highlight a few points where I believe we, or at least I, learnt something useful for this purpose during this conference. A significant point was on the role of small farms. It is clear that, because our profession, notably thanks to TW Schultz, was right to point out several decades ago that small farms are efficient, we may have oversimplified the case for small farms. Recent experience in Eastern Europe, in Russia in particular, shows that very large farms are able to survive and prosper, while there has been little development of family farms as they exist in Western Europe and even North America. The debates and even controversies on this topic at this conference finally led to a clear conclusion, which is a little more sophisticated than perhaps we would have formulated a few years ago: small farms are often efficient indeed, but small farmers may be very poor; and with economic development, it is good that many small farms disappear. In addition, institutions are very important in determining farm size, something we may have tended to neglect in the past.

The second important lesson we learnt here is that there is hope for Africa. The success stories, which were presented in the special session on African agriculture, are powerful testimonies in this regard. Being relatively familiar with several of them (cassava, cotton, and maize in particular), I know that these successes are indeed very significant. I was much interested to note that in most of these stories, overall success was made possible by prior successes in technology development. Agricultural research played a key role. Why is it then that agricultural research does not receive more attention and more support from policy makers and from the donor community?

We, agricultural economists, may be partly to blame for this. Of course, we often point out the high rates of return to agricultural research investments and conclude that most societies, particularly in sub-Saharan Africa, underinvest in agriculture. We may be right, but we certainly have not been very effective communicators or persuasive policy advisers. Could it be because our analyses are not perceived as socially relevant? Another development in our profession, reflected at this conference, would lead to a similar conclusion. I want to refer to work on total factor productivity (TFP) and on production frontier analysis. I am aware that TFP is an intellectually attractive concept to relate change in total production to change in total production costs. But is such a concept relevant for agricultural situations where many factors of production are directly provided to the firm by the family, so to speak, without any market transaction, thus leaving us without solid reference prices to compute an index of total factor use? As for production frontier analysis, it appears very appealing in that it seems to have direct policy implications for the choice between research and extension investments. After 10 years of direct involvement with those issues at the World Bank, I can, however, testify that the choice between research and extension does not figure among the most difficult policy questions. In most situations both are needed. The main problems have to do with very poor institutions for both research and

extension. How can these institutions be improved? Because of the multiple and ubiquitous problems of perverse incentives, it would seem that economists would have much to contribute to the improvement of research and extension institutions. Yet I did not hear much on this topic at this conference. Perhaps, in this field of technological change we have not been very good at using our static concepts to explain dynamic processes. A counterexample may be provided by the analysis of poverty issues, where the distinction between chronic and transitory poverty—clearly a truly dynamic dimension—was shown to be critical for policy analysis and policy advice.

International trade issues and related domestic policy issues did not play a prominent role in our program. I would like, however, to highlight a point that was made regarding the use of producer support estimate (PSE) indicators in the policy debates. We know that this indicator, promoted by one of ours, Tim Josling, was devised to give an order of magnitude of the support received by farmers because of market interventions, adding together public budget costs and costs borne by consumers through higher prices. Thus, we know, or should have known and said loudly, that PSE indicators are not good measures of trade distortions. Yet, we have let politicians, civil servants, and journalists widely use numbers based on PSE calculations, mainly those of the OECD, to denounce the "subsidies" received by farmers in OECD countries. I suspect this is because, as a profession, we do not like those subsidies. But it remains the case that these numbers are often misleading in the way they are used in international negotiations about trade policy reform. They did contribute, for example, to the recent failure of the WTO Cancun meeting.

Regarding the policy debate, which is a central piece of the policy process, the paper on the media and the information market by Swinnen, McCluskey, and Francken was probably the most interesting of the conference for me. It provided a beautiful explanation of why consumers and citizens are flooded with new information and yet are often very poorly informed. Their analysis of the media is valid well beyond the domain of food safety, for which they provide empirical evidence. The paper is very interesting as it properly emphasizes the role of such an important actor in the policy process, and an actor that had curiously not been subjected to much rigorous economic analytical work in the past. In this respect, I would suggest that it would be worthwhile extending the static approach used and putting it in a dynamic political economy context. I am convinced this would improve our understanding of the public policy process and thereby enhance the social relevance of our professional work.

Michel Petit

In the few minutes that I have for what are necessarily brief remarks, I want to focus on a few points that in my view were some of the highlights of this conference. They were chosen considering their analytical interest and their policy relevance today, and according to what, I anticipate, will be influential in shaping the research agenda of the profession in the near future. Of course, there were other attractive topics discussed, and one person's view of the conference's program is inherently a narrow one, considering the large number of parallel sessions in the program—one does not know what one is missing.

The four issues are the following: (i) The Bruce Gardner "shock" in his Elmhirst Lecture on the direction of causality between agricultural growth, overall growth, and impact on rural and farm incomes. (ii) The debate on the fundamental changes occurring in food systems and their policy implications, particularly for small farmers in developing countries. (iii) The "small farm is beautiful" debate. (iv) The analysis on rural viability and the need for a more territorial focus, in contrast to a traditional sectoral focus.

Before discussing these four points, one suggestion on the structure of the program for future conferences. I would like to see more panel sessions with invited speakers, and the encouragement of greater debate within panels, in addition to some time for questions from the floor.

Briefly on the four issues:

1. Agriculture or economy-wide growth as the engine of rural income growth

Bruce Gardner's thesis in his Elmhirst lecture would appear to contradict most of the empirical work showing agriculture as a major engine of income growth

(and thus rural poverty reduction). Am I right in interpreting Gardner's conclusions as contradicting, for example, Peter Timmer's conclusions in his survey article published in 2002 entitled "Agriculture and Economic Development"?[1] For the United States at least, Gardner concludes "I found real income growth in the non-farm sector to be more fundamentally important in increasing low farm incomes than any specifically agricultural variable." Discussing the experience in East Asia, Gardner concludes that in spite of the strong growth in agricultural productivity experienced in that region, it is apparent that the strong performance of agriculture per se would not have generated a sustainable increase in rural household income. "What is necessary is real average income growth in the economy as a whole." More generally for the developing world, and based on his cross-country regressions, his analysis is consistent with the view that growth in the economy-wide demand for labor is the most critical factor, under which "a growing real wage is a sufficient condition for rural income growth."

Agricultural economists, in my view, have been rather fundamentalist in their belief in the direction of causality—traditionally we look to agriculture as the main engine. Gardner has challenged this view. Is Gardner's work applicable mainly to developed countries? It certainly fits with his own analysis for the United States in his recent book entitled *American Agriculture in the Twentieth Century*.[2] Is it applicable to middle- and low-income countries? The thesis and empirical analysis in his Elmhirst Lecture are important and very provocative, and—rightly so in my view—it will be strongly challenged as it applies to developing countries. I anticipate that it will generate great interest among researchers.

2. On the fundamental changes affecting food systems. A 'shock wave' as presented by T. Reardon

Changes occurring on both the international and the domestic front would seem to have accelerated during

---

[1] Chapter 29 in *Handbook of Agricultural Development*, 2002, B. Gardner and G. Rausser (eds.), North Holland, Amsterdam.

[2] Gardner, B. (2002). *American Agriculture in the Twentieth Century: How It Flourished and What It Cost.* Cambridge, MA: Harvard University Press.

these last decades. This has been associated with trends toward vertical integration and the expansion of information technology. And it has been influenced by globalization, and in developing countries it has been linked directly to domestic economic reforms.

On the *international* front, I would highlight three major issues: (a) declining world prices for farm commodities, (b) WTO negotiations, and (c) regional and bilateral agreements, including what has happened in the EU. On the *domestic* front, which was the novelty in Reardon's paper, we observe the results of trade liberalization, deregulation, privatization and foreign direct investment, all part of the economic reforms. And we observe the effects of the economics of vertical integration.

What I found very interesting in Reardon's study is what he calls the supermarket revolution during the 1990s, and the perhaps related increasing concentration in agro-processing observed both in rich and middle-income countries. This seems to be a long-term structural trend, deeply affecting the conditions under which farmers produce and sell. The impact of these changes is not scale neutral. Specifically, as a result of these forces, small farmers appear to be facing an enormously difficult challenge. The question of the future of small farming in a globalized world would not have been raised 10 years ago. And I expect this will be a fruitful topic for future research. This leads to the third issue of scale and development more generally.

3. The "small farm is beautiful" debate

Ten years ago, whoever challenged the notion that the small farm is beautiful faced a hostile reaction, and was often labeled reactionary. Today, the debate is less ideological, and this calls attention to what seems to be a profound underlying structural change, with important policy implications. It appears that farming is becoming an increasingly more complex business, likely to favor more modern and somewhat larger farm operations. This is due to more open trade regimes for imports, the presence of the so-called supermarket revolution and the increasing concentration in agro-processing, and the increasing capital intensity in farming (Mundlak).[3]

In this conference, we heard presentations by Koester, Otsuka,[4] Hazell, Maxwell, and Fan and Chang-Kang, all addressing this apparent dilemma between the preference for small is beautiful and the structural changes in farming that seem to favor medium and larger-sized firms. The typical statement—of the existence of an inverse relation between farm size and efficiency—that a few years ago was dogma, now no longer closes the debate. It is partly an issue of technical efficiency. It is also an issue of whether or not, below a certain production scale, farms can generate a "livable income." It may also be the case that now it takes larger production units to generate sufficient income to stay in farming, given opportunity costs determined elsewhere in the economy. The prevalence of off-farm employment opportunities in some areas (such as in the United States) has been an essential component of the small farm adjustment issue. But off-farm employment develops selectively, only in certain areas—those with better infrastructure and closer to markets. To expect significant off-farm employment opportunities is not realistic in many areas in poor countries. The debate during the conference on this issue showed strong disagreements, but we all learned much from it. For example, Hazell argued "as countries get richer, farmers get bigger." And he called for not neglecting small farmers: too fast a transition can be a problem. Particularly for middle-income countries, the issue of farm adjustment is currently very complex and politically sensitive. Research on this issue will generate great interest on the part of policy makers.

4. Rural viability and the need for a more territorial approach rather than an overly sectoral focus

Our profession has been biased toward viewing the sector in isolation. If our aim is to revitalize the rural economy, should public investments be targeted at favorable areas, where synergisms and agglomeration economies might generate self-sustained growth? Are there strong, inherent disadvantages in areas with low density of economic activity? What should be the strategy toward less-favored areas? Have we considered the

[3] Mundlak Y. (2001). *Agriculture and Economic Growth*, Harvard University Press: Cambridge, MA.

[4] Kei Otsuka led the discussion on the second plenary session relating to small farms.

possible negative externalities of an excessively rapid urbanization as part of the equation?[5]

It seems to me that surrounding the broad theme of territorial development we have many critical and challenging questions confronting us. Territorial development and the interaction between urban and rural growth are areas of study to which our profession has not contributed significantly beyond platitudes. In my view, our research focus has not been drawn to this issue due to its inherent interdisciplinary nature—it combines urban economics, agricultural economics, geography, and political economy. Although this was not a topic covered by this conference, I believe there were some presentations that are relevant to this theme, which I believe will become more important in the future. I would like to highlight the presentations by Simon Maxwell, who emphasized the need for a multi-sector and territorial approach, and by Alain de Janvry, who stressed a territorial view of rural development.

In closing, I found the conference both interesting and challenging, and an excellent way of catching up with the profession's interests. It was also an opportunity to meet with many colleagues that I do not have the opportunity to see except in the IAAE conferences. My brief comments here present what are, in my view, some of the highlights of the conference. The four areas I mention are themes that should—and I believe will—attract the special attention of researchers in the future.

Alberto Valdés

This conference of the IAAE once more took place on the African continent as the continent and the host country, South Africa, are facing major challenges. In Africa as a whole the challenges include issues such as the implementation of the NEPAD philosophy, good governance, and enhancing the role of private finance—all with the major aim of improving growth and alleviating poverty. In South Africa the challenges are similar, with a strong emphasis on issues related

to land reform, rural poverty, and Black Economic Empowerment—all largely a function of the country's legacy.

Agricultural issues on the continent remain largely focused on rural poverty, food security, the general failure to alleviate poverty, and the stagnation of the rural economy. However, countries now also have to deal with large and growing cities. Increasing inequality also remains a major worry for most countries asking therefore for more well-designed policies to avoid increasing inequality. The argument by Joachim von Braun that we need to unpack the distributional effects of policy should therefore be taken seriously.

The conference raised a number of issues and debates of specific relevance for African agriculture. Questions were raised about the role of agriculture in economic development. Can agricultural and rural growth reduce poverty? Bruce Gardner was rather pessimistic about agriculture's role, but there were also other papers showing the positive contribution of agriculture to GDP growth. The debate nevertheless challenged our standard orthodoxy.

At the same time some doubts were raised about the future of smallholders. Do they stand a chance in an environment characterized by (i) growing industrialization of agriculture, (ii) the growing role of supermarkets, (iii) a stricter application of quality and food safety rules and regulations (SPS, EUREPGAP, etc), and (iv) the failure of current WTO negotiations to reduce developed country subsidization and restrictive market access.

We have also heard that transaction costs are the real culprit inhibiting smallholders' performance. In this respect a number of issues are relevant: (i) How can public good provision in rural areas reduce transaction costs? (ii) More should be explored about the potential role of agribusiness and institutional arrangements such as contract farming in creating market opportunities for smallholders. (iii) There is a new role for small village level cooperatives and other local institutions in reducing transaction costs.

The real issues for African agriculture are set in a context of market failure, missing markets, and high transaction costs with, therefore, a limited opportunity for the application of neoclassical economic models of behavior. There is thus a great need for the application of the New Institutional Economic theory

[5] For discussion on this theme, see the analysis by J. Vernon Henderson, "Urban Primacy, External Costs, and Quality of Life," *Resource and Energy Economics*, 24, 2002, pp. 95–106.

to the problems of Africa. However, it was disappointing to see how few papers at the conference addressed the problems of developing countries from this theoretical perspective. There remains a need to shift paradigms in the agricultural economic discipline if we as agricultural economists want to remain relevant and not to become extinct. I maintain that the lack of an appropriate theoretical framework is the reason why we do poor research, design poor strategy, and why we have limited success with implementation.

It is therefore time to challenge the consensus and to think of a more theoretical framework so that our proceedings at these conferences do not merely reflect orthodoxy but challenge our thinking. I trust that 2006 will live up to this task.

Johann Kirsten

Conference Program

INTERNATIONAL EXECUTIVE COMMITTEE

President IAAE	Joachim von Braun *IFPRI, Washington DC, USA*
President Elect IAAE	Prabhu Pingali *FAO, Rome, Italy*
Past President IAAE	Douglas D. Hedley *Agriculture and Agri-Food Canada*
Secretary-Treasurer IAAE	Walter J. Armbruster *Farm Foundation, USA*
Vice President Program	David Colman *University of Manchester, UK*
Member	Roley Piggott *University of New England, Australia*
Member	Johan van Zyl *Santam (Pty) Ltd, South Africa*
Member	Ling Zhu *Chinese Academy of Social Sciences*
Editor IAAE's Journal	Stephan von Cramon Taubadel *Georg-August University, Germany*

CONFERENCE PROGRAM COMMITTEE

Program Co-ordinator	David Colman *University of Manchester, United Kingdom*
Contributed Paper Organizers	Michele Veeman and Terry Veeman *University of Alberta, Canada*
Poster Paper Organizer	Keijiro Otsuka *GRIPS/FASID Joint Graduate Program, Japan*
Mini-symposia and Discussion Paper Organizer	Wallace E. Tyner *Purdue University, United States of America*
Computer Presentation Organizer	Peter Wehrheim *University of Bonn, Germany*

NATIONAL STEERING COMMITTEE

Bongiwe Njobe	National Dept of Agriculture (Chairperson)
Johan van Zyl	Santam (Pty) Ltd
Johan van Rooyen	South African Wine and Brandy Company
Tamás Fenyes	Vista University
Moraka Makhura	Development Bank of Southern Africa
Gerhard Backeberg	Water Research Commission

Gerhard Coetzee	Ebony Consulting International
Bertus van Heerden	Standard Bank of South Africa
André Louw	ABSA Bank
Johann Kirsten	University of Pretoria
Charles Machethe	University of Pretoria
Nick Vink	University of Stellenbosch
Dirk Troskie	Department of Agriculture, Western Cape
André Jooste	University of the Free State
Herman van Schalkwyk	University of the Free State
Gerald Ortmann	University of Natal
Ruvimbo Chimedza	University of Zimbabwe
T.B. Seleka	University of Botswana
Bankies Malan	Landbank
Graham Moor	Cane Growers Association

SOUTH AFRICAN ORGANIZING COMMITTEE

Gerhard Coetzee	Ebony Consulting International (Pty) Ltd and University of Pretoria (Co-Chair)
Johann Kirsten	University of Pretoria (Co-chair)
Melanie Campbell	Event Dynamics (Pty) Ltd

PROGRAM

SATURDAY 16 AUGUST 2003
THE DAY AT A GLANCE

08h00 – 09h00	Workshops and Conference Registration	ICC Foyer
09h00 – 18h00	Learning Workshop 1: Analytical and Empirical Tools for Poverty Research	Meeting Room 12AB
09h00 – 17h00	Learning Workshop 2: Food Security Measurement in a Developing World Context with a Focus on Africa	Meeting Room 21ABC
09h00 – 17h30	Learning Workshop 3: Water Reforms, Institutions' Performance, Allocation, Pricing, and Resource Accounting	Meeting Room 22ABC

SATURDAY 16 AUGUST 2003

Learning Workshop 1: Analytical and Empirical Tools for Poverty Research

Venue: **Meeting Room 12AB**

Organizers: Chris Barrett (Cornell University) and Csaba Csaki (World Bank)

08h00 – 09h00   *Workshop/Conference Registration*
09h00 – 09h05   Introductory Remarks (Chris Barrett and Csaba Csaki)

09h05 – 10h05	**Michael Carter (University of Wisconsin-Madison)**
	Poverty dynamics: An overview of theory and empirical methods using panel data.
10h05 – 11h05	**Chris Barrett (Cornell University)**
	Integrating quantitative and qualitative poverty analysis tools.
11h05 – 11h30	*Refreshment break*
11h30 – 12h30	**Steve Younger (Cornell University)**
	Welfare comparisons across different measures: concepts and methods.
12h30 – 13h45	*Lunch (Meeting Room 12CDE)*
13h45 – 14h45	**Berk Ozler (World Bank)**
	Poverty mapping: integrating survey and census data to generate more spatially comprehensive poverty assessments.
14h45 – 15h45	**Stefan Dercon (University of Oxford)**
	Dynamic vulnerability analysis using panel data.
15h45 – 16h45	*Refreshment break*
16h45 – 17h45	**Jock Anderson (World Bank), Gershon Feder (World Bank), Peter Hazell (IFPRI), Kei Otsuka (FASID, Japan), Tom Reardon (Michigan State University)**
	Panel discussion on current thinking on poverty reduction policy and rural development.
17h45 – 18h00	Closing Remarks (Csaba Csaki and Chris Barrett)

Learning Workshop 2: Food Security Measurement in a Developing World Context with a Focus on Africa

Venue:	**Meeting Room 21 ABC**
Organizers:	Susan Offutt (Economic Research Service, USDA)
08h00 – 09h00	*Workshop/Conference Registration*
09h00 – 09h15	Introductory Remarks (Susan Offutt)
09h15 – 10h15	**Why Do We Measure Food Insecurity?**
	Moderator—Jim Blaylock (Economic Research Service, USDA)
	William Meyers (Iowa State University)
	Understanding Food Security
	Johann Kirsten (University of Pretoria) and Nick Vink (University of Stellenbosch)
	Food Security and Prices in South Africa
10h15 – 10h30	*Refreshment break*
10h30 – 12h00	**Measuring Food Insecurity**
	Moderator—Nicole Ballenger (Economic Research Service, USDA)
	Jorge Mernies (Food and Agriculture Organization)
	FAO Food Balance Sheets.
	Mark Nord (Economic Research Service, USDA)
	The UNIVERSITYS. Household Food Security Survey Module.
	Lisa Smith (International Food Policy Research Institute)
	Measuring Food Insecurity in Rural and Urban Areas.
	Walter Odhiambo (University of Nairobi)
	Discussant(s)

12h00 – 13h00	*Lunch (Meeting Room 21DEFG)*
13h00 – 14h45	**Food Insecurity in Africa.**

Moderator—John Dunmore (Economic Research Service, USDA)

Simon Maxwell (Overseas Development Institute)
An Overview
Shahla Shapouri (Economic Research Service, USDA)
A Cross-Country Comparison with Food Balance Sheets.
David Sahn (Cornell University)
A Multi-Dimensional Look At Well-Being.
Siméon Nanama (Cornell University)
Understanding and Measuring Household Food Insecurity in
    Northern Burkina Faso.
Mark Nord (Economic Research Service, USDA)
Discussant(s)

14h45 – 15h00	*Refreshment break*
16h00 – 17h30	**A Discussion of Ways to Improve Data Collection.**

Moderator—Mary Bohman (Economic Research Service, USDA)

Vincent Ngendakuma (Food and Agriculture Organization)
Challenges faced in capacity building.
Eloi Ouedraogo, Expert en Statistique Agricole (AFRISTAT)
The data collection experience at the regional level.
**John Mngodo (Strategic Grain Reserve, Tanzania Ministry of
    Agriculture and Food Security)**
The data collection experiences at the country level.
Larry Sivers (National Agricultural Statistics Service, USDA)
Outside technical assistance in data collection.

Bentry Chaura, Senior Statistician SADC, Zimbabwe)
Discussant(s)

16h30 – 17h00	Concluding Remarks—Joaquim Von Braun (International Food Policy Research Institute)

Learning Workshop 3: Water Reforms, Institutions' Performance, Allocation, Pricing, and Resource Accounting

Venue:	**Meeting Room 22ABC**
Organizers:	David Zilberman, Ujjayant Chakravorty, R. Maria Saleth, Ariel Dinar, and Rashid Hassan
08h00 – 09h00	*Workshop/Conference Registration*
09h00 – 09h45	*Water Reforms*

**David Zilberman (University of California Berkeley),
    Alberto Garrido (Universidad Politécnica de Madrid, Spain)**
The Economics of Water Reform.

09h45 – 10h30	*Water Institutions*

R. Maria Saleth (IWMI) and Ariel Dinar (World Bank)
Water Institutions and Sector Performance.
Ariel Dinar (World Bank) and R. Maria Saleth (IWMI)
Water Institutions Health Index.

10h30 – 10h45	*Refreshment break*
10h45 – 12h15	**Water Allocation and Pricing**

**Ujjayant Chakravorty (Emory University), Yacov Tsur
(Hebrew University of Jerusalem) and Renan Goetz
(University of Girona)**
Spatial, Dynamic and Risk Considerations in Water Allocation and
Pricing.

12h15 – 13h30	*Lunch (Meeting Room 22DEFG)*

Marie Livingston (University of Northern Colorado)
Evaluating Institutional Change: Methodological Issues at Micro and
Meso Levels. Luncheon Speaker.

13h30 – 15h00	**Water Resource Accounting**

Alessandra Alfieri (United Nations Statistical Division)
Managing Water Resources: The Natural Resource Accounting
Approach.

Rashid Hassan (University of Pretoria)
Water Resource Accounts for Namibia, Botswana and South Africa:
A comparative assessment of water allocation and use.

Glenn-Marie Lange (New York University)
Regional Resource Accounts for Water in Southern Africa: The Case
of the Orange River Catchments.

15h00 – 16h00	**Water Institutions (B)**

Jennifer McKay (University of Southern Australia)
Progress in Institutional Reforms in the Water Sector of Australia.

Robert Hearne (North Dakota State U)
Progress in Institutional Reforms in the Water Sector of Mexico.

**Rachid Doukkali (Institute Agronomique et Veterinaire
Hassan II)**
Progress in Institutional Reforms in the Water Sector of Morocco.

Madar Samad (IWMI)
Progress in Institutional Reforms in the Water Sector of Sri Lanka.

16h00 – 16h15	*Refreshment break*
16h15 – 17h30	**Resource Accounting and Country Institutional Experience:** **Regional Plans for Africa**

A round table chaired by Doug Merrey and Madar Samad (IWMI)

Gerhard Backeberg (Water Research Commission)
Progress in Institutional Reforms in the Water Sector of South Africa.
Piet Heyns (Namibia Ministry of Agriculture)
Progress in Institutional Reforms in the Water Sector of Namibia.
Rashid Hassan (University of Pretoria)
Institutionalizing Water Resource Accounting Work within the
SADC Region.
Glenn-Marie Lange (New York University)
Regional Water Management Strategies for Africa.

SUNDAY 17 AUGUST 2003		
THE DAY AT A GLANCE		

08h30 – 10h00	Conference registration	ICC Foyer
10h00 – 11h00	Opening Ceremony	Hall 2CDE
11h00 – 11h30	Refreshment Break	Hall 3A
11h30 – 13h00	IAAE President's Address and Elmhirst Lecture	Hall 2CDE
13h00 – 14h00	Lunch	Hall 3BC
14h00 – 15h30	Plenary Session 1:	Hall 2CDE
	1. Chris Barrett	
	2. Simon Maxwell	
	3. Alain de Janvry	
15h30 – 16h00	Refreshment Break	Hall 3A
16h00 – 17h30	Contributed papers session 1:	
	11. Economic Impacts of AIDS/Premature Mortality	Hall 2CDE
	12. Valuation Methods for Agricultural Producst	Meeting Room 11AB
	13. The Economics and Politics of Agricultural Policy Reform	Meeting Room 11CD
	14. Assets and Asset Markets in Development	Meeting Room 12AB
	15. Conceptual Tools in Resource and Environmental Management	Meeting Room 12CD
	16. Agricultural Production Issues in Transition Economics	Meeting Room 22DEF
	17. Conceptual and Empirical Issues in Quantitative Policy Analysis	Meeting Room 22ABC
	18. Economic Adjustment to Trade Liberalization	Meeting Room 21DEF
	19. Assessing the Impacts of Adopting Genetically Modified Crops	Meeting Room 21ABC
18h00 – 19h00	Busses to depart for the City Hall	
19h00 – 21h00	Welcome Reception	City Hall, Durban

SUNDAY 17 AUGUST 2003		

08h30 – 18h00	Registration and Travel desk opening hours	
10h00 – 11h00	**Opening Ceremony**	**Hall 2CDE**
11h00 – 11h30	*Refreshment break*	
11h30 – 13h00	IAAE President's Address and Elmhirst Lecture	Hall 2CDE
13h00 – 14h00	*Lunch*	Hall 3BC
14h00 – 15h30	**Plenary Session 1**	**Hall 2CDE**
	Chairperson: Akin Adesina	
	Discussant: Deryke Belshaw	
	Speakers: **Chris Barrett**	
	Rural Poverty Dynamics: Development Policy	
	Implications	
	Simon Maxwell	
	Can we rescue rural development before it's too late?	
	Alain de Janvry	
	Achieving Success in Rural Development: Toward	
	Implementation of an Integral Approach	
15h30 – 16h00	*Refreshment break*	

16h00 – 17h30   **Contributed papers session 1**

11   **Economic Impacts of AIDS/Premature Mortality**   **HALL2CDE**
     Chairperson: Peter Hazell
     Investing in Hope: AIDS, Life Expectancy, and Human Capital
       Accumulation
     Rui Huang, Lilyan, E. Fulginiti & E. Wesley Peterson (USA)
     Prime-Age Adult Morbidity and Mortality in Rural Rwanda: Which
       Households are Affected and What are Their Strategies for
       Adjustment?
     *Cynthia Donovan, Linda Bailey, Edson Mpyisi, & Michael Weber
       (USA)*
     Measuring the Impacts of Prime-Age Adult Death on Rural
       Households in Kenya
     Takashi Yamano, T.S. Jayne & Melody McNeil (Japan)

12   **Valuation Methods for Agricultural Products**   **Meeting Room 11AB**
     Chairperson: Ellen Goddard
     An Experimental Approach to Valuing New Differentiated Products
     *Leigh J. Maynard, Jason , G. Hartell, A. Lee Meyer & Jianqiang
       Hao (USA)*
     SC-X: Calibrating Stated Choice Surveys with Experimental Auction
       Markets
     Frode Alfnes & Kyrre Rickertsen (Norway)
     Measuring Quantity-Constrained and Maximum Prices Consumers
       are Willing to Pay for Quality Improvements: The Case of Organic
       Beef Meat
     Alessandro Corsi & Silvia Novelli (Italy)

13   **The Economics and Politics of Agricultural Policy Reform**   **Meeting Room 11CD**
     Chairperson: Lars Brink
     Policy Dependency and Reform: Economic Gains versus Political
       Pains
     David R. Harvey (UK)
     Effects of EU Dairy Policy Reform for Dutch Agriculture and
       Economy; Applying an Agricultural Programming/Mixed
       Input-Output Model
     Jack Peerlings & John Helming (The Netherlands)
     EU Enlargement: A New Dimension
     P. Salamon, M. Brockmeier, C.A. Herok & O. van Ledebur (Germany)

14   **Assets and Asset Markets in Development**   **Meeting Room 12AB**
     Chairperson: Alain de Janvry
     Cattle as Assets: Assessment of Non-Market Benefits from Cattle in
       Smallholder Kenyan Crop-Livestock Systems
     Emily A. Ouma, Steven J. Staal & Gideon A. Obare (Kenya)
     Land Markets in Uganda: Incidence, Impact, and Evolution Over Time
     Paul Mpuga & Klaus Deininger (USA)
     Land Rental Markets as an Alternative to Government Reallocation?
       Equity and Efficiency Considerations in the Chinese Land Tenure
       System
     Songqing Jin & Klaus Deininger (USA)

15 **Conceptual Tools in Resource and Environmental Management** **Meeting Room 12CD**
Chairperson: Klaus Frohberg
DEA Based Procurement Design in Natural Resource Management
Peter Bogetoft & Kurt Nielsen (Denmark)
Materials Balance Based Modelling of Environmental Efficiency
Ludwig Lauwers & Guido Van Huylenbroeck (Belgium)
Optimal Economic Length of Leys: A Dynamic Programming
    Approach
*Agnar Hegrenes, Gudbrand Lien & Anders Ringgard Kristensen
    (Norway)*

16 **Agricultural Production Issues in Transition Economics** **Meeting Room 22DEF**
Chairperson: Csaba Forgacs
Liquidity and Productivity in Russian Agriculture: Farm-data
    Evidence
Irina Bezlepkina & Alfons Oude Lansink (The Netherlands)
Technical Efficiency of Polish Farms: Estimation According to
    Specialisation and Lessons from Confidence Intervals
*Laure Latruffe, Kelvin Balcombe, Sofia Davidova & Katarzyna
    Zawalinska (France)*
Courts and Contract Enforcement in Transition Agriculture- Theory
    and Evidence from Poland
Volker Beckmann & Silke Boger (Germany)

17 **Conceptual and Empirical Issues in Quantitative Policy Analysis** **Meeting Room 22ABC**
Chairperson: Luca Salvatici
Structural Efficiency and Income Effects of Direct Payments: An
    Agent-Based Analysis of Alternative Payment Schemes for the
    German Region Hohenlohe
Kathrin Happe & Alfons Balmann (Germany)
Testing for Efficiency: A Policy Analysis with Probability
    Distributions
*Klaus Salhofer, Erwin Schmid, Gerhard Streicher & Friedrich
    Schneider (Austria)*
The Calibration of Incomplete Demand Systems in Quantitative
    Policy Analysis
Sophie Drogue, Jean Christophe Bureau & John C. Beghin, (France)

18 **Economic Adjustment to Trade Liberalization** **Meeting Room 21DEF**
Chairperson: Harry de Gorter
The Implications of World Trade Liberalization on Trade and Food
    Security: A Case Study of Sudan
Imad E. E. Abdel Karim & Dieter Kirschke (Germany)
The Nature of Distortions to Agricultural Incentives in China and
    Implications of WTO Accession
Scott Rozelle & Jikun Huang (USA)
A Case Against the Simultaneous Use of Market Access
    Restrictions, Domestic Support, and Export Subsidies
Roberto J. Garcia (Norway)

19 **Assessing the Impacts of Adopting Genetically Modified Crops**   **Meeting Room 21ABC**
Chairperson: Matin Qaim
Can GM-Technologies Help African Smallholders? The Impact of Bt
   Cotton in the Makhathini Flats of KwaZulu-Natal
Lindie Jenkins nee Beyers & Colin Thirtle (South Africa)
Assessing the Potential Impact of Bt Maize in Kenya Using a GIS
   Based Model
Hugo de Groote, Bill Overholt, James Ouma Okuro & Stephen Mugo
   (Kenya)
Simulating the Effects of Adoption of Genetically Modified
   Soybeans in the U.S.
Denis A. Nadolnyak & Ian M. Sheldon (USA)

19h00 – 21h00   **Welcome Reception**
Hosted by His Worship the Ethekwini Municipality Mayor,
   Councillor Obed Mlaba, at the Durban City Hall.

MONDAY 18 AUGUST 2003
THE DAY AT A GLANCE

08h30 – 10h30	Plenary Session 2:	Hall 2CDE
	1. Ulrich Koester	
	2. Dan Sumner	
	3. Peter Hazell	
	4. Shenggen Fan	
10h30 – 11h00	Refreshment Break	Hall 3A
11h00 – 12h30	Poster Session 1:	Hall 3A
12h30 – 13h30	Lunch	Hall 3BC
13h30 – 15h30	Invited Panel Session 1:	
	1. Agricultural success stories from SSA	Hall 2CDE
	2. Agricultural Development Retrospective	Meeting Room 11AB
	3. Socially responsible Agri-business	Meeting Room 11CD
	4. The Right for Food	Meeting Room 12AB
	5. Efficiency versus Equity in Land Reform	Meeting Room 12CD
	6. Water Policy and Agricultural Production at Global and River Basin Levels	Meeting Room 22DEF
	7. Agricultural R&D in the Developing World	Meeting Room 22ABC
	8. The Rise of Supermarkets in Developing Countries: Effects on Agri-food systems and the rural poor	Meeting Room 21DEF
15h30 – 16h00	Refreshment Break	Hall 3A
16h00 – 17h20	Discussion Groups 1	Meeting Room 11E
		Meeting Room 12E
		Meeting Room 22G
		Meeting Room 21G

Mini-symposia 1	Hall 2CDE
	Meeting Room 11AB
	Meeting Room 11CD
	Meeting Room 12AB
	Meeting Room 12CD
	Meeting Room 22DEF
	Meeting Room 22ABC
	Meeting Room 21DEF
	Meeting Room 21ABC

17h30 – 19h00	Contributed papers session 2:	
	21. Agriculture Production and Productivity in Sub-Saharan Africa	Hall 2CDE
	22. Risk Management in Agricultural Production and Marketing	Meeting Room 11AB
	23. Price and Marketing Issues in Developed Nations	Meeting Room 11CD
	24. Valuation of Attributes: Hedonic and Conjoint Analysis	Meeting Room 12AB
	25. Efficiency Analysis in Agricultural Production	Meeting Room 12CD
	26. Labor Adjustment and Employment in Transition	Meeting Room 22DEF
	27. Mitigation of Environmental Damage: GHG Emissions/Nitrates	Meeting Room 22ABC
	28. Adoption of Land-Enhancing Technologies	Meeting Room 21DEF
	29. Production and Trade under Alternative policy Regimes	Meeting Room 21ABC
19h00	Council Meeting 1	IAAE Office

MONDAY 18 AUGUST 2003

08h30 – 18h00	Registration and Travel desk opening hours	
08h30 – 10h30	**Plenary Session 2**	**Hall 2CDE**
	Chairperson: Lieb Nieuwoudt	
	Discussant: Keijiro Otsuka	
	Speakers: **Ulrich Koester**	

A Revival of Large Farms in Eastern Europe? How important are institutions?

Dan Sumner

Size relationships and policy in UNIVERSITYS. agriculture, with emphasis on the dairy industry

Peter Hazell

Ensuring a Future for Small Farms in the Developing World

Shenggen Fan

Farm size and efficiency: Asian experience

10h30 – 11h00	*Refreshment break*	
11h00 – 12h30	**Poster Session 1**	**Hall 3A**

Poverty Reduction and Economic Development
Group: 1-1-1   Chairperson: Madhusudan Bhattarai
Environmental Stewardship including Natural Resource Management
Group: 2-1-1   Chairperson: Eric Jessup
Group: 2-1-2   Chairperson: Tom P. Tomich

Efficiency in Food and Farming Systems
Group: 3-1-1   Chairperson: Eberhard Schulze
Group: 3-1-2   Chairperson: Matin Qaim
Group: 3-1-3   Chairperson: Valentina Hartarska
Group: 3-1-4   Chairperson: M. F. Viljoen
Group: 3-1-5   Chairperson: Timothy J. Dalton
Consumer Safety and Food Security
Group: 4-1-1   Chairperson: Wojciech J Florkowski
Group: 4-1-2   Chairperson: D. K. Grover
Policy Analysis
Group: 5-1-1   Chairperson: Claudio Soregaroli
Group: 5-1-2   Chairperson: Allessandro Olper
Market and Trade Analysis
Group: 6-1-1   Chairperson: Christian Fischer
Group: 6-1-2   Chairperson: Sven Anders

12h30 – 13h30   *Lunch*
13h30 – 15h30   **Invited Panel Session 1**

Agricultural success stories from SSA   **Hall 2CDE**
Chairperson: Mandivamba Rukuni (Kellogg Foundation)
Speakers: Felix Nweke
             The cassava green revolution in Nigeria
             Margaret Ngigi
             Smallholder dairying in Kenya
             Josue Dione
             Smallholder cotton in Mali
             Stan Heri
             Horticulture success for smallholders in Zimbabwe
             Joseph Rusike
             The maize green revolution in east and southern Afirca

Agricultural Development Retrospective   **Meeting Room 11AB**
Chairperson: Bob Evenson (University of Yale)
Discussant: Stefan Dercon
Speakers: Douglas Gollin (Williams College, USA)
             *The Agricultural Producer and Agricultural
                Productivity*
             Duncan Thomas (UCLA; RAND Corporation)
             *The Agricultural Household and the Agricultural
                Community*
             Terry Roe (University of Minnesota)
             *The Political Economy of Support for Public Goods
                and Public Policy*

Socially responsible Agri-business   **Meeting Room 11CD**
Organizers: Mike Cook (IAMA) and Johan van Rooyen (South
                African Wine and Brandy Company)
Chairperson: Jan van Roekel (CEO, Cokon)
Speakers: Mumeka Mwanamwamba-Wright (Zambia)
             Michael L. Cook (University of Missouri)
             Johan van Rooyen (President and CEO South African
                Wine and Brandy Company)

The Right for Food **Meeting Room 12AB**
Chairperson: Lawrence Haddad (IFPRI)
Speakers: Jean Ziegler (UN Special Rapporteur for the Right to
         Food, Charlotte Mcclain (South African
         Commission for Human Rights, South Africa)
         The Right to Food—Implications for Africa and NEPAD
         Daniel W. Bromley (University of Wisconsin-Madison)
         An Economist's Perspective on the Right to Food

Efficiency versus Equity in Land Reform **Meeting Room 12CD**
Chairperson: Clem Tisdell
Discussant: Mike Lyne
Speaker: Danie Cilliers (ABSA—Agricultural Bank of South
         Africa)
         Sophia Davidova (Imperial College at Wye)
         Innocent Maatshe (University of Zimbabwe)
         John Sender (Leiden University)
         Unconvincing Arguments for Land Reform

Water Policy and Agricultural Production at Global and River **Meeting Room 22DEF**
  Basin Levels
Organizer: Mark Rosegrant (CGIAR)
Chairperson: Consuelo Varela-Ortega
Speakers: Mark Rosegrant (IFPRI), Ximing Cai (IFPRI) and
         Sarah Cline (IFPRI)
         Alternative Policies for Global Irrigation and Water:
           Impacts on Food Supply, Demand and Trade
         Claudia Ringler (IFPRI) and collaborators
         Water Allocation Policies for the Dong Nai River
           Basin, Vietnam: An Integrated Perspective
         Charles Rodgers (IFPRI) and collaborators
         Water Allocation and pricing strategies in the Brantas
           Basin, East Java, Indonesia
         Maria Iskandarani, Stefanie Engel and Maria del Pilar
         Useche (University of Bonn)
         Household Water Security in the Ghanaian Volta
           Basin: Why do people not rely on better domestic
           water supplies?

Agricultural R&D in the Developing World **Meeting Room 22ABC**
Chairperson: Philip Pardey (University of Minnesota)
Speakers: Nienke Beintema (IFPR and ISNAR)
         A New Look at African Agricultural Research
         Frikkie Liebenberg (ARC Pretoria)
         Agricultural R&D in South Africa
         Ruben Echeverria (IADB Washington D.C.)
         Reshaping Agricultural R&D in Latin America
         Suresh Pal (NCAP Delhi)
         Agricultural Research System Reforms and
           Management of Intellectual Property in India.
         Bruce Gardner (University of Maryland)
         International Roles for Agricultural Research

The Rise of Supermarkets in Developing Countries: Effects on   **Meeting Room 21 DEF**
agri-food systems and the rural poor
Organizer: Tom Reardon
Chairperson: Herman van Schalkwyk
Speakers: Thomas Reardon (Michigan State University)
                Overview
                Liesbeth Dries and Jo Swinnen (University of Leuven)
                The Rise of Supermarkets in Central and Eastern
                    Europe.
                Thomas Reardon, and Julio Berdegue (RIMISP, Chile)
                Supermarkets in Latin America and Asia; effects on
                    produce markets and farmers.
                Dave Weatherspoon and Thomas Reardon (Michigan
                    State University)
                Supermarkets in South Africa and implications for the
                    rural poor.
                Jean-Marie Codron and Zouhair Bouhsina (INRA
                    France) and Fatiha Fort (ENSA Montpellier)
                Supermarkets in Morocco and Turkey: fresh produce
                    procurement systems.
15h30 – 16h00 *Refreshment break*
16h00 – 17h20 **Discussion Groups 1**

Livestock Research Need for Developing Countries   **Meeting Room 11E**
Chris Delgado (IFPRI) and Guillaume Duteurte (CIRAD)
Capacity Building for Policy Analysis in Developing Countries **Meeting Room 12E**
Rachid Doukkali (IAV-Hassan II, Morocco) and Tancréde Voituriez
    (CIRAD)
Policies to Promote Integrated Pest Management in Africa   **Meeting Room 22G**
Hermann Waibel (Univ. of Hanover)
Agent Based Modeling in Agricultural and Resource Economics   **Meeting Room 21G**
Alfons Balmann (Institute of Agricultural Development in Central
    and Eastern Europe)

Mini-symposia 1

Poverty Impacts and Policy Options of Non-Farm Rural   **Hall 2CDE**
    Employment
Gertrud Buchenrieder (Univ. of Hohenheim)
The Environmental Role of Agriculture in Developing Countries: **Meeting Room 11AB**
    Economic Valuation of Selected Environmental Externalities
    of Alternative Farming Patterns
Frédéric Dévé (Roles of Agriculture project, FAO)
Problems of Outreach in Rural Finance Risk-reducing   **Meeting Room 11CD**
    Mechanisms for Rural Finance Understanding and Improving
    the Impact of Financial Services
Richard Meyer (Ohio State University)
Progress with Biotechnology and GM Crops in Developing   **Meeting Room 12AB**
    Countries
Colin Thirtle (Imperial College London)

Soil Fertility and Food Security for the Poor in Southern Africa: Technical, Policy, and Institutional Challenges	**Meeting Room 12CD**

Mulugetta Mekuria (CIMMYT, Southern Africa)

Market Research for the Development of Commercialized Agriculture in Sub-Saharan Africa	**Meeting Room 22DEF**

Jim Gockowski, Victor Manyong, Patrick Kormawa, Steffen Abele, and Shaun Ferris (IITA)

Evaluation and Rural Development Programs	**Meeting Room 22ABC**

Benjamin Davis (ESA) and Sudhanshu Handa (Inter-American Development Bank)

Economic Evaluation of New Institutional Mechanisms for Funding and Delivery of Agricultural Extension in Developing Countries	**Meeting Room 21DEF**

Derek Byerlee and David Nelson (World Bank)

Is OECD Agricultural Support Stunting Agricultural Development and Poverty Alleviation in Developing Countries?	**Meeting Room 21ABC**

William Meyers (Iowa State University)

17h30 – 19h00   **Contributed papers sessions 2**

21   **Agriculture Production and Productivity in Sub-Saharan Africa**   **HALL2CDE**
Chairperson: Johann Kirsten
Explaining The Failure of Agricultural Production in Sub-Saharan Africa
Guy Blaise Nkamleu & Jim Gockowski (Cote d'Ivoire)
Institutions and Agricultural Productivity in Sub-Sahara Africa
Bingxin Yu, Lilyan E. Fulginiti & Richard K. Perrin (USA)
Cotton Sector Policies and Performance in Sub-Saharan Africa: Lessons Behind the Numbers in Mozambique and Zambia
Duncan Boughton, David L. Tschirley, Afonso Osorio Ofico, Higino M. Marrule & Ballard Zulu (USA)

22   **Risk Management in Agricultural Production and Marketing**   **Meeting Room 11AB**
Chairperson: Alexander Sarris
Is China's Agricultural Futures Market Efficient?
H. Holly Wang & Bingfan Ke (USA)
The Effects of Spot Water Markets on the Economic Risk Derived from Variable Water Supply
Javier Calatrava & Alberto Garrido (Spain)
Weather Derivatives: Efficient Risk Management Tools
Timothy J. Richards & Kurt K. Klein (USA)

23   **Price and Marketing Issues in Developed Nations**   **Meeting Room 11CD**
Chairperson: Alessandro Corsi
The Impact of Food Scares on Price Transmission in Inter-Related Markets
Tim Lloyd, Steve McCorriston, Wyn Morgan & Tony Rayner (UK)
Retail Sales: Do They Mean Reduced Expenditures? German Grocery Evidences
Jens-Peter, Loy & Robert D. Weaver (Germany)

Price Variability or Rigidity in the Food Retailing Sector?
Theoretical Analysis and Evidence from German Scanner Data
Roland Herrmann & Anke Moser (Germany)

24   Valuation of Attributes: Hedonic and Conjoint Analysis   **Meeting Room 12AB**
Chairperson: Euan Fleming
A Hedonic Model of Rice Traits: Economic Values from Farmers in
West Africa
Timothy J. Dalton (USA)
New World Wine in the New World: A Hedonic Analysis of
Reputation and Quality Signaling
Gunter Schamel (Germany)
Consumer Willingness to Pay for Multiple Attributes of Organic
Rice: A Case Study in the Philippines
Shihomi Ara (USA)

25   Efficiency Analysis in Agricultural Production   **Meeting Room 12CD**
Chairperson: Tim Coelli
Profit Efficiency Among Bangladeshi Rice Farmers
Sanzidur Rahman (United Kingdom)
Farm Size and the Determinants of Productive Efficiency in the
Brazilian Center-West
Steven M. Helfand (USA)
Measuring the Impact of Ethiopia's New Extension Program on the
Productive Efficiency of Farmers
Arega D. Alene & Rashid M. Hassan (South Africa)

26   Labor Adjustment and Employment in Transition Economies   **Meeting Room 22DEF**
Chairperson: Klaus Deininger
Agricultural Adjustment and the Diversification of Farm Households
in Central Europe
Hannah Chaplin, Sophia Davidova & Matthew Gorton (UK)
Intertemporal Analysis of Employment Decisions on Agricultural
Holdings in Slovenia
Luka Juvancic & Emil Erjavec (Slovenia)
Human Capital and Labor Flows Out of the Agricultural Sector:
Evidence from Slovenia
Liesbeth Dries, Stefan Bojnec & Johan F. M. Swinnen (Belgium)

27   Mitigation of Environmental Damage: GHG Emissions/Nitrates   **Meeting Room 22ABC**
Chairperson: Ross Vani
Mitigation of Greenhouse Gas Emissions: The Impacts on a
Developed Country Highly *Dependent on Agriculture*
Anita Wreford & Caroline Saunders (New Zealand)
Non-CO2 Greenhouse Gas Emissions from Agriculture: Analysing
the Room for Manoeuvre for Mitigation, in Case of Carbon Pricing
Daniel Deybe & Abigail Fallot (France)
Marginal Abatement Costs for Reducing Leaching of Nitrates in
Croatian Farming Systems
*John Sumelius, Zoran Grgic, Milan Mesic & Ramona Franic
(Finland)*

28   Adoption of Land-Enhancing Technologies   **Meeting Room 21DEF**
Chairperson: John Pender
Productivity and Land Enhancing Technologies in Northern
   Ethiopia: Health, Public Investments, and Sequential Adoption
Lire Ersado, Gregory S. Amacher & Jeffrey Alwang (USA)
The Determinants of Adoption of Sustainable Agriculture
   Technologies: Evidence from the Hillsides of Honduras
David R. Lee & Peter Arellanes (USA)
The Blending of Participatory Research and Quantitative Methods:
   Wealth Status, Gender and the Adoption of Improved Fallows in
   Zambia
*Steven Franzel, Donald Phiri, Paramu Mafongoya, Isaac Jere,
   Roza Katanga & Stanslous Phiri (Kenya)*

29   Production and Trade Under Alternative Policy Regimes   **Meeting Room 21ABC**
Chairperson: David R. Harvey
World Cereals Markets Under Alternative Common Agricultural
   Policy Reforms
C. Benjamin, C. Gueguen, M. Houee & A. Hess-Miglioretti (France)
Reforming the CAP: A Partial Equilibrium Analysis of the MTR
   Proposals
Julian Binfield, Patrick Westhoff & Robert Young (USA)
Trade Effects of Dairy Pricing Arrangements
*Pavel Vavra, Jesus Anton-Lopez, Nobunori Kuga & Joe Dewbre
   (France)*

19h00   **Council Meeting 1**   **IAAE Office**

TUESDAY 19 AUGUST 2003
THE DAY AT A GLANCE

08h30 – 10h30	Plenary Session 3:	Hall 2CDE
	1. Jean Kinsey	
	2. Betsey Kuhn	
	3. Johan Swinnen	
	4. Vittorio Santaniello	
10h30 – 11h00	Refreshment Break	Hall 3A
11h00 – 12h30	Poster Session 2	Hall 3A
11h30 – 12h50	Computer Session (Consecutive part 1):	
	Multi Agent Systems	Meeting Room 11AB
	Tools for Spatial Analysis in Agricultural Economics	Meeting Room 11CD
12h30 – 13h30	Lunch	Hall 3BC
13h30 – 15h30	Contributed papers session 3:	
	31. Health and Schooling Issues in Development	Hall 2CDE
	32. Food Quality and Safety	Meeting Room 11AB
	33. Productivity, Technological change and Growth	Meeting Room 11CD
	34. Agricultural Trade Policy Reform	Meeting Room 12AB
	35. Production, Employment and Efficiency in LDCs	Meeting Room 12CD
	36. Economic Issues in Transition and Developing Economics	Meeting Room 22DEF
	37. The Economics of Land and Environmental Degradation	Meeting Room 22ABC
	38. Water, Irrigation and Agriculture	Meeting Room 21DEF

15h30 – 16h00	Refreshment Break	Hall 3A
16h00 – 17h20	Organized Session by the European Association of Agricultural Economists	Hall 2CDE
	Computer Sessions (Consecutive part 2)	
	Multi Agent Models	Meeting Room 11AB
	Tools for Spatial Analysis in Agricultural Economics	Meeting Room 11CD
	Tools for Agricultural Trade Modeling	Meeting Room 12AB
	Selected issues	Meeting Room 12CD
17h30 – 19h00	Discussion Groups 2	Meeting Room 11E
		Meeting Room 12E
		Meeting Room 22G
		Meeting Room 21G
	Mini-symposia 2	Hall 2CDE
		Meeting Room 11AB
		Meeting Room 11CD
		Meeting Room 12AB
		Meeting Room 12CD
		Meeting Room 22DEF
		Meeting Room 22ABC
		Meeting Room 21DEF
		Meeting Room 21ABC

TUESDAY 19 AUGUST 2003

08h30 – 18h00	Registration and Travel desk opening hours	
08h30 – 10h30	**Plenary Session 3**	**Hall 2CDE**
	Chairperson: Isaac Minde	
	Discussant: Lawrence Haddad	
	Speakers: **Jean Kinsey**	
	Will Food Safety Jeopardize Food Security?	
	Betsey Kuhn	
	Measuring Food Security: A New Approach	
	Johan Swinnen	
	Food Safety, the Media, and the Information Market	
	Vittorio Santaniello	
	Biotechnology: Implications for Food Security	
10h30 – 11h00	*Refreshment break*	
11h00 – 12h30	**Poster Session 2**	**Hall 3A**

Poverty Reduction and Economic Development
Group: 1-2-1   Chairperson: Channing Arndt
Environmental Stewardship including Natural Resource Management
Group: 2-2-1   Chairperson: Konrad Hagedorn
Group: 2-2-2   Chairperson: Ben N. Okumu
Efficiency in Food and Farming Systems
Group: 3-2-1   Chairperson: Raushan Bokusheva
Group: 3-2-2   Chairperson: Takeshi Sakurai
Group: 3-2-3   Chairperson: Melinda Smale
Group: 3-2-4   Chairperson: Els Wynen
Group: 3-2-5   Chairperson: Winfried Manig

Consumer Safety and Food Security
Group: 4-2-1   Chairperson: Maria Raquel Ventura-Lucas
Group: 4-2-2   Chairperson: Kevin Z. Chen
Policy Analysis
Group: 5-2-1   Chairperson: Glenn C. W. Ames
Market and Trade Analysis
Group: 6-2-1   Chairperson: Søren Kjeldsen-Kragh
Group: 6-2-2   Chairperson: S. Jayne
Group: 6-2-3   Chairperson: Guenter Peter

11h30 – 12h50   **Computer Sessions**

<u>**Session No. 1 (consecutive, part 1): Multi Agent Systems**</u>   **Meeting Room 11AB**

Thomas Berger (Centre for Development Research, Germany)
An overview of Multi-Agent Systems Applications in
　Agro-Ecological Development Research
**Christophe Le Page, Patrick d'Aquino, François Bousquet
（CIRAD-Montpellier, France)**
Aiding Policy and Land-Use Management by Linking Role Playing
　Games, GIS and MAS
**Thomas Berger, Pepijn Schreinemachers (Centre for
　Development Research, Germany)**
Agent-based bio-economic simulations—empirical applications to
　developing countries

<u>**Session No. 2 (consecutive, part 1): Tools for Spatial Analyses in**</u>   **Meeting Room 11CD**
<u>**Agricultural Economics**</u>

**Michael Epprecht (IFPRI, Vietnam office) and Daniel Müller
（Institute of Rural Development, University of Göttingen,
Germany)**
GIS and Spatial Analytical Techniques for Agricultural Economists
　I: Linking poverty and the landscape: Using GIS to explore the
　relationship between poverty and the local environment
　(Combination of Spatially Explicit GIS with Population Census
　Data)

12h30 – 13h30   *Lunch*
13h30 – 15h30   **Contributed papers session 3**

31   **Health and Schooling Issues in Development**   **HALLCDE**
　　　Chairperson: Lawrence Haddad
　　　Child Growth, Shocks, and Food Aid in Rural Ethiopia
　　　Luc Christiaensen, Takashi Yamano & Harold Alderman (USA)
　　　Natural Resource Collection Work and Children's Schooling in
　　　　Malawi
　　　Flora J. Nankhuni & Jill L. Findeis (USA)
　　　The Role of Schooling in the Alleviation of Rural Poverty in Ethiopia
　　　Tassew Woldehanna (Ethiopia)
　　　The Impact of Micronutrients on Labor Productivity: Evidence from
　　　　Rural India
　　　Katinka Weinberger (Taiwan)

32 Food Quality and Safety   **Meeting Room 11AB**
Chairperson: Mary Bohman

Public Regulation as a Substitute for Trust in Quality Food
    Markets.What if the Trust Substitute Cannot Be Fully Trusted?
Giovanni Anania & Rosanna Nistico (Italy)

Consumers' Resistance to GM-Foods: The Role of Information in an
    Uncertain Environment
*Wallace E. Huffman, Matt Rousu, Jason F. Shogren & Abebayehu
    Tegene (USA)*

Socioeconomic Determinants of Consumer Food Safety Awareness
    and Perceptions in the United States: Pesticide and Antibiotic
    Residues in Food
Steven T. Yen, K. L. Jensen, & C.-T. J. Lin (USA)

Nutrient Effects on Consumer Demand: A Panel Data Approach
Ana Maria Angulo, B. Dhehibi & J.M. Gil (Spain)

33 Productivity, Technological Change, and Growth   **Meeting Room 11CD**
Chairperson: Colin Thirtle

An Application of the Stochastic Latent Variable Approach to the
    Correction of Sector Level TFP Calculations in the Face of Biased
    Technological Change
Alastair Bailey, Xavier Irz & Kelvin Balcombe (UK)

Fundamental and Induced Biases in Technological Change in Central
    Canadian Agriculture
J. Stephen Clark, K. K. Klein & W. A. Kerr (Canada)

Dual Technological Development in Botswana Agriculture: A
    Stochastic Input Distance Function Approach
Xavier Irz & David Hadley (UK)

Quality Improvement and Sustainable Growth in Japanese
    Agriculture
Shunji Oniki (Japan)

34 Agricultural Trade Policy Reform   **Meeting Room 12AB**
Chairperson: Bill Kerr

On Export Rivalry and the Greening of Agriculture—The Role of
    Eco-labels
Nancy H. Chau, Arnab K. Basu & Ulrike Grote (USA)

Agricultural Markets Liberalization and the Doha Round
*John Beghin, Stephane De Cara, Jacinto Fabiosa, Holger Matthey,
    Murat Isik & Cheng Fang (France)*

May the Pro-poor Impacts of Trade Liberalization Vanish Because of
    Imperfect Information?
*J.M. Boussard, F. Gerard, M.G. Piketty, A.K. Christensen
    & T. Voituriez (France)*

Making Sense of Agricultural Trade Policy Reform
David Vanzetti & Ralf Peters (Switzerland)

35   Production, Employment and Efficiency in LDCs   **Meeting Room 12CD**
      Chairperson: Gerald Ortmann
      Access to Information and Factor Market Participation: Adjustments
          of Land and Labor Margins of Agricultural Households in
          Bangladesh
      Shyamal K. Chowdhury (Germany)
      Labor Market Liberalization, Employment, and Gender in Rural
          China
      Linxiu Zhang, Alan de Brauw & Scott Rozelle (USA)
      Diversification Economies and Specialisation Efficiencies in a Mixed
          Food and Coffee Smallholder Farming System in Papua New
          Guinea
      Tim Coelli & Euan Fleming (Australia)
      Productivity and Efficiency in Chinese Agriculture: A Distance
          Function Approach
      Bernhard Bruemmer, Thomas Glauben & Wencong C. Lu

36   Economic Issues in Transition and Developing Economies   **Meeting Room 22DEF**
      Chairperson: Ken Thomson
      Transition and Food Consumption
      William Liefert, Bryan Lohmar & Eugenia Serova (USA)
      The Dynamics of Agri-Food Trade Patterns-The Hungarian Case
      Imre Ferto & L J Hubbard (Hungary)
      Human Capital and the Agrarian Structure in Transition Economies:
          Micro Evidence from Romania
      Marian Rizov (Belgium)
      Rural Taxation and Government Regulation in China
      Ran Tao, Justin Yifu Lin, Mingxing Liu & Qi Zhang (PR China)

37   The Economics of Land and Environmental Degradation   **Meeting Room 22ABC**
      Chairperson: Simeon Ehui
      Strategies to Increase Agricultural Productivity and Reduce Land
          Degradation: Evidence from Uganda
      *John Pender, Ephraim Nkonya, Pamela Jagger & Dick Sserunkuuma
          (USA)*
      Modelling Land Degradation in Low-input Agriculture: The
          "Population Pressure Hypothesis" Revisited
      Unai Pascual & Edward Barbier (UK)
      Spatial Analysis of Soil Fertility Management Using Integrated
          Household and GIS Data from Smallholder Kenyan Farms
      *S. J. Staal, D. Romney, I. Baltenweck, M. Waithaka, H. Muriuki
          & L. Njoroge (Kenya)*
      Social Welfare and Environmental Degradation in Agriculture: The
          Case of Ecuador
      E. Segarra, D. de la Torre Ugarte, J. Malaga & G. W. Williams (USA)

38	**Water, Irrigation, and Agriculture**	**Meeting Room 21DEF**

Chairperson: K.N. Ninan

The Equity Consequences of Public Irrigation Investments: The Case
  of Surface Irrigation Subsidies in India
Mona Sur & Dina Umali-Deininger (USA)

Integrating Agricultural Policies and Water Policies Under Water
  Supply and Climate Uncertainty
*Patricia Mejias, Consuelo Varela-Ortega & Guillermo Flichman
  (Spain)*

The Cost of Meeting Equity: Opportunity Cost of Irrigation in the
  Fish-Sundays Scheme of South Africa
Beatrice I. Conradie & D.L. Hoag (South Africa)

Multi-criteria Analysis of Factors Use Level: The Case of Water for
  Irrigation
Jose A. Gomez-Limon, Laura Riesgo & Manuel Arriaza (Spain)

15h30 – 16h00   *Refreshment break*

16h00 – 17h20   **Organized Session by the European Association of Agricultural**   **Hall 2CDE**
      Economists

Organizer: Giovanni Anania

**Reform of the agricultural and rural development policies of the
  European Union and its implications for world agriculture.**

Speakers: Sophia Davidova, Chair
          Introduction
          Jo Swinnen
          The reform process: what happened, why, what's ahead
          Jerzy Wilkin
          *The EU enlargement: how will it happen, what it means,
            what is its implications*
          Giovanni Anania
          *EU agricultural and rural development policy reform and
            WTO negotiations*
          Søren E. Frandsen
          *EU agricultural and trade policies: consequences for
            developing countries*

Discussion openers: Alberto Valdes
                      Wally Tyner

16h00 – 17h20   **Computer Sessions**

Session 1 (consecutive, part 2): Multi Agent Models   **Meeting Room 11AB**

**Kathrin Happe(University of Hohenheim, Germany), Alfons
Balmann, and Konrad Kellermann (Institute of Agricultural
Development in Central and Eastern Europe, Halle/Saale,
Germany)**
AgriPoliS – Agricultural Policy Simulator

**Konrad Kellermann, Alfons Balmann (Institute of Agricultural
Development in Central and Eastern Europe, Halle/Saale,
Germany and Kathrin Happe (University of Hohenheim,
Germany)**
PlayAgriPoliS – a Policy Simulation Game

**Stefano Farolfi, (University of Pretoria, South Africa) and
Martine Antona (Cirad – Montpellier, France)**
Using MAS for the analysis of environmental management policies

<u>Session 2 (consecutive, part 2): Tools for Spatial Analyses in Agricultural Economics</u>   **Meeting Room 11CD**

Darla Munroe (Dpt. of Geography and Earth Sciences, University of North Carolina, USA) and Daniel Müller (Institute of Rural Development, University of Göttingen, Germany)
GIS and Spatial Analytical Techniques for Agricultural Economists II: Linking the Household to the Landscape: Application of High-resolution GIS Data with a Household Survey

<u>Session 3: Tools for Agricultural Trade Modeling</u>   **Meeting Room 12AB**

Hans Loefgren and Peter Wobst (IFPRI, Washington, DC, USA)
IFPRI's standard General Equilibrium Model for Agricultural Policy Analysis
Ramesh Shrama and Daneswar Poonyth (FAO, Economics and Social Dept., Rome, Italy)
The Agricultural Trade Policy Simulation Model (ATPSM) of the FAO

<u>Session 4: Selected issues</u>   **Meeting Room 12CD**

Paul Gibson (USDA, Economic Research Service, USDA, USA) and Edward Gillin (FAO, Rome, Italy)
Analysing Market Access with the Agricultural Market Access Database
Gracian Chimwaza, TEFAL, African Office, Zimbabwe
The Essential Electronic Agricultural Library
Ecio F. Costa (Universidade de Pernambuco, Brazil), Jack E. Houston, and Gene M. Pesti (University of Georgia. Athens, GA, USA)
Interactive Broiler Profit Maximizing Model: A Decision Maker's Computer Guide to Profitable Broiler Production and Processing

17h30 – 19h00 | Discussion Groups 2 |

Livestock Research Need for Developing Countries   **Meeting Room 11E**
Chris Delgado (IFPRI) and Guillaume Duteurte (CIRAD)
Capacity Building for Policy Analysis in Developing Countries   **Meeting Room 12E**
Rachid Doukkali (IAV-Hassan II, Morocco) and Tancréde Voituriez (CIRAD)
Policies to Promote Integrated Pest Management in Africa   **Meeting Room 22G**
Hermann Waibel (Univ. of Hanover)
Agent Based Modeling in Agricultural and Resource Economics   **Meeting Room 21G**
Alfons Balmann (Institute of Agricultural Development in Central and Eastern Europe)

Mini-symposia 2

Poverty Impacts and Policy Options of Non-Farm Rural Employment	**Hall 2CDE**
Gertrud Buchenrieder (Univ. of Hohenheim)	
The Potential of Carbon Sequestration through Land Use Change to Contribute to Poverty Alleviation: Comparative Micro-Economic Evidence	**Meeting Room 11AB**
Leslie Lipper (FAO)	
Problems of Outreach in Rural Finance Risk-reducing Mechanisms for Rural Finance Understanding and Improving the Impact of Financial Services	**Meeting Room 11CD**
Richard Meyer (Ohio State University)	
Crop Biotechnology and Developing Countries: Understanding Constraints to Full Acquisition and Use	**Meeting Room 12AB**
Aziz Elbehri and Paul Helsey (USDA)	
Soil Fertility and Food Security for the Poor in Southern Africa: Technical, Policy, and Institutional Challenges	**Meeting Room 12CD**
Mulugetta Mekuria (CIMMYT, Southern Africa)	
Market Research for the Development of Commercialized Agriculture in Sub-Saharan Africa	**Meeting Room 22DEF**
Jim Gockowski, Victor Manyong, Patrick Kormawa, Steffen Abele, and Shaun Ferris (IITA)	
Evaluation and Rural Development Programs	**Meeting Room 22ABC**
Benjamin Davis (ESA) and Sudhanshu Handa (Inter-American Development Bank)	
Economic Evaluation of New Institutional Mechanisms for Funding and Delivery of Agricultural Extension in Developing Countries	**Meeting Room 21DEF**
Derek Byerlee and David Nelson (World Bank)	
Is OECD Agricultural Support Stunting Agricultural Development and Poverty Alleviation in Developing Countries?	**Meeting Room 21ABC**
William Meyers (Iowa State University)	

WEDNESDAY 20 AUGUST 2003 THE DAY AT A GLANCE

08h30 – 10h30	African Session	Hall 2CDE
10h30 – 11h00	Refreshment Break	Hall 3A
11h00 – 12h30	African Session Continues	Hall 2CDE
12h30 – 13h30	Lunch	Hall 3BC
13h30 – 18h00	Technical Tours: *Please refer to the separate Technical Tours brochure*	
19h00 – 22h30	Networking Event	Joe Kool's

WEDNESDAY 20 AUGUST 2003

08h30 – 16h00	Registration and Travel desk opening hours
08h30	**African Session**
	Chairperson: Ms Bongiwe Njobe
08h30 – 09h00	Agricultural Economics and Agricultural Policy for NEPAD **Hall 2CDE**
	Prof Wiseman Nkuhlu, Chairman, NEPAD
09h00 – 10h30	Panel discussion on agricultural economic and agricultural policy **Hall 2CDE**
	issues in Africa
	Panellists: *Mandi Rukuni (Southern Africa)*
	Chris Ackello-Ogutu and Isaac Minde (Eastern Africa)
	George Abulu (West Africa)
	Mohammed Moussaiu (North Africa) to be confirmed
10h30 – 11h00	*Refreshment break*
11h00	**African Session continues**
11h00 – 11h30	Training Agricultural Economists in Africa for Africa. The
	Collaborative Masters Program in Agricultural and Applied
	Economics in Eastern and Southern Africa (COMAAE-ESA
	Initiative).
	Prof Willis Oluoch-Kosura and Jeffrey Fine
11h30 – 12h30	Founding meeting of the *African Agricultural Economics Association*
	Chairperson: Dr P. Anandajayasekeram
12h30 – 13h30	*Lunch*
13h30 – 18h00	**Technical Tours**
	Please refer to the separate Technical Tours brochure
19h00 – 22h30	**Networking Event** **Joe Kool's**

THURSDAY 21 AUGUST 2003
THE DAY AT A GLANCE

08h30 – 10h30	Best Contributed and Poster Papers	Hall 2CDE
10h30 – 11h00	Refreshment Break	Hall 3A
11h00 – 12h30	Poster Session 3:	Hall 3A
12h30 – 13h30	Lunch	Hall 3BC
13h30 – 15h30	Invited Panel Session 2:	
	1. Livestock industrialization: trends, causes, impacts and policy options.	Hall 2CDE
	2. Reappraising Food Aid	Meeting Room 11AB
	3. The Millennium Round so far	Meeting Room 11CD
	4. Economics of Food safety in Developing Countries	Meeting Room 12AB
	5. The September 2002 World Summit on Sustainable Development: Agricultural Perspectives	Meeting Room 12CD
15h30 – 16h00	Refreshment Break	Hall 3A

16h00 – 17h20	Contributed papers session 4:	
	41. Economic Impacts of Biotechnology	Hall 2CDE
	42. Price and Marketing Issues	Meeting Room 11AB
	43. Multifunctionality: Maintaining Environmental Benefits	Meeting Room 11CD
	44. Economic and Rural Development in Latin America	Meeting Room 12AB
	45. CGE Analysis: African Applications	Meeting Room 12CD
	46. Economic Analysis of Pests and Disease	Meeting Room 22DEF
	47. Rural Credit, Savings and Micro-Insurance	Meeting Room 22ABC
	48. Cereal Production and Policy Issues in Developing Countries	Meeting Room 21DEF
	49. Bioeconomic Models and Game Theory in Resource Management	Meeting Room 21ABC
17h30 – 19h00	Council Meeting 2	IAAE Office
19h00 – 22h30	Conference Dinner	Hall 1A

THURSDAY 21 AUGUST 2003

08h30 – 18h00	Registration and Travel desk opening hours
08h30 – 10h30	**Best Contributed and Poster Papers**
10h30 – 11h00	*Refreshment break*
11h00 – 12h30	**Poster Session 3**

Poverty Reduction and Economic Development
Group: 1-3-1   Chairperson: Carlo del Ninno
Environmental Stewardship including Natural Resource Management
Group: 2-3-1   Chairperson: Ernst-August Nuppenau
Efficiency in Food and Farming Systems
Group: 3-3-1   Chairperson: Hugo de Groote
Group: 3-3-2   Chairperson: Takashi Yamano
Group: 3-3-3   Chairperson: Philip Lund
Group: 3-3-4   Chairperson: Cheryl R. Doss
Consumer Safety and Food Security
Group: 4-3-1   Chairperson: Chung L. Huang
Group: 4-3-2   Chairperson: M. G. Chandrakanth
Policy Analysis
Group: 5-3-1   Chairperson: Kei Kajisa
Group: 5-3-2   Chairperson: Manfred Zeller
Market and Trade Analysis
Group: 6-3-1   Chairperson: Michelle Veeman
Group: 6-3-2   Chairperson: Brian J. Revell
Group: 6-3-3   Chairperson: Colin Poulton

12h30 – 13h30	*Lunch*
13h30 – 15h30	**Invited Panel Session 2**

Livestock industrialization; trends, causes, impacts and policy options   Hall 2CDE
Organizer: Chris Delgado (IFPRI)
Chairperson: Nick Vink

Speakers: Christopher Delgado (IFPRI), Clare Garrod (FAO) and
     Simeon Ehui (ILRI)
     Unpacking "economies of scale" in livestock
     production in developing countries: efficiency,
     transaction costs and policy distortions
     Achilles Costales (UNIVERSITY of Philippines),
     M. Lucila Lapar (ILRI), Viroj NaRanong (TDRI.
     Thailand), Nipon Paopongsakorn (TDRI) and
     Christopher Delgado
     Economies of scale in hog production in
     Southeast Asia
     Mohamad Jabbar (ILRI), Fakhrul Islam BSMRAU,
     Bangladesh), Rajesh Mehta (New Delhi), and Vijay
     Paul Sharma (IIM, India)
     Economies of scale in poultry, eggs and dairy in South
     Asia
     Steve Staal (ILRI), John Omiti (IPAR, Nairobi), and
     Christopher Delgado
     Economies of scale in eggs and dairy in East Africa
     Clare Narrod and Geraldo Sant'Ana de Carmargo
     Barros (UNIVERSITY of Sao Paolo)
     Economies of scale in hogs and poultry in Brazil

Reappraising Food Aid   **Meeting Room 11AB**
Organizers: Werner Kiene and Linda Young
Chairperson: Ashok Gulati (CGIAR)
Speakers: Shahla Shapouri (USDA-ERS)
     Past developments, future needs, and the new
     environment for food aid
     Linda M. Young (Montana S. UNIVERSITY) and
     Philip C. Abbott (Purdue UNIVERSITY)
     How well has food aid addressed need: an evaluation
     of the last decade
     Werner Kiene (UNWFP)
     Food aid and the international development agenda:
     focusing on people

The Millenium Round so far   **Meeting Room 11CD**
Chairperson: Harald von Witzke
Speakers: Lynn Kennedy (Louisiana State University)
     Alberto Valdes (Chile)
     Ulrich Koester (Kiel)
     Anastassios Haniotis (EU Commission)
     Mercedita Sombilla-Agcaoili (IRRI)

Economics of Food safety in Developing Countries   **Meeting Room 12AB**
Organizer: Spencer Henson
Chairperson: Giovanni Anania

Speakers: Spencer Henson (University of Guelph) and Laurian
            Unnevehr (University of Illinois)
            *An overview of economic issues associated with food
            safety in developing countries*
            Ivy Drafor (University of Cape Coast, Ghana)
            *Pesticide use and consumer and worker safety:
            Experiences from Kenya and Ghana.*
            Richard Abila (ESRF, Tanzania)
            *Food safety standards and trade: The case of fish
            exports from the Kenyan shore of Lake Victoria.*
            Steve Staal (ILRI, Kenya)
            *An economic assessment of milk safety in Kenya,
            Ghana and Tanzania.*

The September 2002 World Summit on Sustainable   **Meeting Room 12CD**
Development: Agricultural Perspectives
Chairperson: Susan Offutt (ERS/USDA)
Discussants: Betsey Kuhn (ERS/USDA) and Nick Vink (University
            of Stellenbosch)
Speakers: Susan Offutt (ERS/USDA)
            *The World Summit on Sustainable Development: what
            happened and what's ahead.*
            Terry Roe (University of Minnesota)
            Agricultural Trade Perspective
            Susan Capalbo (Montana State University)
            Resource and Productivity Perspective
            Jean Kinsey (University of Minnesota)
            Food Security Perspective

15h30 – 16h00   *Refreshment break*
16h00 – 17h20   **Contributed papers session 4**

41   **Economic Impacts of Biotechnology**   **HALL2CDE**
     Chairperson: David Schimmelpfennig
     International Diffusion of Gains from Biotechnology and the
        European Union's Common Agricultural Policy
     Hans van Meijl & Frank van Tongeren (The Netherlands)
     Biotechnology Boosts to Crop Productivity in China: Trade and
        Welfare Implications
     *Jikun Huang, Ruifa Hu, Hans van Meijl & Frank van Tongeren
        (China)*
     Socio-economic Impact of Biotechnology Applications: Some
        Lessons from the Pilot Tissue Culture (tc) Banana Production
        Promotion Project in Kenya, 1997–2002
     *Stephen G. Mbogoh, Florence M. Wambugu & Sam Wakhusama
        (Kenya)*

42   **Price and Marketing Issues**   **Meeting Room 11AB**
     Chairperson: Stephan von Cramon-Taubadel
     Price Linkages in the International Wheat Market
     Atanu Ghoshray & Tim Lloyd (UK)

Measuring Market Integration in the Presence of Transaction Costs:
A Threshold Vector Error Correction Approach
Jochen Meyer (Germany)
Efficiency of the Italian Agri-food Industry: An Analysis of
"Districts" Effect
Christina Brasili & Elisa Ricci Maccarini (Italy)

43   Multifunctionality: Maintaining Environmental Benefits   **Meeting Room 11CD**
Chairperson: Ernst-August Nuppenau
Multifunctional Agriculture and the Preservation of Environmental
Benefits
Werner Hediger & Bernard Lehmann (Switzerland)
Estimation of Costs of Maintaining Landscape Elements by the
Example of Southern Germany
Jochen Kantelhardt, Elisabeth Osinski & Martin Kapfer (Germany)
Multifunctionality and Non-Agricultural Supply of Public Goods
Tristan Le Cotty & Inra Lameta (France)

44   Economic and Rural Development in Latin America   **Meeting Room 12AB**
Chairperson: Ismail Shariff
The Importance of Social Capital in Colombian Rural
Agro-Enterprises
Nancy L. Johnson, Ruth Suarez & Mark Lundy (Columbia)
Non-Farm Rural Activities (NFRA) in a Peasant Economy: The Case
of North Peruvian Sierra
Jackeline Velazco (United Kingdom)
Social Exchange Rates, MERCOSUR and Economic Development
Leo da Rocha Ferreira (Brazil)

45   CGE Analysis: African Applications   **Meeting Room 12CD**
Chairperson: Terry Roe
Copper Crisis and Agricultural Renaissance in Zambia: An
Economy-wide Analysis
Hans Lofgren, James Thurlow & Sherman Robinson (USA)
Technical Change, Market Incentives and Rural Incomes: A CGE
Analysis of Uganda's Agriculture
Paul Dorosh, Moataz El-Said & Hans Lofgren (USA)
Transgenic Cotton and Crop Productivity: A General Equilibrium
Analysis for West and Central Africa
Aziz Elbehri & Steve MacDonald (USA)

46   Economic Analysis of Pests and Disease   **Meeting Room 22DEF**
Chairperson: Derek Byerlee
Economic Policy Analysis of an Invasive Pest: A Case Study of
Colorado Potato Beetle in Finland
Jukka Peltola & Jaakko Heikkila (Finland)
How Labor Organization May Affect Technology Adoption: An
Analytical Framework Analysing the Case of Integrated Pest
Management
Justus Wesseler & Volker Beckmann (The Netherlands)

Economic Analysis of the Impact of Adopting Herd Health Control
Programs on Smallholder Dairy Farms in Central Thailand
David C. Hall, Simeon K. Ehui & Barry I. Shapiro (Kenya)

47   **Rural Credit, Savings, and Micro-Insurance**   **Meeting Room 22ABC**
Chairperson: Nick Vink
Interlinked Credit and Farm Intensification: Evidence from Kenya
T. S. Jayne, Takashi Yamano & James Nyoro (USA)
The Potential for Financial Savings in Rural Mozambican
Households
*Oliveira Amimo, Donald W. Larson, Mauricio Bittencourt &
Douglas H. Graham (USA)*
Are the Poor too Poor to Demand Health Insurance?
Rajeev Ahuja & Johannes Juetting (India)

48   **Cereal Production and Policy Issues in Developing Countries**   **Meeting Room 21DEF**
Chairperson: Timothy O. Williams
Yield Response in Pakistan Agriculture: A Cointegration Approach
Khalid Mushtaq & P.J. Dawson (Pakistan)
Is Increased Instability in Cereal Production in Ethiopia Caused by
Policy Changes?
*Zerihun Gudeta Alemu, Klopper Oosthuizen & H.D. van Schalkwyk
(South Africa)*
Comparing Economic Determinants of Interspecific Cereal Diversity
on Farms in the Ethiopian Highlands
S. Benin, B. Gebremedhin, M. Smale, J. Pender & S. Ehui (Ethiopia)

49   **Bioeconomic Models and Game Theory in Resource Management**   **Meeting Room 21ABC**
Chairperson: Kurt Klein
Analyzing Negotiation Approaches in Natural Resource
Management- A Case Study of Crop-Livestock Conflicts in Sri
Lanka
Regina Birner (Germany)
Trade-off Between Economic Efficiency and Contamination by
Coffee Processing: A Bioeconomic Model at the Watershed Level
in Honduras
*Bruno Barbier, Robert R. Hearne, Jose Manuel Gonzalez, Andy
Nelson & Orlando Mejia Castaneda (France)*
Poverty, Resource Scarcity, and Incentives for Soil and Water
Conservation: Analysis of Interactions with a Bio-economic
Model
Bekele Shiferaw & Stein Holden (India)

17h30 – 19h00   **Council Meeting 2**   **IAAE office**
19h00 – 22h30   **Conference Dinner**   **Hall 1A**

FRIDAY 22 AUGUST 2003		
THE DAY AT A GLANCE		

08h30 – 10h30   Plenary Session 5:   Hall 2CDE
                         1. Dan Bromley
                         2. Ramon Lopez
                         3. Simeon Ehui and John Pender
                         4. Agnes Rola and Ian Coxhead

10h30 – 11h00   Refreshment Break   Hall 3A
11h00 – 12h30   Discussion Groups 3   Meeting Room 11E
                                                                   Meeting Room 12E
                                                                   Meeting Room 22G
                                                                    Meeting Room 21G

                Mini-symposia 3   Hall 2CDE
                                                                   Meeting Room 11AB
                                                                    Meeting Room 11CD
                                                                   Meeting Room 12AB
                                                                    Meeting Room 12CD
                                                                   Meeting Room 22DEF
                                                                    Meeting Room 22ABC
                                                                    Meeting Room 21DEF
                                                                   Meeting Room 21ABC

12h30 – 13h30   Lunch   Hall 3BC
13h30 – 15h30   Contributed papers session 5:
                         51. Economic Issues Relating to Biodiversity   Hall 2CDE
                         52. Agricultural Research and Development Issues   Meeting Room 11AB
                         53. Trade, Development and Liberalization   Meeting Room 11CD
                         54. Modeling Farmers' response under varying policy regimes   Meeting Room 12AB
                         55. Demand and Marketing Issues   Meeting Room 12CD
                         56. Poverty Issues and Analysis   Meeting Room 22DEF
                         57. Marketing and Policy Issues in Developing Countries   Meeting Room 22ABC
15h30 – 16h00   Refreshment Break   Hall 3A
16h00 – 17h20   Conference Synopsis   Hall 2CDE
17h30 – 19h00   Closing Reception   Hall 1A

FRIDAY 22 AUGUST 2003	

08h30 – 17h00   Registration and Travel desk opening hours
08h30 – 10h30   **Plenary Session 5**   **Hall 2CDE**
                Chairperson: Terry Veeman
                Speakers: **Dan Bromley**
                                  The poverty of sustainable development: saving
                                    economics from platitudes
                                Ramon Lopez
                                  Agriculture and/or Environment; Political economy
                                  and public policy

Simeon Ehui and John Pender
Resource degradation and poverty in Africa: Pathways
out of the spiral
Agnes Rola and Ian Coxhead
Economic Development and Environmental
Management in a Watershed Context

10h30 – 11h00	*Refreshment break*
11h00 – 12h30	**Discussion Groups 3**

Livestock Research Need for Developing Countries   Meeting Room 11E
Chris Delgado (IFPRI) and Guillaume Duteurte (CIRAD)
Capacity Building for Policy Analysis in Developing Countries   Meeting Room 12E
Rachid Doukkali (IAV-Hassan II, Morocco) and Tancréde Voituriez
  (CIRAD)
Policies to Promote Integrated Pest Management in Africa   Meeting Room 22G
Hermann Waibel (Univ. of Hanover)
Agent Based Modeling in Agricultural and Resource Economics   Meeting Room 21G
Alfons Balmann (Institute of Agricultural Development in Central
  and Eastern Europe)

Mini-symposia 3

The Role of Agricultural Growth in Reducing Poverty   Hall 2CDE
Frédéric Dévé (Roles of Agriculture project, FAO)
The Potential of Carbon Sequestration through Land Use   Meeting Room 11AB
  Change to Contribute to Poverty Alleviation: Comparative
  Micro-Economic Evidence
Leslie Lipper (FAO)
Problems of Outreach in Rural Finance Risk-reducing   Meeting Room 11CD
  Mechanisms for Rural Finance Understanding and Improving
  the Impact of Financial Services
Richard Meyer (Ohio State University)
Crop Biotechnology and Developing Countries: Understanding   Meeting Room 12AB
  Constraints to Full Acquisition and Use
Aziz Elbehri and Paul Helsey (USDA)
Transition in Agriculture in South Africa   Meeting Room 12CD
Kristy Cook (AGRILINK II Project)
Macroeconomic Impacts of Agricultural Trade and Policy   Meeting Room 22DEF
  Reform in Sub-Saharan Africa
Scott McDonald (Univ. of Sheffeld)
Impact Assessment in the CGIAR: The Tried, the True, and the   Meeting Room 22ABC
  New
Meredith Soule (USAID)
Impacts of Investment in Less-Favored Areas   Meeting Room 21DEF
John Pender (IFPRI)
Improving the Economics in Livestock Health Economics   Meeting Room 21ABC
Tom Randolph (ILRI)

12h30 – 13h30	*Lunch*

13h30 – 15h30   **Contributed papers session 5**

51   **Economic Issues Relating to Biodiversity**   **HALL2CDE**
      Chairperson: Consuelo Varela-Ortega
      The Economics of Biodiversity Conservation: A Study in a Coffee
          Growing Region of India
      K.N. Ninan & Jyothis Sathyapalan (India)
      Assessing Economic Returns from Farmers' Rights
      C.S. Srinivasan (UK)
      Biodiversity versus Transgenic Sugar Beets: The One Euro Question
      Matty Demont, Justus Wesseler & Eric Tollens (Belgium)
      Crop Diversity as the Derived Outcome of Farmers' "Survival First"
          Motives in Ethiopia: What Role for On-farm Conservation of
          Sorghum Genetic Resources?
      Edilegnaw Wale & Detlef Virchow (Germany)

52   **Agricultural Research and Development Issues**   **Meeting Room 11AB**
      Chairperson: Robert Evenson
      The Impact of Research Led Agricultural Productivity Growth on
          Poverty Reduction in Africa, Asia, and Latin America
      Colin Thirtle, David Hadley & Lin Lin (UK)
      International R & D Spillovers and Productivity Growth in the
          Agricultural Sector
      Luciano Gutierrez & M. M. Gutierrez (Italy)
      Optimizing the Allocation of Agricultural R&D Funding: Is
          Win-Win Targeting Possible?
      Johannes Roseboom, Paul Diederen & Arie Kuyvenhoven (The
      Netherlands)
      Estimating Returns from Past Investments into Beef Cattle Genetics
          RD&E in Australia
      G. R. Griffith, R. J. Farquharson, S. A. Barwick, R. G. Banks, & W. E.
      Homes (Australia)

53   **Trade, Development, and Liberalization**   **Meeting Room 11CD**
      Chairperson: Harald von Witzke
      The Effects of Trade Uncertainty on Chrysanthemum Trade Between
          Taiwan and Japan
      Rhung-Jieh Woo & Hsin-Yeh Tsai (Taiwan)
      Public Investment and China's Grain Production Competitiveness
          Under WTO
      Jing Zhu (China)
      Impacts of FTAA and MERCOEURO on Agribusiness in the
          MERCOSUL Countries
      Erly C. Teixeira & Luiz A. Cypriano (Brazil)

54   **Modelling Farmers' Response Under Varying Policy Regimes**   **Meeting Room 12AB**
      Chairperson: Jack Peerlings
      Price Incentives, Non-Price Factors, and Agricultural Production in
          Sub-Saharan Africa: A Cointegration Analysis
      Rainer Thiele (Germany)

Do Counter-Cyclical Payments in FSRI Act Create Incentives to
   Produce?
Jesus Anton & Chantal Lemouel (France)
Modeling Farmers' Response to a Decoupled Subsidy via
   Multi-Attribute Utility Theory and E-V Analysis
M. Arriaza & J.A. Gomez-Limon (Spain)
Farm Household Decisions Under Various Tax Policies:
   Comparative Static Results and Evidence from Household Data
Arne Henningsen, Thomas Glauben & Christian H.C.A. Henning,
   (Germany)

55   Demand and Marketing Issues   **Meeting Room 12CD**
   Chairperson: Kyrre Rickertsen
   Institutional Innovation to Increase Farmers' Revenue: A Case Study
      of Small Scale Farming in Sheep in Transkei Region, South Africa
   Marijke D'Haese, Wim Verbeke, Guido Van Huylenbroeck, Johann
      Kirsten & Luc D'Haese (Belgium)
   Testing Symmetry and Homogeneity in the AIDS with Cointegrated
      Data Using Full-modified Estimation and the Bootstrap
   R. Tiffen & K. Balcombe (United Kingdom)
   Projecting World Food Demand Using Alternative Demand System
   Wusheng Yu, Thomas Hertel, Paul Preckel & James Eales (Denmark)
   The "New Economy" and Efficiency in Food Market System: A
      Complement or a Battleground Between Economic Classes?
   Chinkook Lee & Gerald Schluter (USA)

56   Poverty Issues and Analysis   **Meeting Room 22DEF**
   Chairperson: Herman van Schalkwyk
   Why Does Poverty Persist in Rural Ethiopia?
   Ayalneh Bogale, Konrad Hagedorn & Benedikt Korf (Ethiopia)
   Development and Poverty Reduction: Do Institutions Matter? A
      Study on the Impact of Local Institutions in Rural India
   Vasant P. Gandhi & Robin Marsh (India)
   An Operational Method for Assessing the Poverty Outreach of
      Development Projects: Results from Case Studies in Africa, Asia,
      and Latin America
   Manfred Zeller, Manohar Sharma, Carla Jane Henry & Cecile
      Lapenu (Germany)
   Theft and Rural Poverty: Results of a Natural Experiment
   Bart Minten & Marcel Fafchamps (Madagascar)

57   Marketing and Policy Issues in Developing Countries   **Meeting Room 22ABC**
   Chairperson: Jean-Marc Boussard
   The Impact of Transport-and Transaction-Cost Reductions on Food
      Markets in Developing Countries: Evidence for Tempered
      Expectations for Burkina Faso
   Arjan Ruijs, C. Schweigman & C. Lutz (The Netherlands)
   The Demand for Commodity Insurance by Developing Country
      Agricultural Producers: Theory and an Application to Cocoa in
      Ghana
   Alexander Sarris (Greece)

Interlinkage in the Rice Market of Ghana: Money-lending Millers
  Enhance Efficiency
Jun Furuya & Takeshi Sakurai (Japan)
How Big is Your Neighborhood? Spatial Implications of Market
  Participation by Smallholder Livestock Producers
*Ma. Lucila A. Lapar, Garth Holloway & Simeon Ehui
  (Philippines)*

15h30 – 16h00	*Refreshment break*	
16h00 – 17h20	**Conference Synopsis**	**Hall 2CDE**
17h30 – 19h00	**Closing Reception**	**Hall 1A**

KEYNOTE SPEAKERS BIOGRAPHIES

Christopher B. Barrett

Chris Barrett is Professor at the Department of Applied Economics and Management, Cornell University. He holds degrees from Princeton (A.B. 1984), Oxford (M.S. 1985) and the University of Wisconsin-Madison (dual Ph.D., 1994). There are three basic, interrelated thrusts to his research program: (i) poverty, hunger, food security, economic policy and the structural transformation of low-income societies, (ii) individual and market behavior under risk and uncertainty, and (iii) the interrelationship between poverty, food security and environmental stress in developing areas. Professor Barrett has published widely and is an associate editor of *Agricultural Economics*, the *American Journal of Agricultural Economics, Environment and Development Economics*, the *Journal of African Economies* and *World Development*. He co-directs Cornell University's African Food Security and Natural Resources Management program and the Rural Livelihoods and Biological Resources workshop.

Simon Maxwell

Simon Maxwell is Director of the Overseas Development Institute in London. He is an economist who has published widely on agriculture, rural development and food security. His career includes ten years overseas, in Kenya, India and Bolivia, first with UNDP and then with the British aid Program, and then fifteen years or so as a Fellow of the Institute of Development Studies at the University of Sussex. He became Director of ODI in 1997. Current research interests include rethinking rural development, food security transitions, and aspects of the new poverty agenda. Simon Maxwell is currently President of the Development Studies Association of the UK and Ireland.

Alain de Janvry

Alain de Janvry is an economist working on international economic development, with expertise principally in Latin America, Sub-Saharan Africa, the Middle-East, and the Indian subcontinent. Fields of work include poverty analysis, rural development, quantitative analysis of development policies, impact analysis of social programs, technological innovations in agriculture, and the management of common property resources. He has worked with many international development agencies, including FAO, IFAD, the World Bank, UNDP, ILO, the CGIAR, and the Inter-American Development Bank as well as foundations such as Ford, Rockefeller and Kellogg. His main objective in teaching, researching, and working with development agencies is the promotion of human welfare, including understanding the determinants of poverty and analyzing successful approach to improve well-being and promote sustainability in resource use. He is a member of the French National Academy of Agriculture and a fellow of the American Agricultural Economic Association.

Ulrich Koester

Ulrich Koester is Professor at the Institute of Agricultural Economics in Kiel. He received his doctoral degree from the University of Göttingen, and became a post-doctorate fellow at the Economics Department of the University of California, Berkeley. For 10 years Prof. Koester was a visiting research fellow for 3–5 months annually at the International Food Policy Research Institute (IFPRI). Among his many publications are *Regional Cooperation to Improve Food Security in Southern and Eastern African Countries*, IFPRI Report No .53, and *Disharmonies in EC and US Agricultural Policy Measures* (Editor), Report for the European Commission.

Daniel A. Sumner

Daniel A. Sumner is the Director of the University of California, Agricultural Issues Center and the Frank H. Buck, Jr. Professor, Department of Agricultural and Resource Economics, University of California, Davis where he has been since 1993. His research centers on the effects of economic and trade policy on agriculture in the United States and Developing countries. Earlier Sumner was on the Economics faculty of North Carolina State University and spent several years on leave in government service as a Senior economist at the President's Council of Economic Advisers and as Assistant Secretary for Economics at the USDA. Sumner's research and writing has won American Agricultural Economic Association awards for Quality of Research, Quality of Communication, and Distinguished Policy Contribution. He was named a fellow of the association in 1999. Sumner was raised on a fruit farm in California. He received his Ph.D. in economics from the University of Chicago in 1978.

Peter Hazell

Peter Hazell joined IFPRI in 1992 as Director of the Environment and Production Technology Division. He was previously a principal economist in the World Bank's Agriculture and Rural Development Department. Prior to that he served at IFPRI as Director of the Agricultural Growth Linkages Program. A British citizen, Hazell obtained his Ph.D. in agricultural economics at Cornell University. He has published widely on agricultural sector modeling, risk management, agricultural growth linkages, property rights systems in Africa, the impact of the green revolution, links between agricultural research and poverty reduction, development strategies for less-favored areas, and sustainable farming practices.

Shenggen Fan

Shenggen Fan is Senior Research Fellow at IFPRI. He leads a research program on rural public investment, growth and poverty reduction. For the past several years, his major efforts have been devoted to analyze the impact of government expenditures on growth and poverty reduction in India, China, Vietnam, and Thailand. He has also been heavily involved in the IFPRI's China program. Prior to IFPRI, he worked for the International Service for National Agricultural Research, and the University of Arkansas. He received his Ph.D. in applied economics from the University of Minnesota.

Jean Kinsey

Jean Kinsey is Professor in the Department of Applied Economics, University of Minnesota, and Co-Director of The Food Industry Center. She obtained her Ph.D. in Agricultural Economics. She has conducted research in and published in leading journals about: consumer credit behavior, changing food consumption patterns, the economics of information and consumers' welfare loss, and the marginal propensity to consume food away from home when wives work part-time and full-time, where people purchase their food, retail food store performance and the retail food industry. She was Chair of the Board of Directors of the Federal Reserve Bank of Minneapolis 1996–1997. She was President of the American Agricultural Economics Association in 2002, and has been President of the American Council on Consumer Interests. She was named distinguished fellow of the American Agricultural Economics Association.

Betsey Kuhn

Betsey Kuhn is the Director of the Food and Rural Economy Division (FRED), Economic Research Service, U.S. Department of Agriculture. The Division is responsible for economic policy analysis of a wide range of issues, including diet and health, food safety, food markets, rural development, and food assistance. She received her M.A. and Ph.D. degrees in agricultural and applied economics from Stanford University. She is the author of many articles on futures markets, food safety, conservation easements, food assistance and farm policy and a book, *Global Environmental Change and Agriculture*.

Johan F. M. Swinnen

Johan F. M. Swinnen is Professor of Development Economics and Food Policy, Katholieke Universiteit Leuven, Belgium and Director of the Research Group on Food Policy, Transition & Development. He is Senior Research Fellow, Centre for European Policy Studies (CEPS), Brussels; consultant to many East European governments and international institutions (World Bank, EBRD, OECD, FAO, European Commission). From 1998–2000, he was Economic Advisor, European Commission, Directorate-General for Economic and Financial Affairs (advising on EU agricultural policy, trade issues, East European reforms and EU enlargement, Balkan reconstruction). He has research expertise and publications on political economy, agricultural and food policy, economic and institutional transition, EU enlargement, economics of the media. He received his Ph.D. in agricultural economics from Cornell University in 1992.

Vittorio Santaniello

Vittorio Santaniello is Professor of Economic Policy at the University of Rome Tor Vergata and Coordinator of the International Consortium on Agricultural Biotechnology Research (ICABR). He teaches agricultural policy, economics of natural resources as well as biotechnology. His research interest includes economics of agricultural biotechnology and intellectual property rights. He has published widely on the topics in journals and books. His recent book entitled *Economic and Social Issues in Agricultural Biotechnology* with R. E. Evenson and David Zilberman in 2002. He was born in Naples, Italy.

Daniel W. Bromley

Daniel W. Bromley is Anderson-Bascom Professor of applied economics at the University of Wisconsin-Madison. He has been editor of the journal *Land Economics* since 1974. He has been a consultant to the Global Environment Facility; the World Bank; the Ford Foundation; the U.S. Agency for International Development; the Asian Development Bank; the Organization for Economic Cooperation and Development; and the Ministry for the Environment in New Zealand. Professor Bromley has written and edited eleven books, among which are: (1) *Economic Interests and Institutions: The Conceptual Foundations of Public Policy*; (2) *Environment and Economy: Property Rights and Public Policy*; (3) *Making the Commons Work: Theory, Practice, and Policy*; (4) *The Handbook of Environmental Economics*; and (5) *Sustaining Development: Environmental Resources in Developing Countries*.

Ramon Lopez

Ramon Lopez is Professor in the Department of Agricultural and Resource Economics, University of Maryland at College Park and is Senior Fellow with the Center for Development Research (ZEF) at the University of Bonn. Professor Lopez is Associate Editor of *Environment and Development Economics* and has been Associate Editor of the *Journal of Environmental Economics and Management*. He obtained his Ph.D. in Economics from the University of British Columbia. He has been an advisor with various governments and a consultant with several international organizations, including the World Bank, Inter-American Development Bank, FAO and others. His research is focused on agriculture and the environment, international trade, and economic development. He is the author of a large number of journal articles and of three books: (a) *The Quality of Growth* (with coauthors); (b) *Rural poverty in Latin America* (with A. Valdes); and (c) *Sustainable Development in Latin America: Financing and Policies Working in Synergy* (with J. Jordan).

Simeon Ehui

A national of Côte d'Ivoire, Simeon Ehui has over fifteen years of professional international agricultural experience in Africa, and most recently in Asia. He is presently Coordinator of the Livestock Policy Analysis Program at the International Livestock Research Institute (ILRI). Ehui has authored or co-authored over 70-policy research papers relevant to national researchers and policy makers in Africa and Asia. His writings have covered various topics including: policies for the sustainable management of land and enhancing food security; policies for improving the competitiveness of smallholder livestock producers; and policies for improving domestic and international market access by smallholder livestock producers in the Africa and Asia. Ehui obtained his PhD degree from Purdue University in 1987.

Dr. John Pender

John Pender is a Senior Research Fellow at IFPRI. He leads IFPRI's research program on strategies for sustainable development of less-favored lands. His research focuses on pathways of development and land management in the East African highlands and the Central American hillsides. Prior to joining IFPRI in 1995, Pender was an assistant professor of economics at Brigham Young University. He received his Ph.D. in agricultural and development economics from Stanford University.

Agnes Rola

Agnes Rola (Ph. D. Agricultural Economics, University of Wisconsin-Madison) is Professor at the Institute of Strategic Planning and Policy Studies and an affiliate Professor at the Department of Agricultural Economics, University of the Philippines Los Banos. Her main research interests are in the areas of sustainable agriculture and natural resource management where she has extensive experience both at the field and policy levels. She has recently been involved in a decade long project on policy analysis for environmental management planning in the Philippines. She teaches courses in Policy Analysis and Production Economics. Currently, she sits on several local and international advisory boards for agricultural research.

GENERAL INFORMATION

LANGUAGE

The language of the Conference will be English.

HOTEL CHECK IN/OUT PROCEDURE

Check in time at the official hotels is 14h00 and check out is 11h00 on the day of departure. Early arrivals will be accommodated where possible and those delegates wishing to check out later, should make arrangements with their respective hotels.
REMEMBER – WHAT YOU SIGN FOR, YOU PAY FOR before leaving your hotel. This includes all incidental expenses such as room service, telephone calls, private drinks, laundry, etc. All room service, including breakfast in your room will be charge to your own "extras" account. EXTRAS can be paid by travellers' cheques, cash or credit card. All major credit cards are accepted.

SHUTTLE SERVICE

Shuttles between the official conference hotels and the ICC have been arranged at no charge to registered delegates. No shuttles will be provided between any hotels not listed below. From Monday 18 August 2003 only delegates wearing conference name badges will be allowed access to coaches. A shuttle schedule has been included in your delegate bag. Please refer to this schedule regarding the times when the shuttles are running between the ICC and the official conference hotels.

OFFICIAL CONFERENCE HOTELS

- Hilton Durban
- Holiday Inn Garden Court North Beach
- Holiday Inn Garden Court Marine Parade
- Holiday Inn Durban Elangeni
- City Lodge Durban

TECHNICAL TOURS: WEDNESDAY 20 AUGUST 2003

The technical tours are included in your registration fee. The tours have limited availability, therefore it is essential to ensure that you have received a voucher for the tour that you have chosen in your name badge

FOREIGN EXCHANGE

There are numerous banks in the vicinity of the International Convention Centre. Please ask at the Registration Desk for more information. There is also a foreign exchange facility available at the International Convention Centre.

CURRENCY AND TIPPING

The South African currency unit is the Rand, denoted by the symbol R. R1 = 100 cents. The international symbol is ZAR.
Tipping for service is a standard practice and a guideline is a 10% gratuity, depending on service and satisfaction.

LOCAL TIME

South Africa is currently two hours ahead of Greenwich Mean Time (GMT)

MESSAGES

The message board will be displayed at the Registration Desk. Please check regularly and once messages are read, please remove them from the board.

INTERNET CAFÉ

An Internet Café will be available for delegates use in the Poster display area between 07h00 and 18h00 daily.

IDENTIFICATION/ENTRY VOUCHERS

You are important! PLEASE WEAR YOUR NAME BADGE AT ALL TIMES. Besides being a vital aid to communication, if badges are not worn, you will be refused admittance to conference sessions and meals. Vouchers for the Networking Evening and the Gala Dinner are included in your name badge pouch—**a verbal identification at the Networking Evening is not sufficient**. You are also required to reserve a seat for the Gala Dinner at the registration desk.

SECURITY

Safety deposit boxes are available at your hotel and we strongly suggest that you keep all items of value in these boxes. As is applicable when visiting any large city in the world, we suggest that you do not display jewelry, cameras etc when walking in the streets.

MEDICAL

Most hotels have a list of doctors, whose names may also be found in the "medical" section of the Durban telephone directory. There is a travel clinic available on the lower level of the International Convention Centre.

SMOKING/CELLULAR TELEPHONES

Please note that all public areas in South Africa are now strictly non-smoking. As a matter of courtesy to other delegates, all cellular telephones should be turned off during the conference sessions.

VAT

Foreign Tourists to South Africa can have the sales tax which they have paid (known as Value Added Tax or VAT) refunded at a port of exit, provided the value of the items purchased exceeds R250.00. To qualify for a refund, visitors must be in possession of a valid passport, the necessary forms, VAT invoices and till slips. Currently the VAT rate is 14% and is levied on most products and services. Foreign delegates will NOT be able to reclaim the VAT included in their registration fees and accommodation charges.

Please note that all prices quoted include VAT. Should the rate be amended, the relevant adjustment will be payable by the delegate.

ENQUIRIES/INFORMATION

The conference registration/information and travel desks will be open daily from 07h30 to 17h30. Please feel free to approach our organizers should you require any assistance.

SOCIAL EVENTS

Registration

Registration for the conference and workshops will commence at 08h00 on Saturday 16 August 2003 in the main foyer of the ICC, DURBAN and will continue daily for the duration of the conference.

Opening Ceremony

The Minister of Agriculture and Land Affairs, Ms. Angela Thoko Didiza will officially open the conference in the plenary venue on Sunday 17 August 2003 at 10h00.
Dress: Business Attire or Traditional Dress

Welcome Reception

This warm South African welcome, hosted by His Worship the Ethekwini Municipality Mayor, Councillor Obed Mlaba, will commence at 19h00 on Sunday 17 August 2003.
Dress: Casual

Networking Evening

This event will be held on Wednesday 20 August 2003 at 19h00 at "Joe Kool's" theme restaurant on the beachfront.
Dress: Casual, bring a jacket to ward off the sea breezes

Conference Dinner

Thursday 21 August 2003 brings us to the last function of the conference. We have planned a fun-filled social event comprising music, cabaret and food which represents the broad spectrum of south African culture and cuisine.
Dress: Jacket & Tie/Traditional Dress

TABLE RESERVATIONS FOR THE GALA DINNER

In order to book a table, gather your colleagues and present your vouchers, found in your name badge pouch, at the conference registration desk from *09h30 on Tuesday 19 August 2003 to 12h00 on Thursday 21 August 2003*. Those delegates who have not reserved a seat by the end of this period, will be seated at the discretion of the organizers. A final seating plan will be available at the entrance to Hall 1AB on Thursday evening.

Lunches

All lunches will be served in Hall 3BC, next to the Poster Display area. Only delegates with lunch vouchers, found in your name badge, would have access to the lunch venue. If you have not booked any lunches, you may still do so at the conference registration desk.

Refreshment breaks

All refreshment breaks will be held in the Poster Display area Hall 3A. Only delegates wearing name badges will be given entry into this area.

NAME BADGE IDENTIFICATION

Workshops:

Learning Workshop on Analytical and Empirical Tools for Poverty Research — Blue insert
Learning Workshop on Food Security Measurement in a Developing Wolrd Context with a Focus on Africa — Green insert
Learning Workshop on Water Reforms, Institutions' Performance, Allocation, Pricing and Resource Accounting — Yellow insert

Conference:

Delegate	White
Accompanying person	Orange
Staff/Crew	Black
Keynote speakers/VIPs	Red
National Committee	Green
Executive Committee	Blue
Day delegates	Turquoise

SECURITY SAFETY TIPS ON THE STREET

- Do not publicize your valuables, e.g. jewellery, camera, etc.
- Use credit cards or if not possible, please carry small amounts of cash.
- At night, avoid isolated dark places.
- If you need any information, a policeman or officer will be glad to assist you.
- If you need a taxi, your hotel or the nearest tourism information office can be recommend a reliable service.
- It is advisable, when going for a walk, to walk in groups of not less than two persons.
- Do not wear your delegate ID card on your person when leaving the ICC or DEC.

IN THE CAR

- Like anywhere else in the world, your safety is strongly dependent on you.
- Always know where you are going.
- Fasten your seatbelt, lock your doors, and only leave your windows open about 5cm.
- Never pick up strangers.
- Do not use your cell phone whilst driving, unless you have a "hands free kit".
- Never display your valuables in the car, e.g. handbags, clothes, cell phones. Lock them in the boot.

ON THE BEACH

- Swim at beaches manned by lifeguards.
- Swim between the beacons.
- Obey instructions of lifeguards.
- Do not bring along valuables to the beach, always leave them behind.
- Use adequate sun protection cream.
- When encountering difficulties in the sea, raise one hand above your head.
- Do not take glass bottles to the beach.
- Avoid mixing alcohol and swimming.

VISITING SITES IN RURAL AREAS

- Establish how to observe the cultural protocol of that area.
- Visit traditional areas via recognized tourism transport.
- Use registered, qualified tour guides.

ACCOMMODATION

- Just like anywhere else in the world, please do not leave your luggage unattended.
- Store valuables in the safety deposit box at reception.
- Keep your room locked, whether you're in or out.
- If someone knocks, check who it is before opening the door.

AT THE AIRPORT

- Always keep your bags where you can see them.
- If you feel uncomfortable with people around you, please go to the nearest security offices.

Name Index

Subject Index